FREE Test Taking Tips Video/DVD Offer

To better serve you, we created videos covering test taking tips that we want to give you for FREE. **These videos cover world-class tips that will help you succeed on your test.**

We just ask that you send us feedback about this product. Please let us know what you thought about it—whether good, bad, or indifferent.

To get your **FREE videos**, you can use the QR code below or email freevideos@studyguideteam.com with "Free Videos" in the subject line and the following information in the body of the email:

      a. The title of your product

      b. Your product rating on a scale of 1-5, with 5 being the highest

      c. Your feedback about the product

If you have any questions or concerns, please don't hesitate to contact us at info@studyguideteam.com.

Thank you!

Chemistry Smart Review

Complete Study Guide Book with Practice Test Questions
[Includes Detailed Answer Explanations]

Joshua Rueda

Interested in buying more than 10 copies of our product? Contact us about bulk discounts:
bulkorders@studyguideteam.com

ISBN 13: 9781637753989
ISBN 10: 1637753985

Table of Contents

Quick Overview

As you draw closer to taking your exam, effective preparation becomes more and more important. Thankfully, you have this study guide to help you get ready. Use this guide to help keep your studying on track and refer to it often.

This study guide contains several key sections that will help you be successful on your exam. The guide contains tips for what you should do the night before and the day of the test. Also included are test-taking tips. Knowing the right information is not always enough. Many well-prepared test takers struggle with exams. These tips will help equip you to accurately read, assess, and answer test questions.

A large part of the guide is devoted to showing you what content to expect on the exam and to helping you better understand that content. In this guide are practice test questions so that you can see how well you have grasped the content. Then, answer explanations are provided so that you can understand why you missed certain questions.

Don't try to cram the night before you take your exam. This is not a wise strategy for a few reasons. First, your retention of the information will be low. Your time would be better used by reviewing information you already know rather than trying to learn a lot of new information. Second, you will likely become stressed as you try to gain a large amount of knowledge in a short amount of time. Third, you will be depriving yourself of sleep. So be sure to go to bed at a reasonable time the night before. Being well-rested helps you focus and remain calm.

Be sure to eat a substantial breakfast the morning of the exam. If you are taking the exam in the afternoon, be sure to have a good lunch as well. Being hungry is distracting and can make it difficult to focus. You have hopefully spent lots of time preparing for the exam. Don't let an empty stomach get in the way of success!

When travelling to the testing center, leave earlier than needed. That way, you have a buffer in case you experience any delays. This will help you remain calm and will keep you from missing your appointment time at the testing center.

Be sure to pace yourself during the exam. Don't try to rush through the exam. There is no need to risk performing poorly on the exam just so you can leave the testing center early. Allow yourself to use all of the allotted time if needed.

Remain positive while taking the exam even if you feel like you are performing poorly. Thinking about the content you should have mastered will not help you perform better on the exam.

Once the exam is complete, take some time to relax. Even if you feel that you need to take the exam again, you will be well served by some down time before you begin studying again. It's often easier to convince yourself to study if you know that it will come with a reward!

Test-Taking Strategies

1. Predicting the Answer

When you feel confident in your preparation for a multiple-choice test, try predicting the answer before reading the answer choices. This is especially useful on questions that test objective factual knowledge. By predicting the answer before reading the available choices, you eliminate the possibility that you will be distracted or led astray by an incorrect answer choice. You will feel more confident in your selection if you read the question, predict the answer, and then find your prediction among the answer choices. After using this strategy, be sure to still read all of the answer choices carefully and completely. If you feel unprepared, you should not attempt to predict the answers. This would be a waste of time and an opportunity for your mind to wander in the wrong direction.

2. Reading the Whole Question

Too often, test takers scan a multiple-choice question, recognize a few familiar words, and immediately jump to the answer choices. Test authors are aware of this common impatience, and they will sometimes prey upon it. For instance, a test author might subtly turn the question into a negative, or he or she might redirect the focus of the question right at the end. The only way to avoid falling into these traps is to read the entirety of the question carefully before reading the answer choices.

3. Looking for Wrong Answers

Long and complicated multiple-choice questions can be intimidating. One way to simplify a difficult multiple-choice question is to eliminate all of the answer choices that are clearly wrong. In most sets of answers, there will be at least one selection that can be dismissed right away. If the test is administered on paper, the test taker could draw a line through it to indicate that it may be ignored; otherwise, the test taker will have to perform this operation mentally or on scratch paper. In either case, once the obviously incorrect answers have been eliminated, the remaining choices may be considered. Sometimes identifying the clearly wrong answers will give the test taker some information about the correct answer. For instance, if one of the remaining answer choices is a direct opposite of one of the eliminated answer choices, it may well be the correct answer. The opposite of obviously wrong is obviously right! Of course, this is not always the case. Some answers are obviously incorrect simply because they are irrelevant to the question being asked. Still, identifying and eliminating some incorrect answer choices is a good way to simplify a multiple-choice question.

4. Don't Overanalyze

Anxious test takers often overanalyze questions. When you are nervous, your brain will often run wild, causing you to make associations and discover clues that don't actually exist. If you feel that this may be a problem for you, do whatever you can to slow down during the test. Try taking a deep breath or counting to ten. As you read and consider the question, restrict yourself to the particular words used by the author. Avoid thought tangents about what the author *really* meant, or what he or she was *trying* to say. The only things that matter on a multiple-choice test are the words that are actually in the question. You must avoid reading too much into a multiple-choice question, or supposing that the writer meant something other than what he or she wrote.

5. No Need for Panic

It is wise to learn as many strategies as possible before taking a multiple-choice test, but it is likely that you will come across a few questions for which you simply don't know the answer. In this situation, avoid panicking. Because most multiple-choice tests include dozens of questions, the relative value of a single wrong answer is small. As much as possible, you should compartmentalize each question on a multiple-choice test. In other words, you should not allow your feelings about one question to affect your success on the others. When you find a question that you either don't understand or don't know how to answer, just take a deep breath and do your best. Read the entire question slowly and carefully. Try rephrasing the question a couple of different ways. Then, read all of the answer choices carefully. After eliminating obviously wrong answers, make a selection and move on to the next question.

6. Confusing Answer Choices

When working on a difficult multiple-choice question, there may be a tendency to focus on the answer choices that are the easiest to understand. Many people, whether consciously or not, gravitate to the answer choices that require the least concentration, knowledge, and memory. This is a mistake. When you come across an answer choice that is confusing, you should give it extra attention. A question might be confusing because you do not know the subject matter to which it refers. If this is the case, don't eliminate the answer before you have affirmatively settled on another. When you come across an answer choice of this type, set it aside as you look at the remaining choices. If you can confidently assert that one of the other choices is correct, you can leave the confusing answer aside. Otherwise, you will need to take a moment to try to better understand the confusing answer choice. Rephrasing is one way to tease out the sense of a confusing answer choice.

7. Your First Instinct

Many people struggle with multiple-choice tests because they overthink the questions. If you have studied sufficiently for the test, you should be prepared to trust your first instinct once you have carefully and completely read the question and all of the answer choices. There is a great deal of research suggesting that the mind can come to the correct conclusion very quickly once it has obtained all of the relevant information. At times, it may seem to you as if your intuition is working faster even than your reasoning mind. This may in fact be true. The knowledge you obtain while studying may be retrieved from your subconscious before you have a chance to work out the associations that support it. Verify your instinct by working out the reasons that it should be trusted.

8. Key Words

Many test takers struggle with multiple-choice questions because they have poor reading comprehension skills. Quickly reading and understanding a multiple-choice question requires a mixture of skill and experience. To help with this, try jotting down a few key words and phrases on a piece of scrap paper. Doing this concentrates the process of reading and forces the mind to weigh the relative importance of the question's parts. In selecting words and phrases to write down, the test taker thinks about the question more deeply and carefully. This is especially true for multiple-choice questions that are preceded by a long prompt.

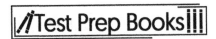

9. Subtle Negatives

One of the oldest tricks in the multiple-choice test writer's book is to subtly reverse the meaning of a question with a word like *not* or *except*. If you are not paying attention to each word in the question, you can easily be led astray by this trick. For instance, a common question format is, "Which of the following is...?" Obviously, if the question instead is, "Which of the following is not...?," then the answer will be quite different. Even worse, the test makers are aware of the potential for this mistake and will include one answer choice that would be correct if the question were not negated or reversed. A test taker who misses the reversal will find what he or she believes to be a correct answer and will be so confident that he or she will fail to reread the question and discover the original error. The only way to avoid this is to practice a wide variety of multiple-choice questions and to pay close attention to each and every word.

10. Reading Every Answer Choice

It may seem obvious, but you should always read every one of the answer choices! Too many test takers fall into the habit of scanning the question and assuming that they understand the question because they recognize a few key words. From there, they pick the first answer choice that answers the question they believe they have read. Test takers who read all of the answer choices might discover that one of the latter answer choices is actually *more* correct. Moreover, reading all of the answer choices can remind you of facts related to the question that can help you arrive at the correct answer. Sometimes, a misstatement or incorrect detail in one of the latter answer choices will trigger your memory of the subject and will enable you to find the right answer. Failing to read all of the answer choices is like not reading all of the items on a restaurant menu: you might miss out on the perfect choice.

11. Spot the Hedges

One of the keys to success on multiple-choice tests is paying close attention to every word. This is never truer than with words like almost, most, some, and sometimes. These words are called "hedges" because they indicate that a statement is not totally true or not true in every place and time. An absolute statement will contain no hedges, but in many subjects, the answers are not always straightforward or absolute. There are always exceptions to the rules in these subjects. For this reason, you should favor those multiple-choice questions that contain hedging language. The presence of qualifying words indicates that the author is taking special care with their words, which is certainly important when composing the right answer. After all, there are many ways to be wrong, but there is only one way to be right! For this reason, it is wise to avoid answers that are absolute when taking a multiple-choice test. An absolute answer is one that says things are either all one way or all another. They often include words like *every*, *always*, *best*, and *never*. If you are taking a multiple-choice test in a subject that doesn't lend itself to absolute answers, be on your guard if you see any of these words.

12. Long Answers

In many subject areas, the answers are not simple. As already mentioned, the right answer often requires hedges. Another common feature of the answers to a complex or subjective question are qualifying clauses, which are groups of words that subtly modify the meaning of the sentence. If the question or answer choice describes a rule to which there are exceptions or the subject matter is complicated, ambiguous, or confusing, the correct answer will require many words in order to be expressed clearly and accurately. In essence, you should not be deterred by answer choices that seem excessively long. Oftentimes, the author of the text will not be able to write the correct answer without

offering some qualifications and modifications. Your job is to read the answer choices thoroughly and completely and to select the one that most accurately and precisely answers the question.

13. Restating to Understand

Sometimes, a question on a multiple-choice test is difficult not because of what it asks but because of how it is written. If this is the case, restate the question or answer choice in different words. This process serves a couple of important purposes. First, it forces you to concentrate on the core of the question. In order to rephrase the question accurately, you have to understand it well. Rephrasing the question will concentrate your mind on the key words and ideas. Second, it will present the information to your mind in a fresh way. This process may trigger your memory and render some useful scrap of information picked up while studying.

14. True Statements

Sometimes an answer choice will be true in itself, but it does not answer the question. This is one of the main reasons why it is essential to read the question carefully and completely before proceeding to the answer choices. Too often, test takers skip ahead to the answer choices and look for true statements. Having found one of these, they are content to select it without reference to the question above. Obviously, this provides an easy way for test makers to play tricks. The savvy test taker will always read the entire question before turning to the answer choices. Then, having settled on a correct answer choice, he or she will refer to the original question and ensure that the selected answer is relevant. The mistake of choosing a correct-but-irrelevant answer choice is especially common on questions related to specific pieces of objective knowledge. A prepared test taker will have a wealth of factual knowledge at their disposal, and should not be careless in its application.

15. No Patterns

One of the more dangerous ideas that circulates about multiple-choice tests is that the correct answers tend to fall into patterns. These erroneous ideas range from a belief that B and C are the most common right answers, to the idea that an unprepared test-taker should answer "A-B-A-C-A-D-A-B-A." It cannot be emphasized enough that pattern-seeking of this type is exactly the WRONG way to approach a multiple-choice test. To begin with, it is highly unlikely that the test maker will plot the correct answers according to some predetermined pattern. The questions are scrambled and delivered in a random order. Furthermore, even if the test maker was following a pattern in the assignation of correct answers, there is no reason why the test taker would know which pattern he or she was using. Any attempt to discern a pattern in the answer choices is a waste of time and a distraction from the real work of taking the test. A test taker would be much better served by extra preparation before the test than by reliance on a pattern in the answers.

FREE Videos/DVD OFFER

Doing well on your exam requires both knowing the test content and understanding how to use that knowledge to do well on the test. We offer completely FREE test taking tip videos. **These videos cover world-class tips that you can use to succeed on your test.**

To get your **FREE videos**, you can use the QR code below or email freevideos@studyguideteam.com with "Free Videos" in the subject line and the following information in the body of the email:

      a. The title of your product

      b. Your product rating on a scale of 1-5, with 5 being the highest

      c. Your feedback about the product

If you have any questions or concerns, please don't hesitate to contact us at info@studyguideteam.com.

Thanks again!

Introduction

Purpose of this Guide

The purpose of this guide is to help students from review and practice concepts in chemistry. The guide includes review material that covers appropriate topics and includes a practice test. The practice test consists of questions that utilize graphs, the periodic table, equations, and formulas.

Alignment Topics

This guide is an overview of chemistry from atomic structure to thermodynamics to acids and bases. The practice questions are similar to questions that are given to students in the common core standards, as they also deal with assessing knowledge through multiple-choice questions and equations.

Skills Assessed

The following points are skills that should be developed by the completion of this guide:

- Learning Atomic Structure and Properties
    - Learning about atomic structure and atomic mass
    - Navigating the Periodic Table of Elements
    - Learning about chemical equations
- Assessing Molecular and Ionic Compound Structure and Properties
    - Learning the different types of chemical bonds
    - Learning how to use Lewis Diagrams
- Discovering Intermolecular Forces and Properties
    - Assessing the properties of solids, liquids, and gases
    - Learning more about the Ideal Gas Law and the Beer-Lambert Law
- Analyzing Chemical Reactions
    - Assessing the difference between chemical and physical changes in matter
- Learning the Applications of Thermodynamics
    - Assessing absolute entropy, kinetic control, and coupled reactions
    - Understanding cell potential
- Understanding Organic and Nuclear Chemistry
    - Assessing structures and properties
    - Learning the different reactions of compounds
    - Understanding radioactivity types
- Evaluating the Uses of Biochemistry and Electrochemistry
    - Understanding DNA and RNA
    - Learning how to balance oxidation-reduction reactions

Atomic Structure and Properties

Atomic Structure

The structure of an atom has 2 major components: the atomic nucleus and the atomic shells (also known as **orbits**). The **nucleus**, which is made up of protons and neutrons, is found in the center of an atom. The 3 major subatomic particles are protons, neutrons, and electrons, which are found in the atomic nucleus and shells.

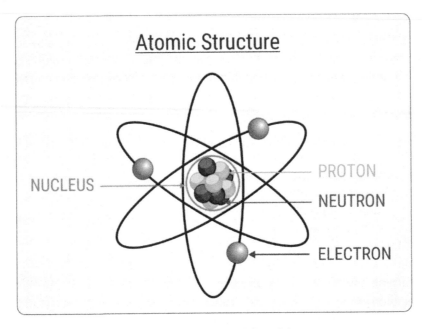

Figure 1. The subatomic particles of the atom

Protons are positively charged particles found in the atomic nucleus. An entirely different element is created when protons are added to or removed from an atom's nucleus. **Neutrons** are also found in the atomic nucleus and are neutral particles, meaning they have no net electrical charge. The addition or removal of neutrons to or from an atom's nucleus does not create a different element; instead it creates a lighter or heavier form of that element, which is called an isotope. **Electrons** are negatively charged particles found orbiting in the atomic shells around the nucleus. A proton or a neutron has nearly 2,000 times the mass of an electron. Table 1 below shows the difference in mass and charge for each subatomic particle.

Table 1. Mass and charge properties of the proton, neutron, and electron				
Subatomic particle	Mass (kg)	Mass (amu)	Charge (Coulomb, C)	Charge (e)
Proton	1.67262×10^{-27}	1.00728	$+1.60218 \times 10^{-19}$	$+1$
Neutron	1.67494×10^{-27}	1.00866	0	0
Electron	9.10939×10^{-31}	0.00055	-1.60218×10^{-19}	-1

The model of the atom in Figure 1 is based off Ernest Rutherford's, Han's Geiger's, and Ernest Marsden's alpha-particle scattering experiments (1911), which bombarded helium atoms toward metal foils. These experiments suggested that a majority of the atom's mass (99.95%) was found at the nucleus, which

contains the positively charged protons. The electrons move around the nucleus and occupy most of the space in the atom. As an analogy, if a ping-pong ball represented the nucleus, then the electrons would move in a space up to 3 miles in diameter from the nucleus. It's important to note that Figure 1 is a simplistic atomic representation that shows electrons confined to fixed orbits and moving around the nucleus.

A more accurate representation of the nuclear model would show that electrons orbit the nucleus in **atomic shells** or **electron clouds**. These electron clouds can vary in shape and size. For example, the first atomic shell (K shell) has a spherical shape and can accommodate 2 electrons. The second atomic shell (L shell) is farther away from the nucleus and consists of a spherical and dumbbell-shaped electron cloud that can hold a maximum of 8 electrons. The third atomic shell (M shell) is similar to the second shell but is farthest away from the nucleus, and it can house a maximum of 8 electrons.

Figure 2 illustrates the first 2 atomic shells closest to the nucleus, which may be composed of several spherical or dumbbell-shaped orbitals, which are spaces where electrons are. The closest shell to the nucleus is called the "1 shell" (also called "K shell"), followed by the "2 shell" (or "L shell"), then the "3 shell" (or "M shell"). The K shell contains only one spherical or "s" orbital, whereas the L shell contains an "s" orbital and 3 dumbbell-shaped "p" orbitals. The numbers preceding "s" or "p" refer to the first shell (K shell) and second shell (L shell). Only 2 electrons are found per s or p orbital. The last column combines the K and L Shells. The second row shows a simple orbit model of the K, L, and combined shells.

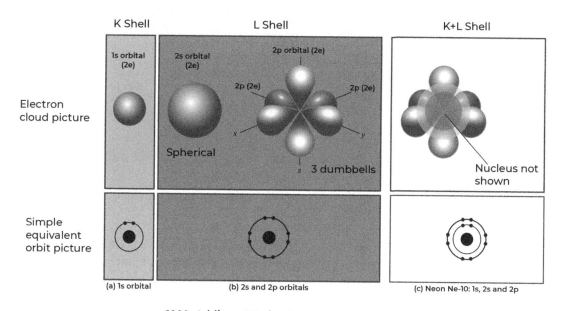

1999, Addison, Wesley, Longman, Inc.

Figure 2. Electron cloud and simple atomic model of the neon atom

The negatively charged electrons orbiting the nucleus are attracted to the positively charged protons in the nucleus via electromagnetic force. The attraction of opposite electrical charges gives rise to chemical bonds, which are how atoms are attached or bonded.

The **atomic number** of an atom is determined by the number of protons in the nucleus. In notation the atomic number is represented by the letter Z. When a substance is composed of atoms that all have the same atomic number, it is called an **element**. In the **periodic table**, elements are arranged by atomic number and grouped by properties. The symbols of the elements in the periodic table are a single letter or a two-letter combination that is usually derived from the element's name.

Figure 3 shows a common chemical symbol represented in some periodic tables. The letter X refers to the specific element. The subscript Z is the element's atomic number. The superscript A is the element's mass number. An atom with a fixed atomic number and mass number is called a **nuclide**. Figure 3 below represents a nuclide.

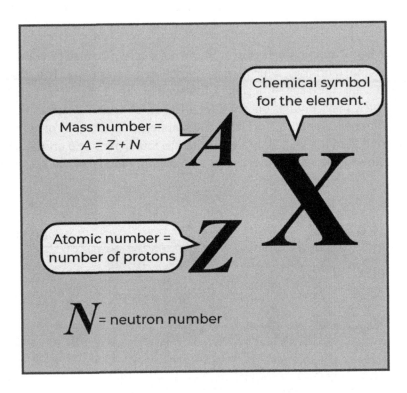

Figure 3. Representation of an element by a symbol X with its associated mass and atomic number

Many of the elements have Latin origins for their names, and their atomic symbols do not match their modern names. For example, iron is derived from the Latin word *ferrum*, so its symbol is Fe, even though it is now called iron. The naming of the elements began with those of natural origin and they have ancient names, often with the ending "-ium." This naming practice has been continued for all elements that have been named since the 1940s. Now the names of new elements must be approved by the International Union of Pure and Applied Chemistry.

An atom's **mass number (A)** is determined by the sum of the total number of protons (Z) and neutrons (N) in the atom. Most nuclei have a net neutral charge, and all atoms of one type have the same atomic number. The neutral charge is due to the equal number of protons and electrons in the atom. Therefore, the atomic number (Z) is often equal to the number of electrons in an atom.

However, there are some atoms of the same type that have a different mass number, due to an imbalance of neutrons. Atoms of this type are called **isotopes**. The atomic number, which is determined

by the number of protons, is the same in all isotopes of a given element, but the mass number, which is determined by adding the protons and neutrons, is different due to the irregular number of neutrons. The **atomic mass** of an element is the weighted average of the naturally occurring atoms of a given element, or the relative abundance of isotopes that might be used in chemistry.

The mass number of chlorine is 35; however, the atomic mass of chlorine is 35.5 amu (atomic mass unit). This discrepancy exists because there are many isotopes occurring in nature. For example, the nucleus could have 36 instead of 35 protons. Given the prevalence of the various isotopes, the average of all of the mass numbers turns out to be 35.5 amu, which is slightly higher than chlorine's number in the periodic table. The nitrogen atom contains 2 types of stable isotopes: nitrogen-14 and nitrogen-15. The nitrogen atom has an average mass number of 14.01 amu. Each number following the name of the element indicates the number of neutrons for each nitrogen isotope. Nitrogen-14 is the most naturally occurring isotope, whereas nitrogen-15 is not. Over 10 radioactive isotopes of nitrogen have also been identified, but these isotopes are unstable or short-lived. The chemical symbols for the 2 stable isotopes can also be designated as:

$$^{14}_{7}N \quad \text{nitrogen-14}$$

$$^{15}_{7}N \quad \text{nitrogen-15}$$

The atomic number of nitrogen is seven ($Z = 7$), $_7N$. The naturally occurring form of nitrogen has 7 protons, 7 neutrons, and 7 electrons. The mass number A would be attributed to the mass of the protons (7) and neutrons (7), which is equal to 14: 7 amu + 7 amu = 14 amu. Therefore, the elemental symbol for nitrogen with 7 neutrons is N-14 or $^{14}_{7}N$, where the left superscript indicates the mass number.

Atomic Mass

For each element found in nature, there is a mixture of isotopes, which will each have a characteristic mass, for example, nitrogen-14 is 14 amu and nitrogen-15 is 15 amu. One **atomic mass unit (amu)** is roughly the mass of one proton or neutron, as shown in Table 1, or one-twelfth the mass of the carbon-12 isotope which contains 6 protons and 6 neutrons: 6 amu + 6 amu = 12 amu. For each element present in the periodic table, the listed mass is an average **atomic mass** because there are other isotopes in addition to the naturally occurring element.

The mass spectrometer is an instrument routinely used to measure the atomic mass of different isotopes. The spectrometer works by comparing the mass of a chosen atom to a standard atom such as the carbon-12 isotope, which has a mass of twelve atomic mass units. A mass spectrometer can identify different isotopes in a given element. First, a gaseous form of the specific element is introduced or injected into a tube in the mass spectrometer. The gaseous element is converted to a positive ion (for example, N to N$^+$), which is accelerated by a negatively charged plate/grid toward an internal magnet. The positive beam of ions is then separated by the magnetic field, from the magnet, based on their mass-to-charge ratio into 2 beams associated with $^{14}_{7}N$ and $^{15}_{7}N$. A detector located at the end of the tube will measure the distance between each beam. A computer connected to the instrument then generates a mass spectrum plot that shows the fractional abundance of the isotope with respect to its mass.

Fractional Abundance

Figure 4 shows a mass spectrum of nitrogen, which indicates the presence of 2 isotopes with masses.

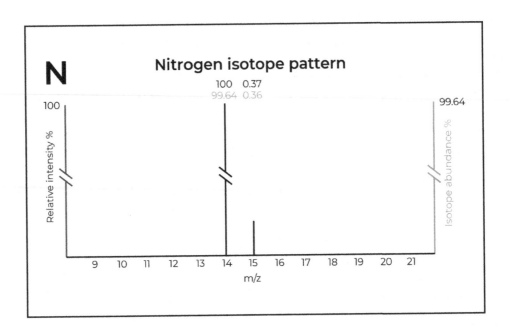

Figure 4. Isotopes of nitrogen: N-14 and N-15. The mass-to-charge ratio is given by m/z

The mass spectrum shows a peak at 14 m/z (14 amu), which corresponds to N-14 with an isotope or fractional abundance equal to 99.64%. The other peak at 15 m/z (15 amu) is N-15, which has one extra neutron compared to N-14 and has a fractional abundance of 0.36%. Table 2 shows the atomic mass and abundance of each isotope.

Table 2. Isotopes of Nitrogen (N)			
Nitrogen isotope	Atomic mass (amu)	Relative intensity %	Fractional abundance
N-14, $^{14}_{7}$N	14.003074005	100	0.9963
N-15, $^{15}_{7}$N	15.000108898	0.37	0.0037

Most of the mass is due to N-14, but the average atomic mass of N will not be exactly 14.0 amu. The atomic mass of N is slightly greater because of N-15. The average atomic mass depends on the intensity or fractional abundance of each isotope. For example, the mass spectrum of N shows 2 isotopes, N-14 (14 amu or m/z) and N-15 (15 amu or m/z), with a fractional abundance of 99.64 and 0.36%. The **fractional abundance** for each isotope is found by dividing the relative intensity of each isotope over the sum of the relative intensities for all isotopes. For example, the fractional abundance of N-14 is:

$$^{14}_{7}\text{N fractional abundance} = \frac{100.0\%}{100.0\% + 0.3700\%} = 0.9963$$

Once the fractional abundance is obtained from the relative intensity provided by the mass spectrometer, the **average atomic mass** of can be calculated. Below are calculations for nitrogen.

$$\text{average atomic mass (amu)} = \sum (\text{atomic mass}) \times (\text{fraction abundance})$$

$$(14.003074005) \times (0.9963) + (15.000108898) \times (0.0037)$$

$$14.01 \text{ amu}$$

The average atomic mass of nitrogen (N) is 14.01 amu, which is an approximate value shown in the periodic table.

The Periodic Table of Elements

The periodic table catalogs all the known elements known, currently 118 (Figure 5). This table is one of the most important references in the science of chemistry. Information in the periodic table includes elements' atomic number, atomic mass, and chemical symbol. The first periodic table was rendered by Mendeleev in the mid-1800s and was arranged in order of increasing atomic mass. The modern periodic table is arranged in order of increasing atomic number and common characteristics.

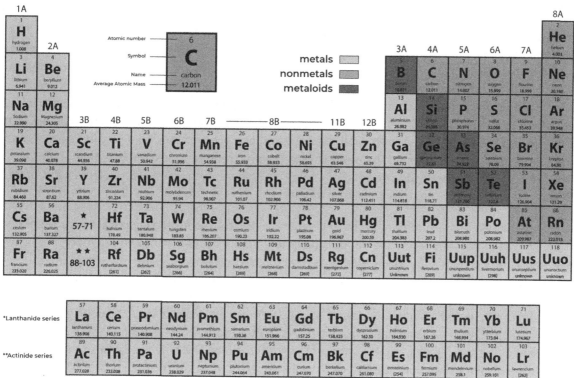

Figure 5. The periodic table of elements

As the atomic number increases, electrons gradually fill the shells of an atom. This is represented in the table from left to right and top to bottom. The horizontal rows, known as **periods**, have different valence electron shell configurations, for example, 1s, 2s, 2p, and 3s. In general, the start of a new period corresponds to the addition of an electron to a new shell. For example, Li: $1s^2\mathbf{2s^1}$ and Na: $1s^2 2s^2 2p^6 \mathbf{3s^1}$. The sum of superscripts is equal to the atomic number. The periodic table contains 7 periods and 18 families. The vertical columns, called **groups** or **families**, are sorted by similar chemical properties and characteristics such as appearance and reactivity. The elements in the periodic table can also be classified into 3 major groups: metals, metalloids, and nonmetals. The elements in each block (for example, metal, transition metal, and nonmetal) of the periodic table are also group by similar electron configurations. **Metals** are concentrated on the left side of the periodic table, and **nonmetals** are concentrated on the right side (Figure 5). **Metalloids** occupy the area between the metals and nonmetals.

Modern periodic tables have the elements arranged according to their valance electron configurations, which also contribute to trends in chemical properties. These properties help to further categorize the elements into blocks or groups, which include metals, nonmetals, transition metals, alkali metals, alkali earth metals, metalloids, lanthanides, actinides, diatomics, post-transition metals, polyatomic nonmetals, and noble gases. For example, the **alkali earth metals** (Li, Na, K, Rb, Cs, Fr) are found in column one, or group IA, and have one loosely bound electron, low electronegativities, and the largest atomic radii within their period. These metals are soft and highly reactive with water. **Noble gases** (for example, He, Ne, Ar, Kr, Xe, Rn) have a full outer electron valence shell of 8 electrons. The elements in this block possess similar characteristics, such as being colorless, odorless, and having low chemical reactivity. In another block, the metals, elements tend to be shiny, highly conductive, and easily form alloys with each other, nonmetals, and noble gases.

Periodic Trends

The elements in the periodic table are arranged by number and grouped by trends in their physical properties and electron configurations. Certain trends are easily described by the arrangement of the periodic table and include the **atomic radius**, which is defined as one half the distance between the nuclei of atoms of the same element. The **atomic radius** increases as elements go from right to left and from top to bottom in the periodic table (Figure 6). Another trend in the periodic table is the **electron affinity energy**, which is the tendency of an atom to attract electrons. This tendency increases from left to right and from bottom to top of the periodic table, which trend is opposite to the that of atomic radius. The **ionization energy**, the amount of energy needed to remove an electron from a gas or ion, follows a trend similar to that of the electron affinity.

Electronegativity is a measurement of the willingness of an atom to form a chemical bond. Elements on the right side and near the top of the periodic table tend to attract electrons and are more electronegative. Consequently, these elements tend to gain a negative charge since there are more electrons than protons. An atom with a charge is referred to as an **ion**, and if an ion has more electrons than protons, it has a negative charge and is called an **anion**. In contrast, elements on the left and near the bottom of the periodic table lose, or give up, one or more electrons in order to bond and are the least electronegative. If an atom has fewer electrons than protons, it has a positive charge and is a called **cation**. Nonmetals typically form anions and metals form cations. The only exceptions to this rule are the noble gases. Since the noble gases have full valence shells (the outermost orbital shell of an atom), they do not tend to lose or gain electrons.

Chemical reactivity is another trend that can be identified by the groupings of the elements in the periodic table. The chemical reactivity of metals decreases from left to right in the table and while going higher in the table. Conversely, nonmetals increase in chemical reactivity from left to right and while

going lower in the table. Again, the noble gases present an exception to these trends because they have very low chemical reactivity.

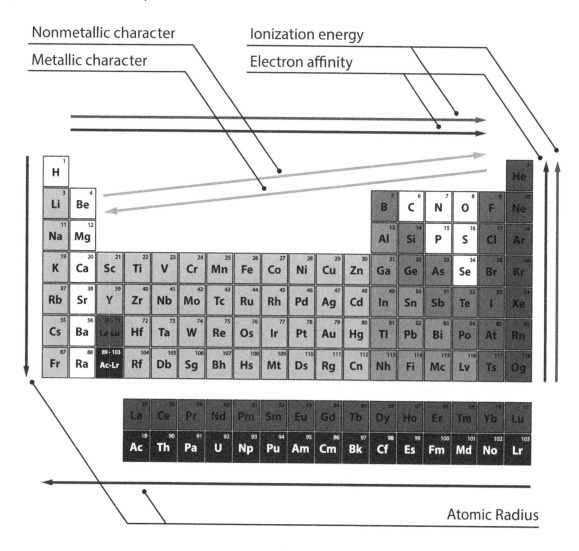

Figure 6. Trends in the periodic table of elements

Chemical Equations

Chemical reactions are represented by **chemical equations**. The equations help to explain how the molecules change during a reaction. For example, when hydrogen gas (H_2) combines with oxygen gas (O_2), two molecules of water are formed. The equation is written as follows, where the plus sign means *reacts with* and the arrow means *produces*:

$$2\,H_2 + O_2 \rightarrow 2\,H_2O$$

Two hydrogen molecules react with an oxygen molecule to produce 2 water molecules. In all chemical equations, the quantity of each element on the reactant side of the equation should equal the quantity of the same element on the product side of the equation. This is due to the law of conservation of matter. If this is true of the equation, the equation is described as balanced. To appropriately label and balance the number of elements on each side of the equation, the coefficient of the element should be multiplied by the subscript next to the element. Coefficients and subscripts are used for quantities larger than one.

The **coefficient** is the number located directly to the left of the element. The **subscript** is the small-sized number directly to the right of the element. In the equation above, on the left side, the coefficient of the hydrogen is 2, and the subscript is also 2, which makes a total of 4 hydrogen atoms. There are 2 oxygen atoms on the left side, so a coefficient of 2 is added in front of water (H_2O) to indicate that there are 2 oxygen atoms. The coefficient multiplied by the subscript in each element of the water molecule also gives 4 hydrogen atoms. This equation is balanced because there are 4 hydrogen atoms and 2 oxygen atoms on each side. The states of the reactants and products can also be written in the equation: gas (g), liquid (l), solid (s), and dissolved in water (aq). If they are included, they are noted in parentheses on the right side of each molecule in the equation.

Molecular and Ionic Compound Structure and Properties

Types of Chemical Bonds

Trends in Electronegativity

The **electronegativity** of an atom is a measure of its ability to attract the electrons of another atom to bond with its own electrons.

One of the ways in which the periodic table is organized is according to the electronegativity of the elements. The electronegativity of the elements increases going from left to right across the table and also increases from the bottom to the top of a group within the table. Fluorine, which is located in the top right corner of the periodic table, is the most electronegative element. Electrons of an atom fill shells around the nucleus. The shells that are closest to the nucleus hold fewer electrons than the ones that are farther away. The shells that are closest to the nucleus are filled first. Moving from left to right across the periodic table, the atoms have an increasing number of electrons in their shells. The atoms in the same group, or column, of the periodic table have the same number of **valence electrons**, which is the number of electrons in the atom's outer most shell.

The shell model of electrons depicts the electron shells as rings around the nucleus. Dots on the ring represent electrons that are filling that shell. For example, hydrogen has only one electron, so it is represented by one dot on one single ring around the nucleus of the atom. Sulfur has 16 electrons in three different shells. Each shell is represented by a ring around the nucleus, and the number of dots in each ring represents the number of electrons filling that shell. The first ring, closest to the nucleus, has two electrons, the next ring has eight electrons, and the third ring has six electrons. For sulfur, the valence electrons are in the outermost shell, which contains six electrons. According to Coulomb's law, the electrons in the outermost shells have less attraction to the nucleus because they are farther away.

Nonpolar Covalent Bonds

Covalent bonds are formed between atoms that have similar electronegativities. These atoms share the electrons equally between them and form a nonpolar molecule. They are usually formed between nonmetals that reside close to each other in the periodic table. Although carbon and hydrogen do not have identical electronegativities, their bond is still considered nonpolar. **Nonpolar covalent bonds** are formed as a balance between the distance between two atoms and the minimization of the potential energy of the atoms.

The length of the bond that is formed is measured as the distance between the two nuclei. It is a distance at which the forces of attraction and repulsion are equal. **Bond energy** is the amount of energy that is needed to break the bond between two atoms. It is a measure of how stable the molecule is compared with the two individual atoms. Remember that in a covalent bond, the electron cloud in the nucleus of one atom is interacting with the cloud of electrons in the nucleus of another atom. This relationship can be visualized graphically. The concave shape of the curve is determined by how much the atoms attract and repulse each other. Hydrogen atoms are often used as examples in these graphs.

If the potential energy was mapped out as a function of distance between the two atoms, something like this would be obtained:

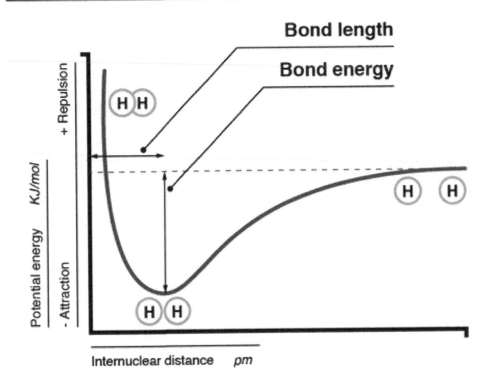

Potential Energy vs. Internuclear Distance Between Two Atoms

The two atoms approach the minimum value of potential energy at the bond energy. The link that corresponds to this is the equilibrium bond length—a happy medium corresponding to a minimum in potential energy. This minimum of energy is achieved when the atomic orbitals are overlapping and so contain two **spin-paired electrons**—when the two atoms' orbitals are overlapping each other. A greater amount of energy is needed to separate those two atoms when they are overlapped.

Polar Covalent Bonds

When two atoms that have nonidentical electronegativities share electrons in a bond, the bond is considered to be **polar covalent**. Because the electronegativities are not equal, their difference is not zero as it is in nonpolar covalent bonds. The atom that has the higher electronegativity retains a partial negative charge, and the other atom retains a partial positive charge. For molecules with two atoms, the partial negative charge is equal in magnitude to the partial positive charge. Even in larger molecules, the sum of all of the partial charges equals the charge of the whole molecule. As the difference in electronegativity increases, larger dipoles are created at the ends of the bond.

Covalent Versus Ionic Bonds

Unlike in covalent bonds, in **ionic bonds**, electrons are transferred from one atom to another atom. They are not shared in between the two atoms. Covalent and ionic bonds represent two ends of the bonding spectrum. Most molecules are formed from bonds that have some characteristics of both types of

bonds, although they may mostly be one or the other. Although for the most part, bonds between nonmetals are covalent, and bonds between a metal and a nonmetal are ionic, there are other factors that can determine what type of bond is formed between atoms of a molecule. It is best to look at the properties of the molecule to determine what type of bond is formed.

Valence Electrons in Metals

Metallic bonds are formed by electrons that move freely through metal. They are the product of the force of attraction between electrons and metal ions. The electrons are shared by many metal cations and act like glue that holds the metallic substance together, similar to the attraction between oppositely-charged atoms in ionic substances, except the electrons are more fluid and float around the bonded metals and form what is called a sea of electrons. The valence electron of one metal atom delocalizes, or detaches from the atom, and forms a molecular orbital with the valence electron shell of the neighboring atom.

This happens for all of the empty spaces in the valence shell of the atom. For example, with sodium atoms, there is one valence electron in its outermost 3s shell. There is room for eight electrons in the 3s shell. So, each sodium atom is surrounded by eight other sodium atoms and molecular orbitals are formed for each space in the 3s shell. The electrons delocalize from their parent atom and form a sea of electrons around and between the atoms. They are free to move around among all of the atoms because the molecular orbitals that are formed cover the whole piece of metal. The molecular structure of the metal is held together by the Coulombic attraction between the sea of electrons and the positive nuclei.

Metallic bonding has characteristics of both covalent and ionic bonding. The electrons disassociate from the atoms, as with ionic bonding, but are also shared among the atoms, as with covalent bonding.

Intermolecular Force and Potential Energy

Graphing the Energy Associated with Making and Breaking Bonds

In **valence bond theory**, electrons are treated in a quantum mechanics fashion. Valence bond theory is good for describing the shapes of covalent compounds. Instead of electrons being pinpointed to an exact location, it's possible to map out a region of space—known as **atomic orbitals**—that the electron may inhabit. In valence bond theory, electrons are treated as excitations of the **electron field**, which exists everywhere. When energy is given to an electron field, electrons are said to exist inside a **wave function**—a mathematical function describing the probability an electron is in a certain place at any given time. Standing waves are created when energy is given to a wave function. Electrons function as standing waves around a nucleus.

There are two things to consider in valence bond theory: the overlapping of the orbitals and the potential energy changes in a molecule as the atoms get closer or further apart. For example, the Lewis structure of H_2 would be $H - H$, but this doesn't describe the strength of the bond. The electron cloud in the nucleus of one atom is interacting with the cloud of electrons in the nucleus of another atom.

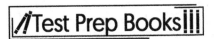

If the potential energy was mapped out as a function of distance between the two atoms, something like this would be obtained:

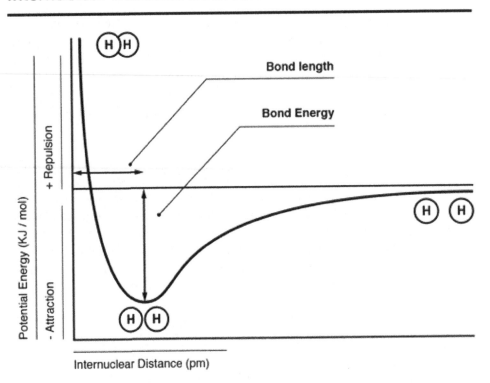

Potential Energy versus Internuclear Distance Between Two Atoms

The two atoms approach this minimum value at a distance known as the **bond energy**. The link that corresponds to this is the equilibrium bond length—a happy medium corresponding to a minimum in potential energy. This minimum of energy is achieved when the atomic orbitals are overlapping and so contain two **spin-paired electrons**—when the two atoms' orbitals are overlapping each other. A greater amount of energy is needed to separate those two atoms when they are overlapped.

The image below shows the potential energy of H_2 with respect to different bond distances as previously discussed. The most stable bond length is found on the minimum of the potential energy

Bond Strength vs. Length
The greater the number of chemical bonds between two nuclei (e.g., single, double, triple), the more extensive the electrostatic interaction between the electron pairs and the nuclei. Consequently, the average bond energy for triple bonds will be higher than double and single bonds. The order of increasing bond strength is single < double < triple.

The more chemical bonds between any two atoms, the shorter and stronger those bonds are because the electrons are held more closely by electrostatic interactions between each atom.

The Attractive Force Between Any Two Ions
There are several factors that influence the strength of the bond between two ions. Recall that Coulomb's law describes the attractive forces between the electrons and protons and the repulsive

forces between two protons or two electrons. The Coulomb force is proportional to the interaction or potential energy (V), which is given as:

$$V = \frac{q_A q_B}{r} \propto Force; \quad Force = -\frac{\Delta V}{\Delta r}$$

The greater the charge of the ions, the greater the force of attraction between them. The shorter the distance between the ions, the greater the force of attraction as well. When the force of attraction has a large magnitude, the melting point of the ionic solid will also be high because a greater amount of heat energy is needed to break the bonds between the ions.

Structure of Ionic Solids

Ionic compounds are those that are held together by **ionic bonds**—also known as **electrostatic forces**—through the transfer of electrons. Ionic substances are neutral overall, but they are balanced out by their **cations**—positively-charged ions and **anions**—negatively-charged ions. Ionic bonds usually form between metals and nonmetals. Most rocks and minerals are ionic compounds, including pyrite (FeS_2).

An example of an ionic compound is sodium chloride, which is held together by ionic bonds, due to sodium (Na^+) having a positive charge and chloride having a negative charge (Cl^-). The sodium ion is positively charged because sodium (Na) only has one electron in its valence shell. Therefore, sodium donates this electron easily to become a positively-charged ion (Na^+). Similarly, chloride (Cl) has seven electrons in its valence shell and accepts an electron easily to become a negatively-charged ion (Cl^-).

Properties of Ionic Solids

Ionic solids are generally formed between metal and nonmetal ions. Metal atoms tend to lose electrons and form cations, and nonmetal ions tend to gain electrons and form anions. The oppositely-charged ions are attracted to each other and bond together with ionic bonds. They form a three-dimensional cubic lattice structure with strong Coulombic attractions between the cations and the anions. These strong Coulombic attractions cause ionic solids to have low vapor pressures, high melting points, and high boiling points. The oppositely-charged ions are neatly arranged so that the forces of repulsion between like-charged ions is minimized. Because of this arrangement, ionic solids are not malleable or ductile.

They are hard and brittle and break apart easily when the like-charged ions from different layers slide over each other. In the solid form, the ions are held tightly in place and cannot move around to transfer electrons and conduct electricity. In liquid or aqueous form, the metal ions are able to move freely and can conduct electricity within their substance. Dissolving a solid substance in water and testing conductivity is one way in which to determine whether the substance is ionic or not. Ionic compounds do not tend to dissolve in nonpolar solvents. The attraction between the ions in the substance are stronger than their attraction to the nonpolar solvent molecules, so the ionic solid tends to stay together. With polar solvents, the polar molecules are attracted to the ions in the solid substance and can break it apart. For aqueous solutions, this process is called **hydration**, with the polar water molecules penetrating the substance and attracting the ions in the solid.

Structure of Metals and Alloys

Understanding Metallic Bonding

Metallic bonding occurs between metal atoms. For example, an aluminum sheet is formed from aluminum atoms metallically bonded together, and a copper wire comprises copper atoms metallically bonded together. Metallic atoms have strong bonds between them, which create high boiling and melting points because it takes more energy to break the strong bonds.

Metallic bonding can be visualized using the electron shell model. In the electron shell model, each electron shell is represented by a ring that circles around the nucleus. With metallic bonding, the outermost shell is always empty and the electrons from that shell are located in the free space between the metal cations. This model depicts the delocalization of the electrons from the parent atoms and the sharing of the electrons in the molecular orbitals that form between the atoms in the metallic structure.

The physical properties of metals can be attributed to the sea of electrons that is formed by metallic bonding. They are good conductors of electricity because of how electrons flow freely in the metallic structure. Once the electrons delocalize, the metal atoms become cations. The cations slide past each other easily and the electrons are always moving, which allows metals to be **malleable** (the ability to bend or form a new shape). The ability for the cations to move away from each other also gives metals **ductility**, which is their ability to stretch or deform under stress. The strength of the metallic bonds also leads to low **volatility** of metals, which means that they do not vaporize easily.

Interstitial Alloys

Metallic alloys are solids that comprise a mixture of metals. The characteristics of an alloy are dependent on the metallic components that it contains. There are two main types of alloys: interstitial and substitutional.

Interstitial alloys are formed from atoms that have different sizes. The smaller atoms fill the interstitial spaces between the larger atoms that form the lattice structure of the solid. Interstitial alloys are less malleable and ductile than pure metal solids. The smaller atoms that fill the interstitial spaces make the solid more rigid and the electrons have less space to move around freely. The solid then becomes less deformable. Steel is an example of an interstitial alloy. It is composed of iron and carbon atoms. Iron cations comprise the lattice, and the smaller carbon atoms fit into the interstitial spaces.

Substitutional Alloys

Another type of alloy is a substitutional alloy. **Substitutional alloys** are formed when metallic atoms that are similar in size are mixed together. Both types of metallic atoms form the lattice structure of the solid. The metal atoms are metallically bonded, which involves some covalent bonding between them and some electron sharing of the delocalized electrons. The strength of the covalent bond and the amount of electron sharing is dependent on the metals that comprise the alloy. Substitutional alloys, like interstitial alloys, are less malleable and ductile than pure metals because of the more rigid structure and balance between the positive kernels of the solid.

The density of substitutional alloys is in between the densities of the individual components. Brass is a metal composed of copper and zinc and is an example of a substitutional alloy. Because alloys are made from metallic ions with metallic bonds and a sea of electrons, they are good conductors of electricity, as with pure metals. Alloys, however, may have surface properties that can be changed through chemical reactions. For example, stainless steel is composed of iron, nickel, and chromium atoms with carbon

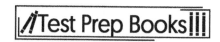

atoms in the interstitial spaces. Oxygen in the air reacts with the chromium to form a chromium oxide layer on the surface that protects the metal alloy from rust and corrosion.

Lewis Diagrams

Lewis Diagrams

Lewis diagrams—also called **electron dot diagrams**—are used to depict how the atoms of a molecule are bonded together covalently and how their electrons are shared in the molecule. The diagram was named after Gilbert Lewis, an American physical chemist who discovered the structure of covalent bonds.

Lewis diagrams show the electrons in the valence shell of an atom. In other words, they show the electrons that are available for covalent bonding. The procedure for drawing Lewis diagrams is as follows:

- Work out the number of available electrons.

- If the compound has more than two atoms, the least electronegative atom goes in the center— remembering that hydrogen (H) must go on the outside.

- Place two electrons between atoms to form a chemical bond—usually represented with a dash.

- Complete the octets of electrons on the outside of the atom. Valence shells need eight electrons, except for hydrogen (H), which only needs two. There are some exceptions to this rule; for example, beryllium (Be) is happy with just four electrons in its valence shell and boron (B) with just six electrons in its valence shell.

- If all of the electrons are used up but the octets still haven't been filled, electrons can be moved from outer atoms, forming double or triple bonds.

An example to consider is carbon dioxide, CO_2. Carbon (C) is in Group 14 of the periodic table, so it has four valence electrons. Oxygen (O) is in Group 16, so it has six valence electrons. The total is ten electrons. The least electronegative atom is carbon, so it should be put in the center of the Lewis diagram, with the two oxygens on either side. Then, two of the ten electrons should be put between the carbon and oxygens to form covalent bonds, and then, the valence shells should be filled up. It is important to note that there aren't enough electrons to fill the octets. However, using two pairs of electrons to form double bonds fills the valence shell and yields the Lewis Dot structure below:

Lewis Diagram

$$\ddot{O} = C = \ddot{O}$$

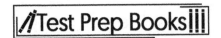

Resonance and Formal Charge

Resonance Structures
In some cases, double bonds can be formed between more than one set of atoms. These molecular structures that have the same atoms with different double bond placement are called **resonance structures**. When resonance structures exist, the Lewis structure is drawn as both configurations with a double-headed arrow in between them. The molecule doesn't switch between the configurations but exists as both within the structure and acts as an average of both.

Using Formal Charge to Determine the Best Lewis Structure
Calculating the formal charge of a Lewis structure can verify the stability of a molecular configuration. First, the number of valence electrons for each atom should be recalled and assigned to the atom. Then, the number of electrons around each atom should be counted and added to half of the atom's bonded electrons. The assigned number should be subtracted from the counted electrons. This should be done for all possible configurations, and the Lewis structure with the lowest formal charge is the one that is preferred.

Limitations of Lewis Structures
The Lewis structure does have some limitations. It cannot predict the geometry of the molecule, including bond angles and length. Additionally, Lewis structures are dependent on the octet rule of filling eight electrons in the valence electron shell and this does not work well when there is an odd number of valence electrons.

VSEPR and Bond Hybridization

The VSEPR Model
The **valence shell electron pair repulsion (VSEPR) theory** can be used to determine the molecular configuration of a structure and how the electrons are paired around the atoms within the molecule. The geometry of the molecule is determined by the valence electrons around the atoms and their tendency to repel each other, as stated in Coulomb's law. The structure will form such that the repulsive forces are minimized around the central atom of the molecule.

Using Lewis Diagrams and VSEPR Theory to Predict Properties of Molecules
Lewis diagrams and VSEPR models can be used together to determine many molecular properties of covalently-bonded molecules.

Molecular Geometry
Bonds and lone pairs of electrons repel each other, which determines the geometry of the molecule.

Bond Angles
The **bond angle** is also determined by these repelling forces. With two bonds, formed from three atoms, the bond angle will be linear at 180°. The central atom is sp hybridized with this bond angle. Four atoms make a trigonal planar molecule with 120° bond angles, and the central atom is sp^2 hybridized. With five atoms, a tetrahedral structure is formed with 109.5° bond angles, and the central atom is sp^3 hybridized.

Relative Bond Energies Based on Bond Order

The **bond order** of a molecule is the number of bonding pairs of electrons between two atoms. Single bonds have a bond order of one, double bonds have a bond order of two, and triple bonds have a bond order of three. The larger the bond order, the stronger the bond, and the greater the bond energy.

The strongest type of covalent bond is a **sigma bond**. It has the highest bond energy. All single bonds are sigma bonds. Double bonds have one sigma bond and one **pi bond**. Triple bonds have one sigma bond and two pi bonds. When a pi bond is present, the bond cannot be rotated, which creates **isomers** and **resonance structures**. In certain structures, such as with rings such as benzene, the electrons need to move to outer more electron shells to complete the pi bonding that exists within the ring. There are three double bonds in benzene. The atomic p-orbitals extend and spread out over the whole molecule to complete the pi bonding. The electrons become delocalized because they are not held between just two atoms anymore. This is another way to describe the resonance of Lewis structures. The pi bonds that create delocalized electrons can also help conduct electricity as with solid metal structures.

Relative Bond Lengths

The **bond length** is determined by the distance between the nuclei of the bonded atoms. The higher the bond order, the shorter the bond length, as the atoms are held closer together. The limiting factor of bond length is the atomic radius of each atom because they can only get as close as touching.

Presence of a Dipole Moment

The molecular geometry caused by the repulsion of electrons may cause a dipole moment in some molecules. For example, with water molecules, the hydrogen atoms each have a slight positive charge and the oxygen atom has a slight negative charge.

Hybridization

The electrons that are shared between the atoms that are bonded form a hybrid orbital between the two atoms. The atoms can also be referred to as **hybridized**.

Molecular Orbital Theory and Hybridization

Atomic orbitals explain the behavior of a single electron or pairs of electrons in an atom. They are regions of space in which the electrons are more likely to spend their time. Every orbital can contain two electrons, and the orbital is at its lowest energy when it has two electrons. One electron spins up, and one spins down. The standard atomic orbitals are known as *s*, *p*, *d*, and *f* orbitals.

- The simplest of the orbitals is the **s orbital**: the inner orbital of any atom or the outer orbital for light molecules, such as hydrogen and helium. The *s* orbital is spherical in shape and can contain two electrons.

- After the first and second *s* orbitals are filled, the **p orbitals** are filled. There are three *p* orbitals, one on each of the x, y, and z axes. Each *p* orbital can contain two electrons for a total of six electrons.

- After the *p* orbitals are filled, the **d orbitals** are filled next. There are five *d* orbitals, for a total of ten electrons.

- The **f orbitals** are next; there are seven *f* orbitals, which can contain a total of fourteen electrons.

Each shell/energy level has an increasing number of subshells available to it:

- The first shell only has the *1s* subshell, so it has two electrons.
- The second shell has the *2s* and *2p* subshells, so it has (2 + 6) eight electrons.
- The third shell has the *3s, 3p* and *3d* subshells, so it has (2 + 6 + 10) 18 electrons.
- The fourth shell has the *4s, 4p, 4d* and *4f* subshells, so it has (2 + 6 + 10 + 14) 32 electrons.

To find the maximum number of electrons per shell, the formula $2n^2$ is used, where n is the shell number. For example, elements in the third period have three subshells—space for up to 18 electrons—but will only have up to eight valence electrons. This is because the *3d* orbitals aren't filled (i.e., the elements from the third period don't completely fill their third shell).

Any element in the periodic table can be written in terms of its electron configuration. For instance, Calcium (Ca), which is in the 4th period on the periodic table and has an atomic number of 20, would be written as $1s^2 2s^2 2p^6 3s^2 3p^6 4s^2$. However, it's important to remember that the transition metals do not follow this rule because quantum energy level rules allow for some of their shells to remain unfilled. For example, the transition metal scandium (Sc), which has an atomic number of 21, has the electron configuration $1s^2 2s^2 2p^6 3s^2 3p^6 3d^1 4s^2$, and the *d* subshell is not filled.

In **molecular orbital theory**, the assumption is that bonding, non-bonding, and anti-bonding orbitals—which have different energies—are formed when atoms are brought together. For N atomic orbitals in a molecule, the assumed result would be N molecular orbitals, which can be described by wave functions.

For example, for a molecule that has two atomic orbitals, two molecular orbitals must be formed: one bonding and one anti-bonding. The molecular orbitals would be separated by a certain energy. A molecule that has three atomic orbitals would form one bonding, one non-bonding, and one anti-bonding molecular orbital. A molecule that has ten atoms would form five anti-bonding and five bonding molecular orbitals.

Sigma bonds (σ bonds) are formed by direct overlapping of atomic orbitals, and they are the strongest type of covalent bond. Sigma bonds are symmetrical around the bond axis. Common sigma bonds—where z is the axis of the bond—are s+s, $p_z + p_z$, s+p_z, and $d_z^2 + d_z^2$.

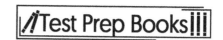

In contrast, **Pi bonds (π bonds)** are usually weaker than sigma bonds and are a type of covalent bond where two ends of one p-orbital overlap the two ends of another p-orbital. D-orbitals can also form pi bonds.

Generally, single bonds are sigma bonds, and multiple bonds consist of one sigma bond plus one pi bond. A **double bond** is one sigma bond plus one pi bond. A **triple bond** consists of one sigma bond and two pi bonds. For example, ethylene has **delta bonds (δ bonds)** that are formed from four ends of one atomic orbital overlapping with four ends of another atomic orbital.

Sometimes, the s, p, d, and f orbitals do not fully explain where an electron will be at any given time. This is where hybrid orbitals come in. **Hybrid orbitals** are combinations of the standard atomic orbitals. For instance, if the s and p orbitals hybridize, instead of being two different kinds of orbitals, they become four identical orbitals. When the s orbital hybridizes with all three p orbitals, it's called sp^3 hybridization, and it forms a tetrahedral shape. This is the type of hybridization that occurs in H_2O. The description used depends on the properties of the compound, including its numbers of lone pairs of electrons.

The Quantum Mechanical (QM) Model

The **quantum mechanical (QM) model** was consistent with the Bohr model and showed that when electrons transition from one orbital to another, the amount of energy that is gained or lost is equal to $\Delta E = nhv \ (n = 1, 2, 3 \dots)$. In addition, the QM model associates each orbital or wave function with a specific energy level, much like how the Bohr model had discrete energy levels associated with each orbital. In contrast, the **Bohr**, or **classical shell model**, described the electron only as a particle. The model was problematic because it required the electron to be fixed to an orbit but couldn't explain why electrons could not transition to other orbitals, e.g., n = 0.5, 1.5, etc.

The QM model resolved this problem by treating the electron not only as a particle but as a wave described by a wave function. The reason an electron is not experimentally observable when transitioning between orbitals or why orbitals with n = 0.5 were not allowed was due to destructive wave inference. The solution to the wave equation provides information on atomic electron structure, such as the number of electrons confined to the different electron subshells and the associated shapes and energy of the different orbitals for multiple-electron atoms. The QM model also explains periodic trends, such as the increasing ionization energy across a row and the stability of the noble gas configurations. For example, the QM model predicts that neon and argon have an energetically stable configuration of subshells that are entirely filled with electrons, thereby making these noble gases unreactive.

Intermolecular Forces and Properties

Intermolecular Forces

Intermolecular forces are the forces of attraction or repulsion between molecules. They are responsible for various properties that are exhibited by some materials, including surface tension, friction, and viscosity. They can be split into two groups:

- **Short-range forces:** when the centers of molecules are separated by three angstroms or less and tend to be repulsive

- **Long-range forces:** also known as van der Waals forces, when the centers of molecules are separated by more than three angstroms and tend to be attractive

The **covalent radius (rcov)** is the length of one half of the bond length when two atoms of the same kind are bonded through a single bond in a neutral molecule. The sum of two covalent radii from atoms that are covalently bonded should, theoretically, be equal to the covalent bond length:

$$rcov(AB) = r(A) + r(B)$$

The van der Waals radius, rv, can be defined as half the distance between the nuclei of two non-bonded atoms of the same element when they are as close as possible to each other without being in the same molecule or being covalently bonded. The covalent radii changes depending on the environment that an atom is in and whether it is single, double, or triple bonded, but average values exist for use in calculations. The van der Waals radii also changes based on the intermolecular forces present, but average values also exist. This information is useful for determining how closely molecules can pack into a solid.

London Dispersion Forces

London dispersion forces are formed only between induced dipoles. These electrostatic forces are also described as **Coulombic interactions**, as **Coulomb's law** states that like charges repel each other and opposite charges attract each other, and it describes how strong the force will be. Although individually, London dispersion forces are very weak, the sum of their forces within a molecule can bring a strong net force to the molecule.

Dispersion Forces and Polarizability

Greater numbers of electrons and larger surface areas increase the number of London dispersion forces that can form from a molecule. As the molecule becomes more polarized and can have greater attraction to nearby atoms and molecules, melting and boiling points increase for the molecule as well. London dispersion forces are also stronger with double and triple pi bonds in a molecule.

Recall that in a chemical bond, the polarity is the result of unsymmetrical electron distribution between two atoms. For example, the C-C bond is a nonpolar covalent bond because the two electrons are shared evenly between each atom. In contrast, an ionic bond such as sodium chloride (NaCl) consists of a metal atom (Na+) and a nonmetal atom (Cl-). The chlorine atom removes an electron from the sodium atom and bears a negative charge. There is no sharing of electrons between Na and Cl, and the ionic bond is held together by electrostatic attractions. The image below shows the continuum of bond polarity.

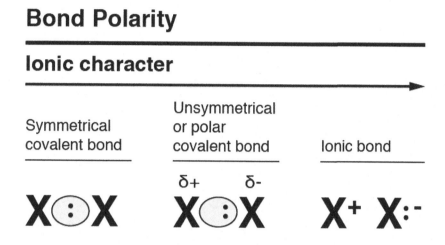

London Dispersion Forces vs. Van der Waals Forces
The term "London dispersion forces" is not synonymous with "Van der Waal's forces," because Van der Waal's forces is a broader term referring to intermolecular attraction. London dispersion forces are only one of the two kinds of Van der Waals forces, the other being dipole-dipole interactions.

Intermolecular Attraction with Dipole Moments
There are several different types of dipole forces that can occur between molecules. **Dipole moments** occur when two molecules are attracted to each other by Coulombic interactions when they are in proximity of one another. Some molecules have permanent dipoles due to the differences in electronegativity of the atoms that they are composed of. These molecules have polar bonds and are called **polar molecules**. **Nonpolar molecules** are neutrally charged on all sides and the electronegativity of their atoms cancel out. **Dipole-dipole interactions** occur between molecules that have permanent dipoles. As they near each other, they arrange themselves so that the negative and positive ends of the molecules interact. That way, the interaction of the molecules is maximized, and the repulsion is minimized. These interactions can be easily visualized through particulate drawings, where the positive and negative ends attract each other for both liquid and solid substances. In liquids, the molecules are constantly moving around, forming these interactions, and then breaking apart and forming them with other molecules. In solids, the molecules will be arranged so that the positive-negative charge interactions line up within the crystalline structure.

Dipole-dipole interactions are stronger than London dispersion forces, but are still considered weak intermolecular forces. They are weaker than both ionic forces and covalent bonds. Polar and nonpolar molecules of similar size have London dispersion forces of equal strength, but the attraction between two polar molecules is greater than that between the nonpolar molecules because of the added dipole-dipole interactions.

Dipole-induced dipole interactions occur between polar and nonpolar molecules. The polar molecule has a permanent dipole, whereas the nonpolar molecule forms an induced dipole as the polar molecule approaches. The stronger the electronegativity of the polar molecule, the greater the strength of the permanent dipole. Molecules with larger dipoles induce larger dipoles in nonpolar molecules and make stronger attractions between them. Dipole-induced dipole interactions are the same type of Coulombic interaction that occurs between two molecules with permanent dipoles.

Temporary, Instantaneous Dipoles

Electrons occupy orbitals that surround the nucleus of an atom. In some molecules, the electrons are concentrated on one side of the molecule, making the molecule polarized. For example, water molecules are polarized because the hydrogen atoms bond to the oxygen molecule to form a V-shape; the hydrogen atoms maintain a slightly positive charge and the oxygen molecule maintains a slightly negative charge. A **dipole** is created in a molecule when the electron distribution becomes distorted, and positively- and negatively-charged areas become concentrated on opposite ends of a molecule, separated by a distance. In nonpolar molecules, the electron distribution can become distorted instantaneously and temporarily.

The electrons move around within their covalent bonds and form induced dipoles. Movement of the electrons within an atom or molecule is dependent on how far the electrons are from the nucleus. Electrons that are farther away move easier than those that are closer to the nucleus. When two molecules near each other, **induced dipoles** are formed as the electrons repel each other and move to opposite sides of the molecule. The positively- and negatively-charged ends then attract each other, as well as other polarized atoms and molecules, and form London dispersion forces between them, which are the weakest of intermolecular forces.

Intermolecular Forces in Polar vs. Nonpolar Molecules

Molecules may contain chemical bonds with atoms similar to one another or with different atoms, which can result in different charge distributions within the molecule resulting in different types of intermolecular forces. Gas molecules such as HCl will have a polarized positive charge on hydrogen and a polarized negative charge on chlorine. The positive end on the molecule will align with the negative end on another HCl molecule forming a **dipole-dipole interaction**. The polar water molecule can induce a dipole on a nonpolar molecule or atom (e.g., CH_3CH_3 or Xe), whereby the negatively polarized oxygen atom in water will polarize another nonpolar molecule or atom, resulting in a **dipole-induced dipole**. **London dispersion** or **induced dipole-induced dipole** forces occur between nonpolar molecules or atoms and are due to the motion of the electrons, e.g., between two He atoms. The electrons on one helium atom can instantaneously align to the left side creating a polarized negative charge on the right

end, which will align with the polarized positive end of the another He atom, as shown in the image below.

Types of Intermolecular Forces

Dipole-induced Dipole

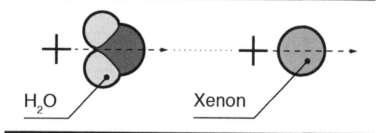

Dispersion Forces

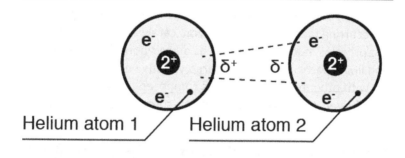

Hydrogen Bonding

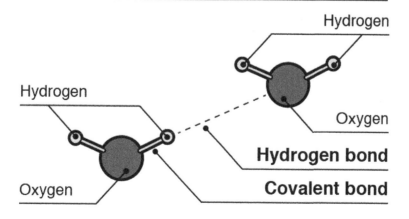

<u>Factors that Increase Dispersion Forces</u>
Polar substances will also have London dispersion forces, which are relatively weaker compared to dipole-dipole interaction forces. However, dispersion forces can become larger as the size of the molecule increases, due to the number of electrons. In hydrocarbons, dispersion forces are greater for

$CH_3(CH_2)_3CH_3$ compared to CH_3CH_3, and higher for a linear five carbon chain compared to a branched five carbon compound. The bulky shape of branched hydrocarbons makes it difficult to form more dispersion interactions.

Hydrogen Bonding

Hydrogen bonds are another type of dipole force. They occur when hydrogen atoms bond to elements in other molecules with high electronegativity and form large dipole-dipole interactions. Hydrogen bonds are the strongest type of the weak intermolecular forces. When hydrogen forms a covalent bond within a molecule, it moves its electron to one side of the nucleus, leaving the positively charged nucleus exposed. The negatively charged side of polar molecules or negatively charged ions can come very close to the hydrogen atom and form a strong Coulombic interaction. It attracts elements—such as fluorine, oxygen, and nitrogen—that are more negatively charged than its own positive charge, and creates exaggerated interactions, so that even small molecules have large interactions. Hydrogen bonds can form between separate molecules as well as different parts of the same molecule. Because hydrogen bonds are exaggerated in strength, they increase the boiling point and viscosity of substances in which they form.

Noncovalent Interactions in Large Biomolecules

Noncovalent interactions refer to intermolecular forces such as hydrogen bonding, dipole-dipole, and London dispersion forces. These intermolecular forces play a vital role in many biological and physiological processes occurring that range from enzyme catalysis to protein folding. Biomolecules such as high molecular weight proteins, nucleic acids such as deoxyribonucleic acids (DNA), large carbohydrate chains, and lipids (fatty acids) will undergo various types of noncovalent interactions between one another or with other molecules. DNA is composed of two strands that coil up like a helix and is held together by hydrogen bonds between nitrogen-containing compounds as illustrated below.

Hydrogen Bonding Between Nitrogen Bases Guanine (G) and Cytosine (C) and the Base Pairs Adenine (A) and Thymine (T)

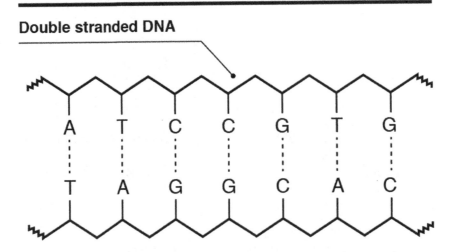

Double stranded DNA

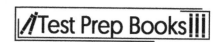

Under certain biological conditions, e.g., a temperature or pH change, the hydrogen bonds will come apart, causing the helical structure to unravel.

Properties of Solids

Intermolecular Forces Affect Properties of Liquids and Solids

The **intermolecular forces** present between molecules in a substance influence the properties of the substance, such as the boiling point, surface tension, capillary action, and vapor pressure. When the intermolecular forces are strong between molecules, the boiling point of the substance increases because it takes more heat energy to break apart the bonds between them.

Boiling Point

The **boiling point** of a substance is the temperature at which the **vapor pressure**, which is the pressure of a gas that is in contact with the liquid or solid form of the same substance, is equivalent to the atmospheric pressure. Substances that have high boiling points have low vapor pressures. Substances with weak intermolecular forces have high vapor pressures and low boiling points.

Surface Tension

Surface tension is determined by the forces that liquid molecules are subjected to at the air-liquid interface. Normally, liquid molecules have forces on them from all three dimensions around them, which cancel each other out and leave no net force on the molecules. The molecules at the surface, however, do not have molecules above them to exert a force, and so they are subjected to a net force that pulls them down into the body of the liquid. An internal pressure is created at the surface, and stronger intermolecular forces are formed. The surface, then, has a stronger interface that can resist external pressure.

Capillary Action

Capillary action occurs by a combination of cohesion and adhesion. The liquid molecules are attracted to each other by cohesion and to their container by adhesion. The combination of the two forces allows liquid molecules to defy gravity and be drawn up the sides of the container. Substances that have similar intermolecular forces between their particles are generally soluble or miscible with each other. The same amount of energy is needed to break the bonds of the original molecules, so this can occur at the same time. The ions and elements can then interact with each other to form new bonds.

Vapor Pressure

The **vapor pressure** of a liquid is defined as the partial pressure of the vapor over the liquid measured at a specific temperature at equilibrium. As temperature increases, molecules move faster and the vapor pressure increases.

Particulate Representations

Particulate representations are visual representations of different molecules and how they change within a chemical reaction. They can be useful for looking at the relative amounts of each element in a reaction and the stoichiometry of the equation. They can also illustrate the relationship of reactants and products at equilibrium. The following image shows particulate representations for the chemical reaction of sulfur dioxide and oxygen gas to form sulfur trioxide, $2SO_2 + O_2 \rightarrow 2SO_3$.

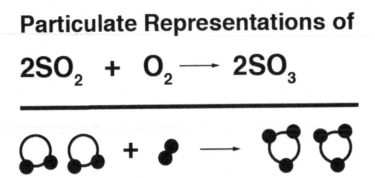

Particulate Representations of
$$2SO_2 + O_2 \longrightarrow 2SO_3$$

The particulate representation makes it easy to see how the elements would have to dissociate and then bond again in a new formation. It also demonstrates that for every two sulfur dioxide molecules, only one oxygen molecule is needed for the reaction to proceed. These diagrams can also show the mixture of reactants and products for reactions that reach equilibrium without complete dissociation of the reactants, when K does not have a large magnitude.

A Particulate Representation of Incomplete Dissociation at Equilibrium

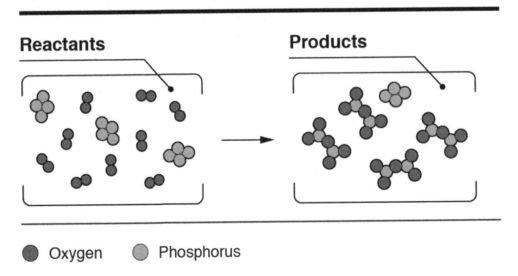

Reactants   Products

● Oxygen   ● Phosphorus

Ionic Solids

Ionic bonds form between two atoms when one atom completely transfers its valence electrons to the other atom, forming two oppositely charged ions. One becomes a **cation**, which has a positive charge, and the other an **anion**, which has a negative charge. This is unlike covalent bonds in which the electrons are shared between the atoms. In an **ionic crystal**, the atoms bond together to form a three-dimensional lattice structure that maximizes the attraction between the ions that are generated and minimizes that repulsion between them. Coulombic attractions bond the cations and anions to each other because they have opposite charges that attract each other. Ionic bonds often form between metals and nonmetals. The metal transfers its valence electrons and becomes a cation, and the nonmetal gains more valence electrons and becomes an anion. Ions are most stable when their outermost electron shell is full. Transferring valence electrons helps ions to achieve a stable noble gas electron configuration. Although the ionic crystal is formed by ions, it will be electrically neutral by the combination of cations and anions.

The strength of an ionic bond is dependent on the charge of the ions and the distance between the ions. The greater the charge of the ion, the stronger the interaction between the anion and cation. This makes the bond stronger. Additionally, when the ions are small in size and can get closer to each other, the bond they form can be stronger.

Ionic compounds are held together in a specific, endlessly-repeating pattern, known as a **lattice**. The structure of the lattice depends on the ions present in the compound. The size of the lattice depends on how many single molecules are present. A single crystal of table salt, for example, is formed by a lattice of trillions of sodium chloride (NaCl) molecules.

Sodium and chloride ions alternate with each other in each of the x, y, and z dimensions to form a cube-like structure:

Sodium Chloride Lattice

6:6 Coordinated

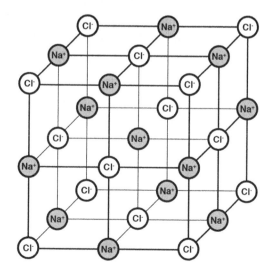

The sodium ion in the middle is joined to six chloride ions. If the chloride ion was centered, it would also be joined to six sodium ions. Therefore, sodium chloride is described as **6:6 coordinated**.

The type of lattice depends on how many attractions are possible between the ions. One sodium ion can only accept six chloride ions before the chloride ions start to repel each other from being too close to one another. If the chloride ions were to repel each other, the entire crystal structure would become unstable.

Different types of lattices are possible depending on the ions present. The compound zincblende (ZnS), for example, has a coordination of four for each ion, so it is said to be **4:4 coordinated.**

Lattice Energies
When two ions with opposite charges are attracted to each other and bond together, energy is released. This is called the **lattice energy**. Single ions need less energy to stay together in a crystal lattice.

The two main factors affecting lattice energy are the charges on the ions and the ionic radii. As the charges on ions increase in magnitude, the bonds become stronger. For example, a bond between a -2 anion and a +3 cation would be much stronger than a bond between a -1 anion and a +1 cation.

Lattice energy (U) is always a positive number and can be calculated as follows:

$$U = k' \frac{(q_1)(q_2)}{r_o}$$

The constant k' depends on the type of lattice and valence electron configurations; $(q_1)\ and\ (q_2)$ are the charges on the ions, and r_o is the internuclear distance.

It is also common to see this equation represented as $\Delta H = (C \times Z^+ \times Z^-)/R_0$, where C is a constant that depends on the type of ionic crystal, Z^+ is the charge of the cation, Z^- is the charge of the anion, and R_0 is the distance between the nuclei.

Two ionic compounds may have the exact same lattice arrangement but have different lattice energies. For example, sodium chloride (NaCl) and magnesium oxide (MgO) have the same crystal structure but different lattice energies. In magnesium oxide, +2 ions are bonded to -2 ions, but in sodium chloride, +1 ions are being attracted to -1 ions. A salt with a metal cation with a +2 charge and a nonmetal anion with a -2 charge will have a lattice energy four times greater than a salt with single charges, if the ions are of similar sizes.

The smaller the distance between the ions, the larger the magnitude of the lattice energy. As the charge of the ions increases, the magnitude of the lattice energy also increases.

Ionic Radii
Ionic radii can be defined as the radius of an atom's ion. When ions are small, the valence shell is closer to the nucleus, and the nucleus has a stronger pull. The nucleus can also affect nearby atoms, having a similar pull. This results in smaller ions having stronger bonds.

Radius/Ratio Effects
The **radius ratio** of any given pair of ions is defined as the ionic radius of the smaller ion divided by the ionic radius of the larger ion. Often, the smaller ion is the cation, and the larger ion is the anion. The lattice structure for crystals of simple 1:1 compounds (NaCl) depends on the radius ratio of the ions that

are present. If the radius of the positive ion is greater than 73% of the negative ion, an 8:8 coordination is possible. If the radius of the positive ion is between 41 – 73% of the negative ion, a 6:6 coordination is possible.

2D and 3D Covalent Network Solids

Covalent network solids form from nonmetal atoms that have strong covalent bonds between them. They can form a two- or three-dimensional network. Diamond, graphite, silicon dioxide, and silicon carbon are all examples of structures that are covalent network solids. The elements found in group 14 of the periodic table can all form four covalent bonds and are good candidates for creating covalent network solids. With covalent bonding, the bond angles are fixed between the atoms, so the solid structure that forms is rigid and firm. The bonds are strong, so the melting point is high as well. Diamonds have a tetrahedral structure that is formed from carbon atoms each bonding to and sharing electrons with four other carbon atoms. A giant covalent structure is formed that is very strong and hard. The melting point of diamond is very high because of the strong bonds between the carbon atoms. These strong bonds also make diamond insoluble to water and organic solvents. The properties of a covalent network solid depend on the atoms that comprise it and how they are bonded together.

Graphite

Graphite is covalent network solid that is made up of two-dimensional sheets that stack on each other. It is composed of carbon atoms that bond to only three other carbon atoms, leaving one free valence electron for each carbon atom. The free valence electron becomes delocalized and moves freely throughout the graphite sheet. The carbon atoms within the sheet are held together through strong covalent bonds, giving it a high melting point. Sheets of graphite are held together by London dispersion forces. The delocalized electrons form opposite dipoles on each side of the sheet that allow them to be bonded together but also to slide easily past each other. The sea of electrons that is created allows electricity to be conducted within the single layers but not between the different layers. Graphite is soft and can be used as a lubricant.

Silicon

Silicon is a covalent network solid that forms a three-dimensional structure that is similar to that of a diamond. It is a **semiconductor**, which means that is has properties that are between being a good conductor and a good insulator but is neither completely. The conductivity of silicon increases with temperature due to its semiconductive nature. The energy gap between the valence band and conduction band of electrons is small in silicon, so with the addition of heat energy, the electrons from the valence band can jump to the conduction band and increase electrical conductivity of the solid.

Silicon can also be doped to turn it into a conductor, which means that a small amount of another element is added to the silicon. Elements in group 15 of the periodic table have one more valence electron than silicon and create an **n-type**, or negative, semiconductor. The extra electrons help move the charge through the material. Elements in group 13 of the periodic table have one less valence electron than silicon and create a **p-type**, or positive, semiconductor. A hole is created in the material where there is an absence of an electron. The hole can then accept an electron and move the current through the material. Junctions of n-type and p-type semiconductors restrict the flow of current to only one direction, which is an important part of electricity flow in electronic devices.

Molecular Solids

Molecular solids are made up of molecules that are formed from nonmetals. The molecules are held together by London dispersion forces, which are weak intermolecular forces. The individual molecules that comprise molecular solids are made from nonmetal atoms that are covalently bonded together. The intramolecular forces are much stronger than the intermolecular forces. Since the molecular solids are held together by weak forces of attraction, they have a low melting point and tend to be soft and not rigid. They also often sublimate directly into the gas phase from the solid phase because of their high volatility. Molecular solids have low densities because the molecules are not packed together tightly and are joined by long intermolecular bonds.

Elements that can easily form dipoles, such as those with electrons that are less tightly bound to the nucleus, have a greater tendency to form London dispersion forces and create molecular solids. Molecular solids are poor electrical conductors because the electrons are shared tightly in covalent bonds and are not delocalized. There is also a large gap between the valence band and conduction band, so it is very hard for any electrons to jump from the valence band to the conduction band. Solid carbon dioxide is an example of a molecular solid that has many commercial applications and is also an example of a sublimating molecular solid. Sucrose is another molecular solid that is a polymer with vital biological applications regarding energy storage.

Metallic Solids
Metallic solids consist of positively-charged metal ions surrounded by a sea of valence electrons that can move around freely. The metal cations are packed together closely in a lattice structure and can be called **positive kernels**.

Because the valence electrons are delocalized from their original atom, metallic solids are good conductors of electricity. The electrons can move throughout the solid rather than being strictly "assigned" to a particular atom, which creates an electric current.

The characteristic of freely-moving electrons also allows metallic solids to be malleable and ductile. The structure of the solid does not change much with deformation because the electrons move freely around the positive cores to keep the repulsive forces to a minimum. The environment around each positive kernel can remain the same even when the shape of the solid changes. Metallic solids are also good conductors of heat. The positively-charged ions are packed closely together, and heat energy can be passed easily between the positive kernels.

Coulombic Forces in Biological Systems
Many biological interactions are determined by the **Coulombic interactions** that occur between molecules. For example, DNA has specific base pairing between adenine and thymine and guanine and cytosine because of the hydrogen bonds that are formed between each pair. Enzymes interact only with specific substrates because of the intermolecular forces that are formed between them. If the interactions are too weak, the resulting molecule will not form properly, and, therefore, not function properly. Proteins form three-dimensional structures based on the hydrophilic and hydrophobic regions of their elements. When interacting with aqueous solutions, the protein can orient itself so that the hydrophilic parts can interact with the water molecules.

Structure Influences Function in Molecules

The enzymes found within our body are efficient biomolecules that catalyze chemical reactions, converting a reactant to a product, and are primarily dependent on noncovalent interactions. An enzyme will contain a small pocket called the **active site** where a small reactant molecule or substrate can dock. The substrate can bind to the inner pocket of the enzyme through hydrogen bonding, dipole-dipole forces, or even London dispersion forces. The active site of the protein is specific to the substrate and may change shape to accommodate the small molecule, forming an enzyme-substrate complex. The intermolecular forces between the enzyme and substrate will stabilize and modify the transition state structure, thereby lowering the reaction barrier and speeding up the rate at which the product is produced. The enzyme DNA polymerase adds a nucleotide to an available strand (as shown in the image below). The amino acid residues found within the polymerase can stabilize the phosphate groups on the nucleotide through hydrogen bonds thereby making the new strand longer. Eventually, the strand containing the new nucleotides will coil and form a double helix structure.

DNA Polymerase Directs One Strand to Its Active Site Where a New Nucleotide Is Added

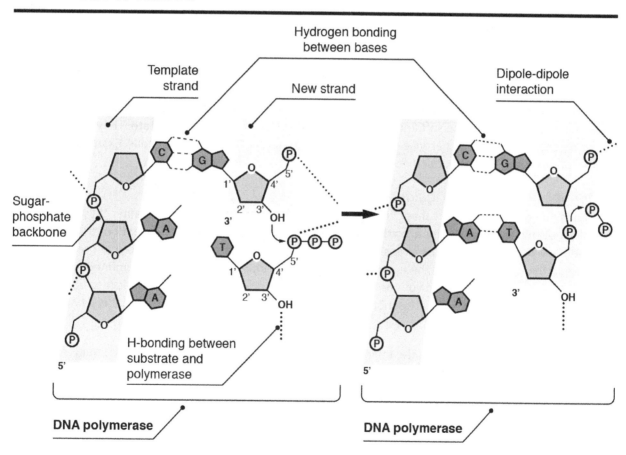

Solids, Liquids, and Gases

Crystalline and Amorphous Solids

The different states of matter—solid, liquid, and gas—each have different properties. In **solids**, the atoms and molecules are packed together closely and cannot move around freely. They each have a fixed position and may only vibrate around that position. Because of this, solids have a fixed three-dimensional shape and volume. The solid is formed from the bonds that link the particles that make up the solid. The hardness of the solid is based on how strong or weak the attraction is between the particles.

There are two subcategories of solids: crystalline solids and amorphous solids. **Crystalline solids** are made up of repeating groups of atoms, ions, or molecules in an ordered structure with a symmetrical pattern. The smallest group of the repeating structure is called a **unit cell**. The pattern of unit cells is repeated throughout the crystalline solid, forming a network, called a **crystal lattice**. Crystalline solids cannot be compressed into a smaller volume. The strength of the bonds between each unit cell is the same, so they have an exact melting point. The bonds between each unit cell will break simultaneously at the same temperature. The symmetry and repeating characteristic of the unit cells also causes cleavage in crystalline solids, which means that when they break, they break along a plane with straight, smooth surfaces. Examples of crystalline solids are table salt, NaCl, sugar, and diamonds.

Unlike crystalline solids, **amorphous solids** are not composed of repeating unit cells and do not have a symmetrical crystal lattice. They are made up of different atoms, ions, and molecules bonded together. Because the particles may be different, amorphous solids do not have a distinct melting point. The bonds between the particles break at different temperatures and, therefore, at different times. These types of solids first become soft and pliable before they transition to a liquid state. They do not exhibit cleavage when they break because of their lack of symmetrical repeating unit cells. Examples of amorphous solids are rubber and gels.

Motions of Particles in Liquids

Liquids do not have a distinct shape; instead, they take on the shape of the container they are in. The molecules that make up liquids are attracted to each other and form brief transient bonds when they collide but also break apart just as easily as the molecules move around. Liquids have more space between them than solids, so the molecules can move around more and are not confined to one fixed position. They still have a fixed volume and cannot be compressed. Liquids are subject to both cohesion and adhesion. **Cohesion** is the attraction between the liquid molecules themselves. It is what causes surface tension.

The stronger the bond between the molecules, the greater the surface tension of the liquid. **Adhesion** is the attraction between the liquid molecules and the container the liquid is in. The liquid molecules may appear to crawl up the side of the container higher than the surface of the liquid due to adhesion properties. Adhesion also causes **capillary action**, which is when the liquid molecules have a stronger attraction to the container than to the other liquid molecules. The **viscosity** of a liquid refers to how well the liquid moves under stress and is also determined by intermolecular attraction. Highly viscous liquids resist moving under stress, such as pouring, due to the strong bonds between the molecules. Low viscosity liquids move freely under stress due to weak bonding.

Volume of Solids vs. Liquids

A solid has a rigid, or set, form and occupies a fixed shape and volume. Solids generally maintain their shape when exposed to outside forces.

Liquids and gases are considered fluids, which have no set shape. Liquids are fluid, yet are distinguished from gases by their incompressibility (incapable of being compressed) and set volume. Like solids, they do have a set volume. Liquids can be transferred from one container to another, but cannot be forced to fill containers of different volumes via compression without causing damage to the container. For example, if one attempts to force a given volume or number of particles of a liquid, such as water, into a fixed container, such as a small water bottle, the container would likely explode from the extra water.

Properties of Gases

In gases, the particles have a large separation and no attractive forces. The constant movement of particles causes them to bump into each other, thus allowing the particles to transfer energy among one another. This bumping and transferring of energy helps explain the transfer of heat and the relationship between pressure, volume, and temperature. The constant motion of molecules in the gaseous phase makes it so that gases lack a definitive volume and shape.

Heating Curves

Each state of matter is considered to be a phase, and changes between phases are represented by **phase diagrams**. These diagrams show the effects of changes in pressure and temperature on matter. The states of matter fall into areas on these charts called **heating curves.**

Heating and cooling curves portray the changes between states of matter for pure substances and when they occur according to temperature changes. The heat energy needed to cause changes between solids and liquids and liquids and gases are delineated through these graphs.

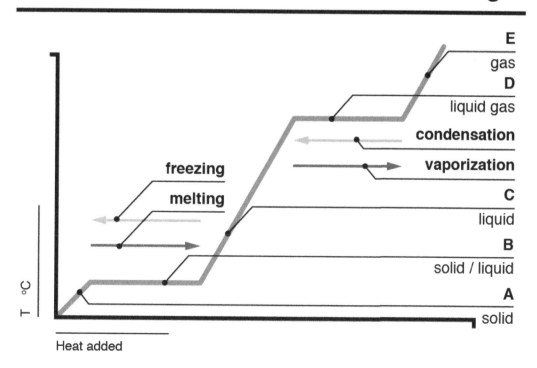

Heat Energy Needed to Cause Phase Changes

Ideal Gas Law

Ideal Gases

The **Ideal Gas Law** states that pressure, volume, and temperature are all related through the equation: $PV = nRT$, where P is pressure, V is volume, n is the amount of the substance in moles, R is the gas constant, and T is temperature.

Through this relationship, volume and pressure are both proportional to temperature, but pressure is inversely proportional to volume. Therefore, if the equation is balanced, and the volume decreases in the system, pressure needs to proportionately increase to keep both sides of the equation balanced. In contrast, if the equation is unbalanced and the pressure increases, then the temperature would also increase, since pressure and temperature are directly proportional.

A whole field of study on particle movement in gas laws has evolved over time to create the **ideal gas law:**

$$PV = nRT$$

Scientists' contributions on advancements in ideal gas laws are referenced in this chart:

Charles' Law	Direct: As temperature increases, volume increases (Pressure and Number of moles held constant)	$\dfrac{V_1}{T_1} = \dfrac{V_2}{T_2}$
Gay Lussac's Law	Direct: As temperature increases, pressure increases (Volume and Number of moles held constant)	$\dfrac{P_1}{T_1} = \dfrac{P_2}{T_2}$
Boyle's Law	Inverse: As pressure increases, volume decreases (Temperature and Number of moles held constant)	$P_1V_1 = P_2V_2$
Avogadro's Law	Direct: As the Number of moles of gas increase, gas volume increases (Pressure and Temperature held constant)	$\dfrac{V_1}{n_1} = \dfrac{V_2}{n_2}$
Combined Law	Combination of Boyle's and Charles' Law	$\dfrac{P_1V_1}{T_1} = \dfrac{P_2V_2}{T_2}$
Joule's Law	The internal energy of a gas depends only on temperature. At a constant temperature, the internal energy is equal to zero.	$\dfrac{U}{T}$

Gas questions can also require stoichiometry, where Avogadro's constants will be useful:

1 mole of a gas at standard temperature and pressure (STP) = 22.4 L (STP is 1 atm and 0°C)

1 mole of any substance = 6.022×10^{23} particles

Example problem:

Given the combustion equation $2C_3H_6 + 9O_2 \rightarrow 6CO_2 + 6H_2O$, how many liters of CO_2 are produced if 173 g of oxygen are involved and the reaction happens at STP? The molar mass of O_2 is 32 g/1 mole

173g O_2	1 mol O_2	6 mol CO_2	22.4 L CO_2	
	32g O_2	9 mol O_2	1 mol CO_2	= 80.7 L

The properties of **gases** vary significantly from both solids and liquids. They do not have a definitive volume or shape, and they are compressible. Gas molecules are constantly moving and spread out as much as they can. Therefore, they fill the entire volume of the container they are in. As the particles move around and collide with the walls of their container, they exert a measurable pressure on the container. Unlike solids and liquids, the gas molecules themselves are not attracted to each other and move independently, continuously, and rapidly in random order. Laws about the behavior of gases are often based on the idea of an **ideal gas**. The molecules of ideal gases have perfectly elastic collisions with each other, where no energy is lost. One mole of an ideal gas has a volume of 22.7 liters at standard temperature and pressure (STP) of 273 K and 1 atm, respectively.

Partial Pressures

When different ideal gases are mixed together, the pressure exerted by each gas is independent of the other gases. **Dalton's law** states that the total pressure of a mixture of gases is equal to the sum of the partial pressures of each individual gas and can be calculated with the equation $P_{Total} = P_A + P_B + P_C$. The ideal gas equation, $PV = nRT$, can be used to mathematically relate the pressure (P), volume (V), number of moles (n), and temperature (T) of ideal gases to each other. R is the ideal gas constant, which is equal to 8.314 J mol^{-1} K^{-1}. The pressure and volume of a gas are inversely proportional. When the pressure of a gas increases, the volume decreases and vice versa.

Therefore, according the **Boyle's law**, the left side of the equation, PV, for a gas is always constant. If one property changes, for example, the pressure increases, the volume would decrease and still be equivalent to the product of the original pressure and volume. **Charles's law** states that the relationship between volume and temperature of a gas are directly proportional. When volume is divided by the temperature in Kelvin, the quotient is constant for each gas. When temperature increases, so does the volume of the gas and vice versa. **Avogadro's law** describes the relationship between moles of a gas and the volume of the gas as being directly proportional. As the number of moles increases, the volume also increases. The relationship is constant, similar to that of volume and temperature, so the changes that occur between moles and volume are of the same proportionate magnitude.

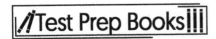

Using Graphical Representations of P, V, and T to Describe Gas Behavior

The relationships between pressure, volume, and temperature in a gas are easily visualized graphically. **Absolute zero** on the Celsius scale occurs when the pressure of a gas equals zero. When pressure versus temperature is graphed, this occurs at the *x*-intercept.

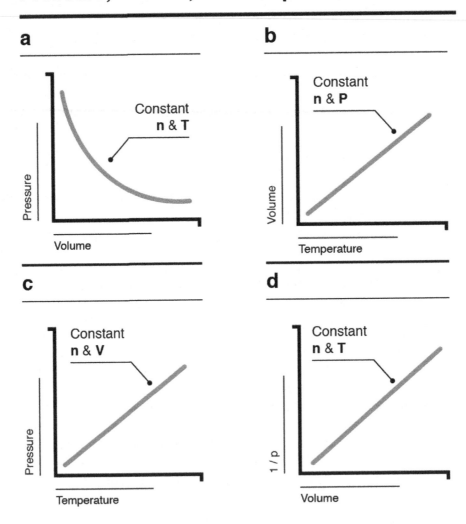

Visualizing the Relationships Between Pressure, Volume, and Temperature in a Gas

Kinetic Molecular Theory

Kinetic Molecular Theory and the Maxwell- Boltzmann Distribution

According to **kinetic molecular theory**:

- Gas particles have **Brownian,** or random, motion (consistent with the second law of thermodynamics)
- Particles travel in a straight line until they hit another object and change direction
- Volume of a gas particle is virtually zero

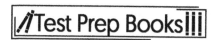

- There are no attractive or repulsive forces between gas particles, and there are no unaccounted chemical or physical interactions between particles
- There are elastic collisions of the gas with its container and other particles: particles don't slow down and are constantly moving
- Movement, or kinetic energy, is a function of temperature. As temperature increases, kinetic energy increases. Average kinetic energy is equal to temperature in **Kelvin**.

The Kelvin Temperature of a Sample

The average kinetic energy of a gaseous molecule or particle is:

$$Average\ kinetic\ energy = \frac{3}{2}RT$$

Thus, the temperature is proportional to the energy of motion for molecules. The gas constant, denoted R, is equal to 8.314 Jmol^{-1} K^{-1}. The temperature, T, is given in Kelvin (K), which is a measure of the absolute thermodynamic temperature with zero being the null point where all molecular motion stops within the classical description. A nonzero temperature implies that the particles or molecules are always moving. Increasing the temperature or doubling it will increase or double the average kinetic energy of the molecules. A temperature of zero Kelvin would indicate that there is no molecular motion. However, in the modern or quantum mechanical description, every substance (e.g., particle, molecule, etc.) has zero-point energy, meaning that substance will still vibrate or move at 0 K. The image below shows the average kinetic energy as a function of temperature.

The Average Kinetic Energy as a Function of Temperature

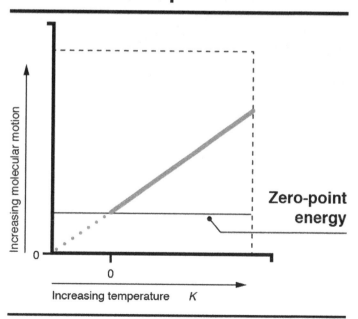

All particles are in motion even at zero Kelvin

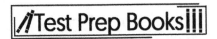

The Maxwell-Boltzmann Distribution and Temperature

According to the Maxwell-Boltzmann distribution, as the temperature of the chemical system increases, there is a greater percentage, or fraction, of molecules that collide such that the energy of that system exceeds the energy needed for a chemical reaction or phase change. The image below shows how the probability distribution of molecules with some kinetic energy widens, or disperses at increased temperatures.

Two Different Maxwell-Boltzmann Distributions at a Temperature of 300 and 500 K

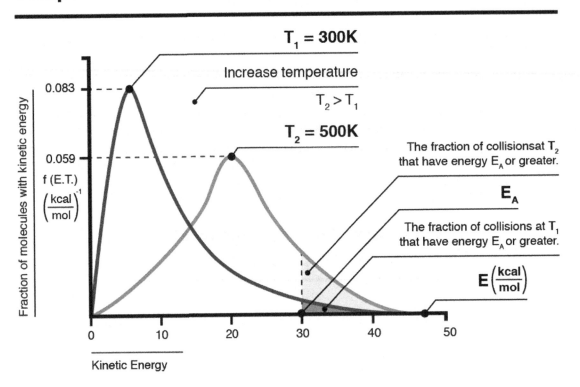

At 300 K, the is a specific fraction of molecules (n_j) that have an energy E greater than the activation energy (E_a). This distribution can also apply to molecules undergoing a phase change (e.g., liquid to gas). At 500 K, the distribution becomes more dispersed, and there are a greater number of molecules (n_i) that have $E > E_a$.

Deviation from the Ideal Gas Law

Real and Ideal Gases

Gases are considered **ideal** if their molecules do not interact with each other. **Real gases** do have intermolecular interactions between their particles. These interactions alter the properties of real gases from those of theorized ideal gases. At low temperatures and high pressures, the interactions between gas molecules increases, and they increasingly repel or attract each other. This can lead to condensation of the gases into liquids. The trends of the graphs for real gases and ideal gases also differ. The

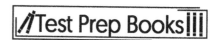

properties that are directly proportional or inversely proportional have more curves and bumps in their shape with real gases.

The ideal gas law makes two key assumptions:

- All gas particles have a volume of zero, even though protons, neutrons, and electrons actually do have mass, albeit very little

- All gas particles have no repulsive and attractive forces so collisions are elastic. As kinetic energy decreases, this becomes less and less true

At small volumes, gas particles are very close together. Therefore, the volume occupies a greater ratio of space, and renders incorrect the collision theory's assumption that gas particles have no mass. At high pressures, the volume that the particles occupy is so small and the particles are so close that there is an increase in intermolecular bonding, rendering the second postulate of collision theory incorrect.

Van der Waals derived the following formula to account for these effects:

$$\left[P + \frac{an^2}{V^2}\right](V - nb) = nRT$$

In this equation, b is a constant equal to the actual volume occupied by 1 mole of the particles. If the volume is very large in comparison to the volume taken up by the actual particles, the volume variable will be consistent with that proposed in the ideal gas law.

Van der Waals made another correction since a real gas' pressure is greater than the ideal equation proposes. Both constants, a, and b, are physical properties of the gas and a is a proportionality factor that corrects for intermolecular forces.

Fugacity is a measure of a real gas' behavior in proportion to what would be expected if the gas behaved like an ideal gas. It involves an assumption that the real and ideal gas have the same chemical potential.

$$\varphi = \frac{f}{P}$$

As pressure decreases, gas particles behave more like ideal gases; so, as pressure decreases, the fugacity coefficient, φ, increases. An ideal gas has a fugacity coefficient of one.

Most gases behave as ideal gases when pressure is relatively low and temperatures are relatively high. When these conditions change, the ideal gases become real gases. Although ideal gas molecules do not have an attraction to each other, as the gases become real gases, this changes and they begin to interact with each other. The larger the molecules and the stronger the interactions and bonds, the greater the deviation from ideal gas behavior.

Solutions and Mixtures

Types of Solutions

A **solution** is a homogenous mixture of more than one substance. A **solute** is another substance that can be dissolved into a substance called a **solvent**. If only a small amount of solute is dissolved in a solvent, the solution formed is said to be **diluted**. If a large amount of solute is dissolved into the solvent, then

the solution is said to be **concentrated**. For example, water from a typical, unfiltered household tap is diluted because it contains other minerals in very small amounts.

Solution Concentration

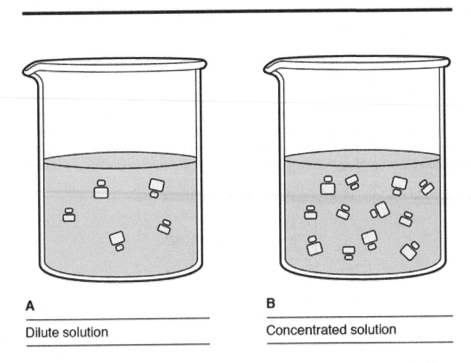

A

Dilute solution

B

Concentrated solution

If more solute is being added to a solvent, but not dissolving, the solution is called **saturated.** For example, when hummingbirds eat sugar-water from feeders, they prefer it as sweet as possible. When trying to dissolve enough sugar (solute) into the water (solvent), there will be a point where the sugar crystals will no longer dissolve into the solution and will remain as whole pieces floating in the water. At this point, the solution is considered saturated and cannot accept more sugar. This level, at which a solvent cannot accept and dissolve any more solute, is called its **saturation point**. In some cases, it is possible to force more solute to be dissolved into a solvent, but this will result in crystallization. The state of a solution on the verge of crystallization, or in the process of crystallization, is called a **supersaturated** solution. This can also occur in a solution that seems stable, but if it is disturbed, the change can begin the crystallization process.

Again, **solutions** are homogeneous mixtures of different components. They are made up of a **solute** and a **solvent** component, where the solute is dissolved in the solvent. All parts of a solution are identical. Solutions can be solids, liquids, or gases. In a liquid solution, the solute that is dissolved into the solvent can be a solid, liquid, or gas. Because of their homogeneous composition, solutions cannot be separated by filtration. They also cannot scatter visible light.

Examples of solutions include gaseous, liquid, and solid solutions. An example of a gaseous solution is air, which contains miscible gases such as nitrogen (N_2), argon (Ar), oxygen (O_2), and trace gases. An example of a solid solution is dental fillings, which may consist of silver (Ag) or different metal alloys. Liquid solutions include soda, which contains carbon dioxide (CO_2) gas dissolved in liquid water, and alcohol, which contains ethanol mixed with liquid water. Remember that the terms *solute* and *solvent* describe the types of components in a solution. The solute is the component in the lesser amount, and the solvent is the component in the greater amount. Ethanol is often mixed with octane to form a

miscible gasoline solution with a ratio of 1:9 ethanol to octane. Ethanol is the solute, while octane is the solvent. In contrast, oil and water are not miscible. It's important to understand the factors that explain solubility, as they explain why solutions are either miscible or immiscible.

Preparing Solutions with Specified Molarity

The strength of a solution can be described in several ways. Although the terms *dilute*, *concentrated*, *saturated*, and *supersaturated* give qualitative descriptions of solutions, a more precise quantitative description needs to be established for the use of chemicals. This holds true especially for mixing strong acids or bases. The method for calculating the concentration of a solution is done through finding its **molarity**. The **molality** of a solution, abbreviated as m, is defined as the moles of solute dissolved in one kilogram of solvent. More commonly, solutions are described in terms of molarity, abbreviated as M, which describes the number of moles of solute dissolved per liter of solution. In the laboratory, use of accurately graduated glassware, including graduated cylinders and volumetric flasks, is imperative for making solutions. In some instances, such as environmental reporting, molarity is measured in **parts per million** (ppm). Parts per million, is the number of milligrams of a substance dissolved in one liter of water.

Concentrated solutions are often diluted with additional solvent to make a solution of the desired strength. First, take a known volume of concentrated solution and calculate the number of moles of solute in it. Next, dilute that amount of concentrated solution to the appropriate volume with solvent to give the desired molarity of the new solution. For example, say you have a 10 M NaCl solution, which has 10 moles NaCl per liter of solution, and need a 1 M NaCl solution, which has 1 mole NaCl per liter of solution. You can take 0.1 L 10 M NaCl, which contains 1 mole NaCl, and dilute it to 1 L total volume by adding 0.9 mL solvent to achieve a 1 M NaCl solution.

To find the molarity, or the amount of solute per unit volume of solution, for a solution, the following formula is used:

$$c = \frac{n}{V}$$

In this formula, *c* is the molarity (or unit moles of solute per volume of solution), *n* is the amount of solute measured in moles, and *V* is the volume of the solution, measured in liters.

Example:

What is the molarity of a solution made by dissolving 2.0 grams of NaCl into enough water to make 100 mL of solution?

To solve this, the number of moles of NaCl needs to be calculated:

First, to find the mass of NaCl, the mass of each of the molecule's atoms is added together as follows:

23.0g (Na) + 35.5g (Cl) = 58.8g NaCl

Next, the given mass of the substance is multiplied by one mole per total mass of the substance:

2.0g NaCl × (1 mol NaCl/58.5g NaCl) = 0.034 mol NaCl

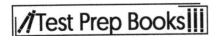

Finally, the moles are divided by the number of liters of the solution to find the molarity:

$$(0.034 \text{ mol NaCl})/(0.100\text{L}) = 0.34 \text{ M NaCl}$$

To prepare a solution of a different concentration, the **mass solute** must be calculated from the molarity of the solution. This is done via the following process:

Example:

How would you prepare 600.0 mL of 1.20 M solution of sodium chloride?

To solve this, the given information needs to be set up:

$$1.20 \text{ M NaCl} = 1.20 \text{ mol NaCl}/1.00 \text{ L of solution}$$

$$0.600 \text{ L solution} \times (1.20 \text{ mol NaCl}/1.00 \text{ L of solution}) = 0.72 \text{ moles NaCl}$$

$$0.72 \text{ moles NaCl} \times (58.5\text{g NaCl}/1 \text{ mol NaCl}) = 42.12 \text{ g NaCl}$$

This means that one must dissolve 42.12 g NaCl in enough water to make 600.0 L of solution.

Representations of Solutions

Particulate Drawings

Physical and chemical processes are often represented by **particle drawings**, such as chemical ball and stick or space-filling models. The image below shows a ball and stick model for the reaction chemical equation of sulfur dioxide, a pollutant, with liquid water to form sulfurous acid. Space-filling models are similar to ball and stick models but only show the atomic spheres, which may overlap with one another.

Ball and Stick Reaction Showing the Formation of Sulfurous Acid

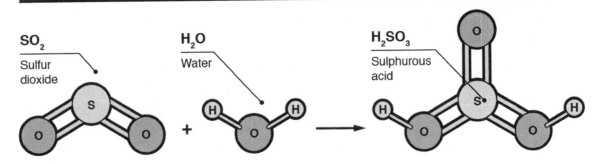

The Effect of a Solute on Its Solvent

Solutes have effects on their solvents, and these effects are called **colligative properties**. A classic example of a colligative property is **freezing point depression** when electrolytes are added to water. This phenomenon happens because water particles have a harder time forming their lattice-like structure from hydrogen-bonding due to the solute molecules getting in the way, preventing water from freezing at its normal temperature.

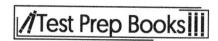

Other colligative properties include:

Vapor pressure lowering: Presence of other solid solutes lowers the surface area of the solvent, thus decreasing solvent's formation into gas

Boiling point elevation: Because vapor pressure decreases with other solutes, it takes more energy to produce gas, so boiling point rises

Osmotic pressure: A concentration gradient depends on the presence of solute, and the more the solute, the more water will move from high concentration to low concentration to reach equilibrium

Freezing point depression (K_f) can be calculated by subtracting the freezing point of the pure solvent minus the freezing point of the solution to find T_f, and multiplying it by the molality (m) of the solution (moles of solute/kg of solvent):

$$T_f = i \cdot K_f \cdot m$$

Note that i is equal to the van't Hoff factor (postulated by Dutch physical and organic chemist J.H. van't Hoff) and is related to disassociation of an ionic solute. It doesn't have a unit, and, for insoluble solids, the factor is one. In the case of ionic NaCl, which disassociates into 1 mole of sodium ion and 1 mole of chlorine, the factor will equal 2. It makes sense that i influences freezing point depression because for every one compound, there are two particles, and if each particle has an effect on freezing point depression, they both must be accounted for.

Boiling point elevation is the same calculation as that of freezing point depression, excepting that the constant is K_b.

Calculating osmotic pressure can be accomplished using the formula π = iMRT, where M is the molar concentration, R is the ideal gas constant, T is temperature in Kelvin, and i is equal to the van't Hoff factor.

Raoult's Law, established by French chemist Francois-Marie Raoult, is used to calculate vapor pressure depression. With solutions containing a non-volatile solute, vapor pressure is equivalent to that of the pure solvent multiplied by its mole fraction. Vapor pressure is dependent on temperature.

The equation is:

$$P_{solution} = X_{solvent} \; x \; P°C_{solvent}$$

$P_{solution}$ = vapor pressure of the solution

$P°C_{solvent}$ = vapor pressure at the temperature stated (physical property)

$X_{solvent}$ = molar fraction of the solvent, which can be calculated by dividing the moles of solvent by the total number of moles in the solution

Raoult's Law assumes that solvent and solute attraction is identical throughout a solution. In reality, as each solvent particle breaks away, the solution becomes more concentrated and forces of attraction increase. The larger the solution, the more valid Raoult's Law is.

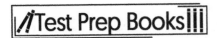

Separation of Solutions and Mixtures Chromatography

Separating Solutions

To separate the components of a solution, the proper technique must be employed that will break the intermolecular bonds between the solute and solvent particles. Two such techniques are chromatography and distillation.

Chromatography

Chromatography separates a solution based on the attraction of the different particles to either the stationary phase or the mobile phase of the experiment. In paper chromatography, the different components of the solution will travel up the paper at different rates based on the intermolecular attractions of the different solution particles to the paper molecules. Solution particles that have a greater affinity for the stationary phase will bond to the paper and not move up the paper. Solution particles that have a greater affinity for the mobile phase will move farther up the paper along with the solution. **Column chromatography** works in a similar manner with a stationary phase and a mobile phase to separate the components of a solution.

The stationary phase of the column is made up of a matrix that has functional groups that interact with the particles of the solution. The solution of interest is loaded onto the top of the column. As it runs through the column, the components of the solution that have weaker attractions to the column matrix will run farther down the column, whereas the components that have a stronger attraction to the column matrix will stick to the column closer to the top where the solution was loaded. As the solution is eluted from the column, the different components will come out at different times, and can be captured at different times, based on how strong their interactions were with the column matrix.

Distillation

The process of **distillation** uses heat to separate the components of a solution. Because different substances have different boiling points, the bonds between certain solute molecules and the solvent molecules will break before others. The molecules with lower boiling points will be distilled first, followed by the ones that have higher boiling points, in order of their boiling point temperatures.

Solubility

Miscibility of Substances

The intermolecular forces between molecules can limit the solubility. Generally, "like dissolves like" and if two substances are similar to one another, they will be **miscible**. Mixing will occur If the attractions between the solute and solvent are more significant than the attractions between solute-solute species and solvent-solvent species. A solution is **immiscible** if the attractions between the solute and solvent components are weak, and if the attractions between solute-solute and solvent-solvent components are strong. Water will not mix with gasoline because of the strong hydrogen bonding network between water molecules. These bonds are stronger than the London dispersion forces that are present between the polar water and nonpolar octane molecules.

Spectroscopy and the Electromagnetic Spectrum

<u>Using Molecular Motion to Study Bonds and Structure</u>

Molecules absorb light energy discretely, which is given by $E = \frac{hc}{\lambda}$, where $v = c / \lambda$. The speed of light is given by c. The frequency v is related to the wavelength λ of light by the following equation:

$$\lambda = \frac{c}{v}; \quad c = 3.00 \times 10^8 \text{ m/s}$$

The frequency (v) or wavelength (λ) of light that is absorbed by a molecule will determine how the atoms vibrate within the molecule. The frequency and wavelength have units of 1/s and meters, e.g., 1 nm = 1.0×10^{-9} m. Most molecular vibrations (e.g., stretching, bending) are associated with the absorption of infrared radiation, which contains wavelengths of light roughly between 1 μm and 1 mm. Molecules containing specific types of chemical groups or bonds (e.g., -O-H, as shown in the image below) have a characteristic absorption wavelength that corresponds to the stretching of a bond or movement between two atoms.

Vibrations and wave-numbers of water

Symmetric stretching	Antisymmetric stretching	Bending (scissors)
3700 cm⁻¹ or 2.7μm	3600 cm⁻¹ or 2.8μm	1750 cm⁻¹ or 5.71μm

Absorption of infrared radiation by different molecules shows the characteristic wavelengths for different vibrations, which provides a fingerprint of the molecules. In contrast, when molecules absorb higher energy ultraviolet/visible (UV) light, the electrons in the atom become excited but still release a specific amount of energy in multiples of hv. The various colors of light that are emitted by an atom provide information on the electronic energy levels, or states, within the atom. The Bohr model relates the frequencies of light from an atom to different electron transitions between electron orbitals, equal to a multiple of hv.

Photoelectric Effect

<u>Energy of a Photon</u>

In the 1900s, Max Planck carried out experiments that involved the heating of everyday solid objects. Every solid object, regardless of its composition, radiates or absorbs electromagnetic radiation with an energy equal to a photon of light, $E = hv$. The spectrum, or frequency, of light that is emitted from an object varies with the temperature. At room temperature, objects radiate infrared light, which is not visible to the human eye. Colors from objects, e.g., a gray nail, are still seen due to the absorption and reflection of light from that object, which is not the result of temperature. As light energy strikes an object, the atoms on the surface of the molecule absorb photons with energy, $E = nh\upsilon$ (n = 1, 2, 3, ...). Temperature increases in an object result in a similar absorption or emission process.

For a given experimental temperature, the light density E(λ, T) with respect to its wavelength (λ) will provide a specific distribution curve. If the temperatures of an object was increased to 700°C or 4700°C (approximately 5000 K), it would emit red or white light. Most hot objects have maximum temperatures below 3000°C, which corresponds to an object emitting a yellow color. Between these temperatures, the colors vary from orange to yellow to blue. Heating a nail with the tip of a propane torch will result in the hottest part of the nail emitting a yellow color; the cooler part of the nail will emit a red color.

The Wavelength (Λ) of Light That Is Produced from a Hot Body Is Temperature Dependent

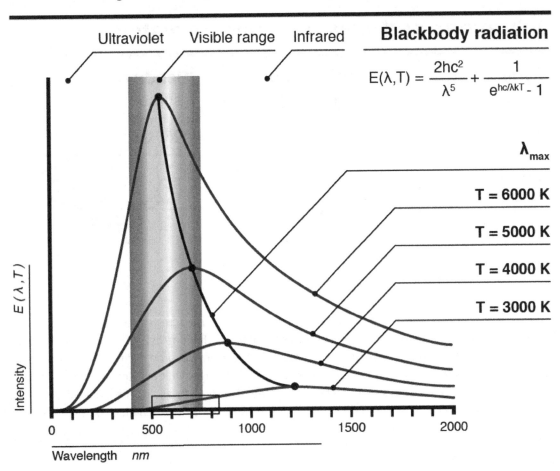

The intensity of the emitted wavelength or frequency of light depends on the temperature and was modeled by Planck with an equation called the **spectrum energy density**, $E(\lambda, T)$. The image above

shows four different curves because there are four different temperatures, and each curve has a maximum emission wavelength, e.g., T = 5,000 K, λ_{max} = 700 nm. For an object that has a temperature of 3000 K (approximately 2700°C), the maximum light emission wavelength is outside the visible range, but the tail end of the curve (red box) indicates that yellow or red light will be seen by an observer.

At temperatures near the sun, 5000 K, the maximum wavelength corresponds to the visible spectrum of light where all visible colors of light are emitted (purple to red). When all visible colors of light are emitted, our eyes observe white light, which explains why light produced from the sun is white. Importantly, Planck was only able to replicate the observed spectrum by incorporating $hc/\lambda = hv = E$ into the $E(\lambda, T)$ function. In other words, the atoms in an object are radiating light or photons with energy equal to hv, which indicated that atoms vibrated in discrete energy states. Therefore, when a molecule is heated, it absorbs energy equal to $E = hv$, causing electrons in the atom to transition to a discrete excited state. During emission, the molecule releases a photon of light with energy, $E = hv$, and the electrons move to a lower energy state.

Beer-Lambert Law

The Beer-Lambert Law
UV spectrophotometers are often used to find the concentration of a photoactive molecule within a solution. For molecules in a solution, the **Beer-Lambert Law** relates to the intensity of molecular absorbance to the molecular concentration:

$$A = \varepsilon lc$$

The term A is the absorbance, which can range from 0 to 1 and is related to the intensity of absorbed light. If A = 0, no light passed through the sample. The term ε is the molar absorptivity coefficient (M/cm), l is the length of the solution cuvette (cm) in which light passes, and c is the concentration of the cell with units of molarity or moles/L. Suppose that wavelength of absorbed light (500 nm) has values of A = 0.75, $l = 1.0\ cm$, and $\varepsilon = 8000\ M/cm$. Then, the concentration of the absorbing molecules in the solution is:

$$c = \frac{A}{\varepsilon l} = \frac{0.75}{(8000\ \text{M/cm})(1.0\ \text{cm})} = 9.4 \times 10^5\ \text{M}$$

A manometer can measure the pressure change for the decomposition of dinitrogen pentoxide to nitrogen dioxide and oxygen.

$$N_2O_5(g) \rightarrow 4NO_2(g) + O_2(g)$$

The partial pressure of dinitrogen pentoxide, taken from manometer readings, is measured in intervals over time. Because pressure is related to the concentration (e.g., n/V = P/RT), the rate of reaction can be calculated. As time increases, the partial pressure of the reactant will decrease.

Color, or the absorption of light energy by a chemical species, can also be used to follow the progress of a reaction. For example, the iron (III) cation (Fe^{3+}) reacts with thiocyanate ion (SCN^-) in solution to produce iron (III) thiocyanate ($Fe(SCN)^{2+}$), a deep red solution.

$$Fe^{3+}(aq) + SCN^-(aq) \rightarrow Fe(SCN)^{2+}(aq)$$

The absorbance spectrum of iron (III) thiocyanate ion contains a peak at about 447 nm, and the concentration of the ion is proportional to the intensity of the absorption peak. **Beers Law**, $A = abc$, where A is the absorbance of the chemical substance (e.g., $Fe(SCN)^{2+}$), "a" is a given constant called the **molar absorptivity**, "b" is the width or distance of the solution container used in the spectroscopic determination, and "c" is the concentration of the absorbing species. During the reaction, the absorbance is measured to find the concentration of $Fe(SCN)^{2+}$ indirectly.

Chemical Reactions

Introduction for Reactions

Chemical vs. Physical Changes

A change in the physical form of matter, but not in its chemical identity, is known as a **physical change.** An example of a physical change is tearing a piece of paper in half. This changes the shape of the matter, but it is still paper.

Conversely, a **chemical change** alters the chemical composition or identity of matter. An example of a chemical change is burning a piece of paper. The heat necessary to burn the paper alters the chemical composition of the paper. This chemical change cannot be easily undone, since it has created at least one form of matter different than the original matter.

Matter primarily exists in three states: gas, liquid, or solid. For completeness, there is one other state of matter called **plasma**, which is seen in lightning, television screens, and neon lights. Plasma is most commonly converted from the gas state at extremely high temperatures. Each of these primary three states of matter has specific characteristics, and by observing a substance, the state of matter can be determined. Substances can also change between the different states when subjected to external influences, such as the environment or other substances.

Changes in states of matter are physical changes because the identity of the matter remains the same but its physical form (and properties) changes. **Gases** have a fixed shape or volume and are easy to compress. They generally expand or compress to fill the volume of the container they are in. **Liquids** have an exact volume but mold to the shape of the container they are in. Liquid molecules are attracted to each other and slide over each other when they are poured into a new container. **Solids** have a definitive volume and shape. The molecules are packed tightly and cannot move within the substance. Visual observations can help to identify the state of matter of a substance.

Making Observations and Recording Data from Physical and Chemical Changes

Chemical substances can undergo chemical and physical changes. Chemical changes involve changes to the molecular structure, whereas physical changes have to do with the appearance of the substance. The **chemical properties** of a substance include flammability, reactivity, oxidation ability, and toxicity. They describe how the substance changes as it becomes a different substance, generally due to contact with another substance. For example, iron bars get rusty from exposure to oxygen, turning the iron into iron oxide. **Physical properties** of a substance include color, odor, density, mass, volume, and fragility. They can mostly be determined with quantitative measurements.

Also, while physical properties can be determined by observation, on a macroscopic level, more in-depth experiments need to be completed to determine the changes in the chemical properties of a substance. Experiments that use known materials or processes with known outcomes reveal the properties of an unknown substance by showing how that substance reacts to the material and process. If a substance is passed through a flame and it catches fire and falls apart, it can be described as flammable. If it stays intact, it is not flammable. Although a physical change occurred, a chemical change was also occurring. Similarly, substances can be tested for reactions with acids, bases, water, and different gases. The more experiments that are done, the more properties and chemical characteristics of the substance can be revealed.

Net Ionic Equations

Chemical Equations Must Be Balanced

Chemical reactions describe the changes that occur within a substance when two or more substances interact with each other and with the environment. They describe how bonds between the atoms within one molecule are broken and then reformed with different configurations to form new, different molecules.

Synthesis reactions involve the combination of smaller molecules to make a larger molecule with different chemical properties. These types of reactions are seen often in laboratory experiments. One example is the combination of sodium and chloride to make sodium chloride, or table salt. The chemical reaction is written as $2Na + Cl_2 \rightarrow 2NaCl$. The equation starts with three molecules but ends up with two molecules. The stoichiometry of this reaction shows you that you need two sodium atoms for every one chloride molecule. The chloride molecule splits into separate atoms and each chloride atom then bonds to one sodium atom.

Decomposition reactions are the opposite of synthesis reactions. They involve the dissolution of larger molecules into smaller molecules or atoms. When baking soda, or sodium bicarbonate, is heated, it breaks into three components: sodium carbonate, water, and carbon dioxide. The reaction is written as follows: $2NaHCO_3 \rightarrow Na_2CO_3 + H_2O + CO_2$. The sodium carbonate is a solid, the water is a liquid, and the carbon dioxide is a gas that is released.

Combustion reactions specifically involve the addition of oxygen gas to a molecule, such as when a substance is burned.

When a chemical reaction is written, it must be balanced for the reaction to be able to work. **Balanced equations** have the same number of each element on the left and right side of the equation. The elements are rearranged from the reaction proceeding, but the count must still be equal on both sides. For example, in the methane combustion reaction $CH_4 + 2O_2 \rightarrow 2H_2O + CO_2$, one methane molecule and two oxygen molecules produce two water molecules and one carbon dioxide molecule. There is one carbon atom, four hydrogen atoms, and four oxygen atoms on each side of the reaction. The coefficient in front of a molecule indicates the quantity of that molecule involved in the reaction.

The **stoichiometry** of a chemical reaction can be calculated based on the quantity of each element that is present on each side of the equation. The number of molecules or moles of each element must be equal on both sides for the equation to be balanced; they all must be accounted for. Take the example of combining nitrogen and fluoride to create nitrogen trifluoride. The chemical reaction is written as follows: $N_2 + F_2 \rightarrow NF_3$. This equation, however, is not balanced because there are two nitrogen atoms on the left and only one on the right. There are, also, two fluoride atoms on the left and three on the right. The stoichiometry needs to be determined so that the reactants are combined at the correct ratio. The minimum number of nitrogen atoms that can be produced is two, so there must be at least two nitrogen trifluoride molecules produced per reactant nitrogen molecule. This would make six fluoride atoms on the right, which is equivalent to three reactant fluoride molecules. The correct equation reads as follows: $N_2 + 3F_2 \rightarrow 2NF_3$.

The Various Forms of Balanced Chemical Equations

Chemical reactions can be written in different forms. **Molecular equations** use elemental formulas with subscripts to indicate the number of each element present in the molecules. The molecular formula for glucose is written as $C_6H_{12}O_6$, which indicates that there are six carbon atoms, twelve hydrogen atoms,

and six oxygen atoms bonded together to form one glucose molecule. The state of matter of the molecule can also be indicated in parentheses following the molecular formula. Solid is indicated as (s), liquid as (l), gaseous as (g), and aqueous as (aq). When mixing hydrogen chloride and sodium hydroxide, the molecular equation is written as follows: $HCl\ (aq) + NaOH\ (aq) \rightarrow NaCl\ (aq) + H2O\ (l)$. **Ionic equations** describe reactions that involve ions within an aqueous solution. Each species is represented by its ionic components, including species that are not involved in the reaction itself, which are called **spectator ions**.

The same reaction of hydrogen chloride and sodium hydroxide is written as an ionic equation as follows: $H^+(aq) + Cl^-(aq) + Na^+(aq) + OH^-\ (aq) \rightarrow Na^+(aq) + Cl^-(aq) + H_2O\ (l)$. **Net ionic equations** are similar to ionic equations except that they do not include the spectator ions. Therefore, the same reaction would be written as a net ionic equation as follows: $H^+(aq) + OH^-(aq) \rightarrow H_2O\ (l)$. The Na^+ and Cl^- ions are present on both sides of the equation, so they are not involved in the reaction and are spectator ions.

Describing Chemical Reactions In Experiments

Chemical reactions can describe many different types of laboratory activities. They can explain what happens when an acid and base mix together, such as with hydrogen chloride and sodium hydroxide $(HCl(aq) + NaOH(aq) \rightarrow NaCl(aq) + H_2O(l))$, to produce a salt and water or when something burns, such as with the combustion of methane $(CH_4 + 2O_2 \rightarrow 2H_2O + CO_2)$ and produces water vapor and carbon dioxide. They can also explain how glucose energy is produced by photosynthesis in plants: $6CO_2 + 6H_2O + light \rightarrow C_6H_{12}O_6 + 6O_2$.

Representations of Reactions

Symbolically Representing Conservation of Matter

The law of conservation of mass explains that mass is always conserved in any chemical or physical process. A chemical equation shows the transformation of an initial reactant to a final product. The chemical process or reaction chemical equation of solid sulfur and iron, which produces solid iron (II) sulfide, is symbolically represented as:

$$\overbrace{Fe(s) + S(s)}^{\text{reactants}} \rightarrow \overbrace{FeS(s)}^{\text{products}}$$

The symbols F and S represent the elemental symbols taken from the periodic table, and the phase notation (s), which follows, indicates these reactants are in the solid phase. The final solid product, FeS, is represented by the combination of F and S, with the metal symbol written before the nonmetal. The forward reaction arrow, $\rightarrow$, means "reacts to" or refers to the progress of a reaction. The reactants, Fe and S, and the product, FeS, both have the same number of atom types, which is consistent with the conservation of mass.

The ionic solid FeS exists in a crystal lattice, meaning there is more than one Fe-S bond, and therefore FeS in the chemical equation represents a formula unit. Matter can be represented by other phase notations, depending on the phase of the reactants or products, and include "(l), (aq), (g)," which denote liquid, aqueous, and gas phases. Not every chemical equation indicates a chemical reaction; however, they may also be used to represent a physical process, such as a change in phase. The dissolution of

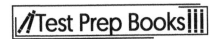

table sugar in water when making sweet tea is an example of solid sucrose breaking apart into its constituent molecules in a solution or aqueous phase:

$$C_{12}H_{22}O_{11}(s) \overset{H_2O}{\rightleftharpoons} C_{12}H_{22}O_{11}(aq)$$

The reaction is reversible if water is evaporated, and the chemical identity of the sugar molecule remains intact while in solution. The states of matter are typically solid, liquid, and gas, but for solution mixtures, the liquid phase can also exist in an aqueous phase. If solid sucrose or table sugar was heated to its melting point of 186°C, the chemical equation can be written to indicate that liquid sucrose form:

$$C_{12}H_{22}O_{11}(s) \rightarrow C_{12}H_{22}O_{11}(l)$$

Physical and Chemical Changes

Chemical vs. Physical Actions

A **chemical reaction process** involves the breaking or forming of a chemical bond whereby a new substance is formed. The dissociation of H_2 is an example of chemical processes. The combustion of methane or a hydrocarbon (e.g., gasoline) with oxygen from the air is a chemical process that produces carbon dioxide and water. Indicators of chemical changes include temperature and color changes, gas evolution, odor formation, heat evolution or absorption, and formation of a solid or new species. In contrast, a **physical process** refers to the changes in intermolecular interactions and includes phase changes. When you fill up an ice tray with water and place it inside the freezer, the liquid water will undergo a phase transition from a liquid to a solid. In the liquid phase, hydrogen bonding forces (not covalent bonds) are continually forming and coming apart. In the solid phase, water expands to maximize the hydrogen bonding forces, which stay fixed in a crystal lattice arrangement. The condensation that you see on the outside of a cold beverage is an example of a physical process where water vapor from the air condenses or forms a liquid on the cold surface of the bottle.

Processes that Blur the Lines of Chemical and Physical

The classification of a physical or chemical change in the context of salts can be ambiguous compared to molecular compounds. Ionic compounds contain **ionic bonds**, which consist of a metal that becomes a **cation**, a positively-charged ion, due to loss of its electron to the nonmetal that becomes an **anion**, a negatively-charged ion. Table salt, or sodium chloride, is an ionic compound and will not exist solely as NaCl, but is found in nature as a crystal structure composed of repeating NaCl units. It can be argued that the dissolution of NaCl in water is a chemical process because the ions break apart and are surrounded or solvated by water molecules.

$$NaCl(s) \rightarrow Na^+(aq) + Cl^-(aq)$$

However, dissolution of NaCl in water, from solid to liquid, may be considered a physical change, which is reversible if the liquid water is evaporated.

Stoichiometry

Coefficients of Balanced Chemical Equations

For chemical reactions to proceed, the correct amount and proportion of reactants must be available. The coefficients in the reaction (of the constituent molecules) indicate the number of that particular molecule needed or produced for the reaction. Many times, a different quantity of each reactant is needed to produce the products. For example, when propane gas is burned, it requires a greater

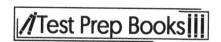

proportion of oxygen molecules than propane molecules to produce the products of water vapor and carbon dioxide. The reaction is written as follows: $C_3H_8 + 5O_2 \rightarrow 4H_2O + 3CO_2$. For every one propane molecule, five oxygen molecules are needed to complete the reaction.

A balanced chemical equation enforces the conservation of atoms, which allows the number of molecules, formula units in an ionic compound, and the number of atoms to be computed from a given mass. If a liter of air contained approximately 0.9800 gram of nitrogen gas (N_2), the number of nitrogen gas molecules is given by the relationship between molar mass and Avogadro's number:

$$0.9800 \text{ g } N_2 \times \frac{1 \text{ mol } N_2}{28.02 \text{ g } N_2} \times \frac{6.02 \times 10^{23} N_2 \text{ molecules}}{1 \text{ mol } N_2} = 2.105 \times 10^{22} N_2 \text{ molecules}$$

If the question asked how many nitrogen atoms, then:

$$2.105 \times 10^{23} N_2 \text{ molecules} \times \frac{2 \text{ N atoms}}{1 \text{ } N_2 \text{ molecule}} = 4.211 \times 10^{22} \text{ N atoms}$$

Therefore, given the mass or moles of any substance in a chemical equation, the quantity of that substance can be found.

A chemical equation represents the reactants and products involved for some chemical or physical process. For instance, consider the evaporation of ethanol from an alcoholic solution or beverage:

$$C_2H_6O(aq) \rightarrow C_2H_6O(g)$$

The **subscripts** shown in the chemical formula indicate the number of atoms for a given molecule. For example, there are two carbon atoms, six hydrogen atoms, and one oxygen atom in ethanol. Both the reactants and products have an equal number of atoms, which is consistent with the conservation of mass.

Suppose that a chemical process was initiated by burning ethanol, an ingredient in gasoline, with oxygen found in air. An unbalanced reaction chemical equation can be written as:

$$C_2H_6O(g) + O_2(g) \rightarrow CO_2(g) + H_2O(g)$$

For the respective molecule, the subscripts indicate the number of atoms within that molecule. There are two molecules each in the reactants and products. However, there are not an equal number of atoms on each side; therefore, the equation must be balanced by adding coefficients in front of each molecule. Placing coefficients will change the number of molecules on each side of the general equation and shouldn't be confused with the subscripts. The subscripts should remain unchanged to keep the identity of the molecule the same. The balanced reaction chemical equation is:

$$1C_2H_6O(g) + 3O_2(g) \rightarrow 2CO_2(g) + 3H_2O(g)$$

The equation can be read as, "1 molecule of ethanol will react with three molecules of oxygen to produce two molecules of carbon dioxide and three molecule of water." The quantity *mole* could be used in place of *molecule*. The total number of each atom will now be the same on each side, e.g., two carbons in ethanol and two carbons in carbon dioxide.

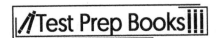

Using Conservation of Matter to Calculate Reactant or Product Masses

The conservation of mass requires that a chemical equation be balanced and therefore makes it possible to calculate the masses of the product given the reactant mass or the mass of the reactant given the mass of the product. An extremely useful relationship when converting a reactant to product mass is:

$$1A \rightarrow 1B$$

$$\text{Mass of A} \xrightarrow[\text{table}]{\text{periodic}} \text{moles of A} \xrightarrow[\text{ratio}]{\text{stoichiometric}} \text{moles of B} \xrightarrow[\text{table}]{\text{periodic}} \text{mass of B}$$

The mass of a substance may be specified in grams. Given the molar mass of A, B, and the stoichiometric relationship between A and B (e.g., one to one), the mass or moles of that substance can be found. If there are 5.00 grams of B and the molar mass of A and B is 24.0 and 12.0 g/mol, the mass of A is:

$$5.00 \text{ g B} \times \left(\frac{1 \text{ mole B}}{12.0 \text{ g B}}\right) \times \left(\frac{1 \text{ mole A}}{1 \text{ mole B}}\right) \times \left(\frac{24.0 \text{ g A}}{1 \text{ mole A}}\right) = 10.0 \text{ g A}$$

However, in some cases, it's not always necessary to perform these lengthy computations if the mass of the products or reactants is known. Suppose that 5.00 grams of iron reacted with an unknown amount of sulfur to produce 7.87 grams of FeS. How many grams of sulfur reacted with iron? Based on the conservation of mass, the mass of sulfur can be calculated as follows: 7.87 g FeS – 5.00 g Fe = 2.87 g S:

$$\overbrace{Fe(s) + S(s)}^{\text{reactants}} \rightarrow \overbrace{FeS(s)}^{\text{products}}$$

The following computation can verify the initial amount of iron that reacts with sulfur:

$$2.87 \text{ g S} \times \left(\frac{1 \text{ mole S}}{32.06 \text{ g S}}\right) \times \left(\frac{1 \text{ mole Fe}}{1 \text{ mole S}}\right) \times \left(\frac{55.85 \text{ g Fe}}{1 \text{ mole Fe}}\right) = 5.00 \text{ g Fe}$$

Using Stoichiometry to Determine the Molar Mass of a Gas

Stoichiometry investigates the quantities of chemicals that are consumed and produced in chemical reactions. Chemical equations are made up of reactants and products; stoichiometry helps elucidate how the changes from reactants to products occur, as well as how to ensure the equation is balanced.

Atoms and molecules are microscopic in size, so it is impossible to run reactions in the laboratory in terms of single or few atoms and molecules. Instead, chemical equations are often thought of in terms of moles. One **mole** is equivalent to 6.02×10^{23} molecules, which is a number that is known as **Avogadro's number**. The coefficient ratio of the equation must stay the same but each molecule quantity can be multiplied by Avogadro's number to find the number of moles needed to run the experiment in a real-life experiment.

Moles can be converted to grams by multiplying the number of moles by the molar mass, or molecular weight, of the molecule. In the acid-base reaction $2HCl + MgOH \rightarrow MgCl_2 + H_2O$, it would be impossible to combine two molecules of hydrogen chloride with one molecule of magnesium hydroxide. The molar mass of hydrogen chloride and magnesium hydroxide are 18 g/mole and 20 g/mole, respectively. The reaction requires twice as much hydrogen chloride as magnesium hydroxide, so in a laboratory, you could combine 36 grams of hydrogen chloride with 20 grams of magnesium hydroxide to run the reaction, or any lesser or greater amount as long as the ratios of the two chemicals remained the same.

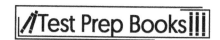

The amount of product that is produced by a chemical reaction is limited by the quantity of the starting reactants. It is important to determine which reactants will be used up and which will be in excess. The reactant that has the smallest amount of substance is called the limiting reactant. The **limiting reactant** is completely consumed by the end of the reaction. The other reactants are called **excess reactants**. Given the hydrogen chloride and magnesium hydroxide reaction, if you have equal amounts of moles of both reactants, the hydrogen chloride will be used up first because twice as many moles are needed to run the reaction compared with the magnesium hydroxide. For example, if you started with six moles of each, once the six moles of hydrogen chloride are used up, three moles of magnesium hydroxide would be remaining because only three moles of magnesium hydroxide would be used up in the reaction.

If you know the quantities of starting materials, you can also determine the product yield of the reaction. Using the same example and mole ratios of reactants and products, for every two moles of hydrogen chloride and one mole of magnesium hydroxide, one mole of magnesium chloride and one mole of water are produced. The **theoretical yield** is the amount of product that should be produced based on the starting materials. If you start with ten moles of hydrogen chloride and five moles of magnesium hydroxide, following the molar ratios, you should end up with five moles of magnesium chloride and five moles of water, which is the theoretical yield. Since the reactants do not always act as they should, the actual amount of resulting product is called the **actual yield**. The actual yield is divided by the theoretical yield and then multiplied by 100 to find the **percent yield** for the reaction. The percent yield is the ratio of the actual yield obtained in the laboratory experiment to the theoretical yield. So, if you actually obtained 4.75 moles of magnesium chloride instead of the theoretical five moles, your percent yield would be 95 percent.

Solution Chemistry

The strength of a solution is often indicated by **molarity**, which is the quantity of moles of a substance dissolved in the solution. A one molar (1 M) solution has one mole of molecules of the substance per liter of solution. For example, a 1 M NaOH solution has one mole of NaOH molecules, or 40 grams of NaOH as determined by molar mass, dissolved in one liter of solution. A 2 M NaOH solution would have two moles of NaOH molecules, or 80 grams of NaOH, dissolved in one liter of solution, making it twice as strong.

When combining two solutions, you can determine the strength of the final solution by determining the number of moles of the substance in the final solution. For example, in the reaction $2 \text{ M HCl} + 2 \text{ M NaOH} \rightarrow \text{NaCl} + H_2O$, a salt water solution is produced. If you start with 1 L of each solution, you will have 2 moles of Cl^- ions and 2 moles of Na^+ ions that will combine to form 2 moles of NaCl molecules. You will also have 2 L total of solution. Take 2 moles and divide by 2 L to determine the number of moles per liter and you find that the final solution is a 1 M NaCl solution.

Titration is a method of determining the concentration of analytes in an unknown solution. If you know the concentration and quantity of analytes in one solution, when the solution is neutralized or all of the known analytes are used up, you can use the volume used of the unknown solution to determine the quantity of unknown analytes that were used. This information can then be used to calculate the concentration of the unknown solution. Using the hydrogen chloride and sodium hydroxide example, if you have one mole of hydroxide ions to neutralize, once the product solution reaches a neutral pH, you know that one mole of chloride ions was used up in the solution Depending on the volume of the chloride solution used, you can determine the strength of the chloride solution. If only 0.25 L was used, divide one mole by 0.25 L and you find that the solution strength was 4 M. But if 2 L were used, you find that the solution was only 0.5 M.

Introduction to Titration

Titrations

Solution stoichiometry deals with quantities of solutes in chemical reactions that occur in solutions. The quantity of a solute in a solution can be calculated by multiplying the molarity of the solution by the volume. Similar to chemical equations involving simple elements, the number of moles of the elements that make up the solute should be equivalent on both sides of the equation. **Molarity** is the concentration of a solution. It is based on the number of moles of solute in one liter of solution and is written as the capital letter M. A 1.0 molar solution, or 1.0 M solution, has one mole of solute per liter of solution. The molarity of a solution can be determined by calculating the number of moles of the solute and dividing it by the volume of the solution in liters. The resulting number is the mol/L or M for molarity of the solution. Alternatively, **percent concentration** can be written as parts of solute per 100 parts of solvent.

When the concentration of a particular solute in a solution is unknown, a **titration** is used to determine that concentration. In a titration, the solution with the unknown solute is combined with a **standard solution**, which is a solution with a known solute concentration. The point at which the unknown solute has completely reacted with the known solute is called the **equivalence point**. Using the known information about the standard solution, including the concentration and volume, and the volume of the unknown solution, the concentration of the unknown solute is determined in a balanced equation. For example, in the case of combining acids and bases, the equivalence point is reached when the resulting solution is neutral. HCl, an acid, combines with NaOH, a base, to form water, which is neutral, and a solution of Cl⁻ ions and Na⁺ ions. Before the equivalence point, there are an unequal number of cations and anions and the solution is not neutral.

A titration in which the end point is determined by formation of a precipitate is called a **precipitation titration**. The most common precipitation titrations are **argentometric** (*Argentum* means silver in Latin). These reactions involve a silver ion with a titrant of silver nitrate ($AgNO_3$). Silver nitrate is a useful titrant because it reacts very quickly. All products in these titrimetric precipitations are silver salts. A very simple argentometric precipitation titration is often performed to determine the amount of chlorine in seawater, as the chlorine anion is present most abundantly. The reaction is as follows

$$AgNO_3 (aq) + NaCl (aq) \rightarrow AgCl (s) + NaNO_3 (aq)$$

Silver chloride is almost completely insoluble in water, so it precipitates out of solution. A selective indicator, potassium chromate (K_2CrO_4), is added to determine when the reaction is complete. When all of the chloride ions from the sample have precipitated, silver ions then react with chromate ions to form Ag_2CrO_4, which is orange in color. The volume of the sample, amount of titrant required to complete the precipitation, molar mass of the reacting species, and the density of the sample provide sufficient information to calculate the concentration of chloride ions.

Types of Chemical Reactions

The chemical changes that occur can be classified by acid-base, redox reactions, or precipitation reactions.

Acid-Base Reactions

Acid-base reactions are also known as **neutralization reactions** because the acid and base dissociate and neutralize each other in solution to form a salt and water. When hydrochloric acid and sodium hydroxide, a base, are mixed, the products that are formed are sodium chloride, a salt, and water.

Electrons in Redox Reactions

Oxidation-reduction reactions, also known as **redox reactions**, are chemical reactions in which electrons are transferred from one molecule to another. The molecule that loses the electrons is oxidized and becomes more positively charged, whereas the molecule that gains the electrons is reduced and becomes more negatively charged. Within a chemical equation, each atom or ion is assigned an oxidation number, which is equal to the total number of electrons that the atom or ion gains or loses when bonding to another atom or ion.

Oxidation Numbers

The sum of the oxidation numbers for each element in a molecule should be equal to the overall charge of the molecule. A neutral molecule should have a sum of zero for oxidation numbers, whereas a molecule with a single negative charge should have a sum of -1 for oxidation numbers, and a positively-charged molecule would have a positive number for the sum of its oxidation numbers.

Redox Reactions in the Energy Production Processes

Energy production uses many redox reactions. The cycles and chains involved in cellular respiration have many electron transfer steps that help break down sugars. Metabolism of fats and proteins also involve the transfer of electrons between components. Combustion of hydrocarbons, such as in the burning of fossil fuels, occurs when hydrocarbons are combined with molecular oxygen and then produce new molecules that incorporate the oxygen atoms. For these reactions to occur, electrons must be transferred from the oxygen atoms, making them redox reactions.

Precipitation Reactions

Precipitation reactions occur when a chemical reaction produces an insoluble salt from the reaction of two soluble salt solutions. An example of a precipitation reaction is the formation of the solid silver chloride when silver nitrate and sodium chloride solutions are mixed.

When a reaction produces a solid, the solid is called a **precipitate.** A precipitation reaction can be used for removing a salt (an ionic compound that results from a neutralization reaction) from a solvent, such as water. For water, this process is called ionization. Therefore, the products of a neutralization reaction (when an acid and base react) are a salt and water. Therefore, the products of a neutralization reaction (when an acid and base react) are a salt and water.

Introduction to Acid-Base Reactions

All liquids have either acidic or basic traits. When hydrogen ions (H^+) are released in a liquid, the solution becomes more acidic. Conversely, when hydroxide ions are released in a liquid (OH^-), it becomes basic. These are the two ions that determine whether a solution is an acid or a base.

The Brønsted-Lowry Concept of Acids and Bases

In the **Brønsted-Lowry Theory** of acids and bases (alkalines), the **acid** is a proton donor (a hydrogen ion), and the **base** is a proton acceptor. Acids and bases can be described using what is known as the pH scale. The pH scale ranges from 0-14. A pH of 7 is **neutral**, while values less than seven are acidic and those greater than seven are alkaline (basic).

Acids and basis can be divided into strong and weak categories. The strength of an acid can be measured using the following equation:

$$K_a = \frac{[H^+]\,[A^-_{weak}]}{[HA_{weak}]}$$

When K_a is large, it is a strong acid (e.g. HCL $K_a = 1 \times 10^3$). An acid with a small K_a is a weak acid (e.g. H_2O $K_a = 1.8 \times 10^{-16}$). The strength of a base can be measured using the following formula:

$$K_b = \frac{[HB^+][OH^-]}{[B]}$$

In Brønsted theory, every acid has a conjugate base, and every base has a conjugate acid.

Water is Amphoteric

Water is made up of two hydrogen atoms bonded to one oxygen atom in a V-shaped molecule. The atoms have unequal sharing of electrons, so water is considered a **polar** molecule. The hydrogen atoms are slightly positively charged and the oxygen atom is slightly negatively charged. In a glass of water where there are many water molecules together, the molecules bump into each other and form weak hydrogen bonds with each other for milliseconds at a time. The polar nature of water allows it to accept protons and donate protons easily to other species, making it an **amphoteric molecule**, which means it can act as an acid or a base.

Water acts like an acid when it donates a proton, or hydrogen ion, to a base, and it acts like a base when it accepts a proton from an acid. The ion that is left after the molecule has donated a proton is called the **conjugate base**. The ion that is formed after it has accepted a new proton is called the **conjugate acid**. For example, when water, acting as an acid, and the base ammonia are combined, the reaction is written as follows: $H_2O + NH_3 \rightarrow HO^- + NH_4^+$. Water starts out as an acid, donates a proton, and becomes hydroxide ions, which are the conjugate base. The ammonia gains a proton and becomes the conjugate acid. When acetic acid and water, acting as a base, are combined, the reaction is written as follows: $CH_3CO_2H + H_2O \rightarrow CH_2CO_2^- + H_3O^+$. The acetic acid loses a proton and becomes the conjugate base, while the water molecules gain a proton and become the conjugate acid.

Water is amphoteric and **self-ionizable** in that it tends to dissociate, or split, into hydrogen ions (H^+) and hydroxyl ions (OH^-) randomly. No other substance on Earth may be found naturally in all three states of matter—liquid, solid, and gas. An amphoteric substance can act as an acid or a base depending on the properties of the solute.

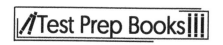

Comparing the Strengths of Conjugate Acid-Base Pairs

Conjugate acid-base pairs can be identified by their positive or negative ionic property and their nonionic counterpart with or without a proton added. Whenever an acid or base is combined with water to produce conjugate acids and bases, two acid-base reactions occur. The acid or base molecules are ionizing as well as the water molecules. The strength of an acid is determined by its ability to dissociate in water and is represented by its K_a value, or **dissociation constant**. Stronger acids dissociate easier in water and release their hydrogen ions faster than weaker acids, giving them a lower pH because of the increased amount of hydrogen ions in the solution. Stronger acids produce weaker conjugate bases, and weaker acids produce stronger conjugate bases.

Oxidation-Reduction (Redox) Reactions

Balancing Redox Equations

Redox reactions involve the exchange of electrons to form new molecules. When hydrogen molecule H_2 and fluoride molecule F_2 are combined, hydrogen fluoride, HF, molecules are formed from hydrogen losing electrons and fluoride gaining electrons. These types of chemical changes are not visually observable but can be determined by analyzing the chemical reactions that are occurring.

Similar to the stoichiometry of chemical equations, the electrons in redox reactions must also be balanced on both sides of the chemical equation. Looking at the oxidation numbers of the atoms within a chemical reaction can assist with making sure the equation is balanced. Redox reactions can be split into half-reactions, which isolate the oxidation and reduction reactions separately. When looking at the half-reactions, the quantity of electrons on the product side of the oxidation reaction must equal the quantity of electrons on the reaction side of the reduction reaction. Take the example of chlorine gas oxidizing iron. The equation is written as follows: $2Fe^{2+} + Cl_2 \rightarrow 2Fe^{3+} + 2Cl^-$. The molecules are different on each side of the equation, and they also have different charges. The equation can be separated into separate reduction and oxidation equations and electrons have to added to each equation, which are not seen in the total equation.

$$\text{Oxidation: } 2Fe^{2+} \rightarrow 2Fe^{3+} + 2e^-$$

$$\text{Reduction: } Cl_2 + 2e^- \rightarrow 2Cl^-$$

Adding up the charges on the left side of the equation, you get a +2 net charge. Adding up the charges on the right side of the equation, you also get a +2 net charge, so the electrons are balanced for the equation. Because the transfer of electrons is not noted in the general chemical equation, the use of oxidation numbers can help determine whether or not a redox reaction has occurred. Once all atoms have been assigned oxidation numbers on both sides of the equation, you can compare the oxidation numbers from the elements on the reactant side to the elements on the product side of the equation. If a redox reaction has occurred, exactly two of the elements will have oxidation numbers that changed between the reactant and product sides. If the oxidation number increased, the element was oxidized. If the oxidation number decreased, the element was reduced. When iodine pentoxide and carbon monoxide are combined, they produce iodine molecule and carbon dioxide. Once oxidation numbers are assigned, you can see that the iodine and carbon atoms have changes in their oxidation numbers, and therefore, the reaction is a redox reaction.

Equation:	$I_2O_5 + 5CO_2 \rightarrow I_2 + 5CO$
Oxidation Numbers:	+5 -2 +2 -2 0 +4 -2

Redox titrations can also help identify redox reactions in the laboratory. These experiments are a type of titration where a solution of a known agent is treated with a reducing agent to produce the ion form of the known agent. Iodine is often used in redox titrations because it is blue in color and when it has been completely reduced to its iodide form, I⁻, the blue color disappears. The electron composition of the reducing agent can be determined based on the volume of solution used to completely reduce the iodine.

Kinetics

Reaction Rates

Rate of a Reaction

The reaction rate of any chemical reaction can be expressed as the change in reactant concentration over time or as the change in product concentration with respect to time. For example, consider the following chemical reaction:

$$aA + bB \rightarrow cC + dD$$

The lowercase letters refer to the **coefficients**, and the capital case letter refers to the different **chemical species**. Suppose that the coefficients have the following values of $a = 1$, $b = 2$, $c = 1$, and $d = 2$. Then the equation would take the following form:

$$A + 2B \rightarrow C + 2D$$

The terms A and B are the **reactants**, and C and D are the **products**. If the rate of reaction is expressed in terms of the reactants, the rate will have a negative sign.

$$Rate\ of\ reaction\ for\ reactants = -\frac{\Delta[A]}{\Delta t} = -\frac{1}{2}\frac{\Delta[B]}{\Delta t}$$

The triangle or **delta symbol** (Δ) refers to a change or difference between the final and initial concentrations. The **brackets** around the species refer to the concentration. The numerator $\Delta[A]$ can be expressed as:

$$\Delta[A] = [A]_{final} - [A]_{initial}$$

Similarly, the denominator Δt can be expressed as:

$$\Delta t = t_{final} - t_{initial}$$

Notice that the rate for *B* has a fraction in front, which is due to the coefficient in the original equation. The rate of reaction of *B* is one-half that of reactant *A*. The following figure shows a graphical representation for the reaction rate of *A*.

Graphical Representation of the Rate of Reaction for A and C

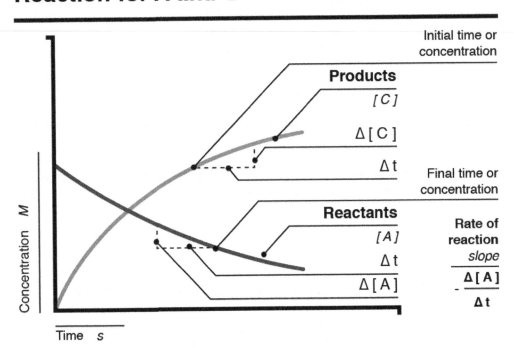

Factors that Affect the Rate of a Reaction

There are a variety of factors that affect the rate of a reaction, such as temperature, surface area of the reactants, and a variety of environmental factors. Reaction concentration is another major factor, except in zero-order processes. In a **zero-order process**, the reaction rate does not depend on the concentration of the reactants and is equal to the rate constant: reaction *rate = k*. In **nonzero order processes**, the rate will depend on many factors. Steel wool does not burn in air; however, when the oxygen or reactant concentration is increased from 20 percent to 100 percent, the material burns and forms a white flame, thereby increasing the rate of reaction.

As the temperature increases, the rate of reaction also increases because more reactant molecules contain enough kinetic energy to surmount the activation energy barrier and form reaction products. Adding a strip of zinc into a solution of hydrochloric acid will take longer to decompose compared to adding the same strip but chopped into pieces. This observation is due to the surface area of zinc, which increases when it's cut into pieces, allowing the rate of reaction to increase. A **catalyst** is a chemical substance or enzyme that allows the rate of reaction to increase without it being consumed and lowers the activation energy needed to form a product. If the surface area of the catalyst increases, so does the rate of the reaction.

Introduction to Rate Law

The Rate Law

For the following reaction with a catalyst or enzyme (E):

$$aA + bB \overset{E}{\rightleftharpoons} cC + dD$$

The rate law can be written as:

$$Rate = k[A]^m[B]^n[E]^p$$

The **rate constant**, denoted k, is a proportionality constant that relates the concentration and rate and varies with temperature. The units for this constant will depend on the determined rate law. The superscripts or exponents (m, n, and p) are found through experiment and are not necessarily integers. Importantly, these exponents cannot be obtained from the coefficients of the balanced equation (a, b, c, and d). Suppose the determine rate law for some chemical reaction is:

$$Rate = k[A]^1[B]^2[E]^0$$

The reaction order for species A, B, and E is one, two, and zero. The overall order is three because $1 + 2 + 0 = 3$. As the value of the exponent for the catalyst is zero, the rate of reaction does not depend on the catalyst concentration and is zeroth order. So, the catalyst is not included in the equation, and the final rate law can be rewritten as:

$$Rate = k[A]^1[B]^2 \ or \ Rate = k[A][B]^2$$

Using Initial Rate to Predict Reaction Progression Over Time

Using the initial rate method, the reaction order can be determined, which requires a series of experiments whereby the initial concentrations of reactant are varied. If there are two experiments, the initial concentration of the reactants will be different, and each experiment will have a different reaction rate. The method allows for the reaction orders to be determined from the rates. The table below shows two experiments with different initial concentrations of N_2O_5.

	Initial $[N_2O_5]$	Initial Rate for N_2O_5 decomposition
Experiment 1	2.0×10^{-2} mol/L	9.6×10^{-6} mol/(L·s)
Experiment 2	4.0×10^{-2} mol/L	1.9×10^{-5} mol/(L·s)

The rate law for the reaction is:

$$Rate_1 = k[N_2O_5]^m$$

Because the concentration is doubled in the second experiment, the second rate law can take on the following form:

$$Rate_2 = k(2[N_2O_5])^m = k2^m[N_2O_5]^m = 2^m k[N_2O_5]^m$$

The rate for Experiment 2 is twice that of Experiment 1, which can be shown as follows:

$$\frac{Rate_2}{Rate_1} = \frac{1.9 \times 10^{-5} \text{ mol/(L·s)}}{9.6 \times 10^{-6} \text{ mol/(L·s)}} = 2.0$$

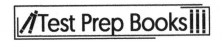

This equation also is written as:

$$\frac{Rate_2}{Rate_1} = \frac{2^m k[N_2O_5]^m}{k[N_2O_5]^m} = 2^m$$

Taking the last two terms, the exponent (m) can be determined:

$$2^m = 2.0$$

The value of "*m*" must equal one which gives a reaction order of one, which is consistent with the image of first-order reactions, as shown previously. The general rate law is:

$$Rate = k[N_2O_5]$$

If the $[N_2O_5]$ is doubled and the rate doubles, then m = 1. Hypothetically, If $[N_2O_5]$ is doubled and the rate quadruples, then m = 2.

The Rate Constant

From the table above, the determined rate law for the decomposition of N_2O_5 allows for the proportionality, or **rate constant**, to be found because the rate and initial concentrations are known. If values are used from experiment one:

$$Rate_1 = k[N_2O_5]$$

$$k = \frac{Rate_1}{[N_2O_5]} = \frac{9.6 \times 10^{-6} \text{ mol/(L} \cdot \text{s})}{2.0 \times 10^{-2} \text{ mol/L}} = 4.8 \times 10^{-4} \text{ s}^{-1}$$

Therefore, the rate constant is an important quantity that can be quantified. The units for the rate constant depend on the overall order of the rate law. Because the overall order is one, the units for the rate constant are s^{-1}. If the order were two, then the units would be $L \cdot mol^{-1} \cdot s^{-1}$.

The Magnitude of Rate Constants

A small change in the reactant concentration or temperature will change the reaction rate, and consequently, the value of the rate constant can vary by several orders of magnitude. One reason for the significant change is that the rate constant has an exponential relationship with temperature. The table below shows the change in the rate constant with respect to temperature, and for fixed values of $Z = 1$ and $p = 1$. The value for R is 8.314 J/(mol $\cdot$ K), and E_a is 50,000 J $\cdot$ mol^{-1}. The table below indicates that for every 10°C change in temperature, the rate constant or reaction rate more than doubles.

Exponential Relationship Between the Rate Constant and the Temperature	
k	**T (K)**
2.71×10^{-10}	273
1.72×10^{-9}	298
9.95×10^{-8}	373
2.44×10^{-3}	1000
4.94×10^{-2}	2000

Temperature Dependence and the Rate Constant

The rate constant, k, can also be expressed as follows:

$$k = Zfp \; ; \; f = e^{-E_a/RT}$$

Where Z is the collision frequency or the number of times molecules or atoms collide, f is the fraction of collisions that have an energy greater than the activation energy barrier E_a, and p is the fraction of molecular collisions that take place when reactant molecules are correctly aligned to one another. Therefore, for a chemical reaction to progress, the molecules or atoms must hit, hit hard, and hit at the right orientation. The collision frequency Z changes slowly with temperature; f can have a greater change because it depends on the temperature (T) and E_a. Based on the exponential relationship between f and T, as the temperature increases so do the fraction of collisions that have an energy exceeding E_a. Consequently, the rate constant k has a temperature dependence, and as k increases the temperature increases.

Concentration Changes Over Time

Using Spectrophotometry to Determine How Concentration Varies With Time

By plotting the concentration of the reactant for different time intervals, either through spectroscopy or by measuring pressure changes, the order for a reactant can be inferred. For example, if there is a linear relationship when plotting the natural logarithm of the reactant concentration versus time, then the reaction is first order. The image below shows a first-order relationship for the decomposition of dinitrogen pentoxide, N_2O_5.

First-Order Reaction: $\ln[N_2O_5]$ Versus Time

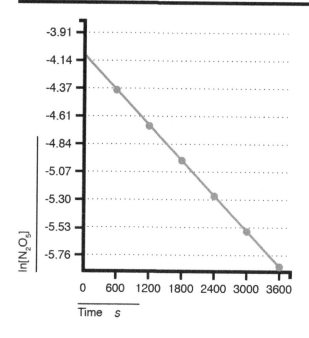

Time	$[N_2O_5]$	$\ln[N_2O_5]$
0	0.0165	- 4.104
600	0.0124	- 4.390
1200	0.0093	- 4.678
1800	0.0071	- 4.948
2400	0.0053	- 5.240
3000	0.0039	- 5.547
3600	0.0029	- 5.843

If the plot of the inverse reactant concentration as a function of time shows a linear relationship, then the reaction is second order. The image below shows a plot of the inverse concentration of iron III thiocyanate ($1/[FeSCN^{2+}]$) vs. time, which is second order for $Fe(SCN)^{2+}$ decomposition. The concentration of $Fe(SCN)^{2+}$ is proportional to the measured absorbance (A) as described by Beer's Law.

Second-Order Reaction: $1 / [FeSCN^{+2}]$ Versus Time

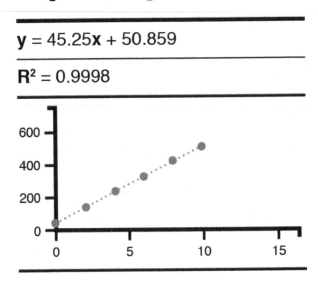

$y = 45.25x + 50.859$

$R^2 = 0.9998$

The Order of an Elementary Reaction

An **elementary reaction** describes the molecular collision of molecules that results in a chemical reaction. Elementary reactions equations are single molecular events and can be described by **molecularity**, which relates the number of reactants, such as molecules, atoms, radicals, or ions on the reactant side. The decomposition of dinitrogen pentoxide (N_2O_5) to its products nitrogen dioxide (NO_2) and oxygen (N_2) is an example of **unimolecular reaction** because it involves only one reactant molecule. The reaction of iron (III) (Fe^{3+}) with thiocyanate ion (SCN^-) is an example of **biomolecular reaction** because it involves two different species on the reactant side.

First Order Reactions

The rate law for an elementary reaction tends to take on a simple form, and the rate is proportional to the product of each reactant concentration. The reaction,

$$A \rightarrow B + C; \quad Rate = k[A]$$

will be first order and as the concentration of A ($[A]$) increases it is more likely to break down into products. The rate is proportional to $[A]$.

Second Order Reactions

A bimolecular reaction such as

$$A + B \rightarrow C + D; \quad Rate = k[A][B]$$

tends to be second order. For the reaction to occur, A and B must combine, and as the concentration of each reactant increases, there is a greater likelihood of a reaction. As the concentration of A and B increase, so do the number of A and B. Therefore, the rate of reaction is proportional to the frequency of collision (Z) of these two reactants and is proportional to the product [A][B].

Half-Life

Half-life (T1/2) describes the time taken for half of the radioactive material to decay. Half-life is linked to the isotope undergoing radioactive decay. For example, the isotope of carbon—carbon-14 (^{14}C)—can undergo beta decay to nitrogen-14 (^{14}N). Carbon-14 has a half-life of 5,730 years. This means that if one started with 100g of carbon-14, then after 5,730 years, 50g of carbon-14 would be left over, and 50g of nitrogen-14 would be produced. After an additional 5,730 years—11,460 years in total—there would be 25g of carbon-14 and 75g of nitrogen-14. Because half-life is constant, it can be used in carbon dating to work out the age of certain organic objects.

The following equation can be used to work out the fraction of parent material that remains after radioactive decay:

$$\text{Fraction remaining} = \frac{1}{2^n} \quad \text{(where n = \# half-lives elapsed)}$$

The first-order rate law for the decomposition of N_2O_5 can be expressed as:

$$Rate = k[N_2O_5] = -\frac{\Delta[N_2O_5]}{\Delta t}$$

Or more generally as:

$$Rate = k[A] = -\frac{\Delta[A]}{\Delta t}; \quad aA \rightarrow reaction\ products$$

based on calculus, this first order expression can be rearranged as:

$$ln\frac{[A]_t}{[A]_0} = -kt \quad \text{or} \quad \overset{y}{\overbrace{ln[A]_t}} = \overset{m}{\overbrace{-k}}\overset{x}{\overbrace{t}} + \overset{b}{\overbrace{ln[A]_0}}$$

The plot in the second-order reaction is represented by the second equation, which has the form y = mx + b, where the rate constant is equal to the negative slope (m). In the first term, the concentration of A at some time t is given by $[A]_t$ and the initial concentration of A at time zero is $[A]_0$. This expression is useful because the ratio $\frac{[A]_t}{[A]_0}$ relates the fraction of reactant A at that is left remaining at time t. If only half of reactant A remains at time t, called $t_{1/2}$ or the **half-life**, then the equation can be written as:

$$ln\frac{\frac{1}{2}[A]_0}{[A]_0} = ln\frac{1}{2} = -kt_{\frac{1}{2}} \quad or \quad 0.693 = kt_{1/2}$$

The negative sign is canceled, and the equation can be simplified too:

$$t_{1/2} = \frac{0.693}{k}$$

The rate constant and half-life are inversely proportional to one another and don't depend on the reactant concentration. The half-life is useful for a carbon-14 radioactive dating method. In the upper atmosphere, carbon-14 is produced from the bombardment of neutrons and nitrogen-14, which combines with oxygen to form carbon dioxide. Plants consume carbon dioxide and maintain an abundance of carbon-14. Animals contain the carbon-14 isotope because they eat plants and have a certain ration of carbon-14 to carbon 12. When dinosaurs were wiped from the planet, the ratio of carbon-14 to carbon-12 decreased through radioactive decay. Give the half-life of carbon-14 (5730 years), and the carbon-14/carbon-12 ratio, the date when dinosaurs became extinct can be found.

Radioactive Decay

Radioactive decay is a nuclear reaction that occurs when the nucleus of an atom spontaneously breaks apart, producing a lighter atom, along with the emission of radiation or particles. Atoms with $Z > 83$ tend to be unstable and undergo radioactive decay. The energy changes for these reactions are much greater than the breaking of chemical bonds. Several types of radioactive decay exist. **Alpha (α) emission** produces ^4_2He nuclei. Radium-236 decays to radon-222 where A decreases by four and Z by two:

$$^{226}_{88}Ra \rightarrow {}^{222}_{86}Rn + {}^4_2He$$

Beta (β^-) rays are high-speed electrons ($^0_{-1}e$) and result in the conversion of a neutron to a proton. The beta decay of uranium-239 produces neptunium-239, where Z increases by one due to proton formation:

$$^{239}_{92}U \rightarrow {}^{239}_{93}Np + {}^0_{-1}e$$

In **positron** or **beta (β^+) emission** ($^0_{+1}e$), particles identical to the mass of an electron are produced but have a positive charge instead. This process results in the conversion of a proton to a neutron. The decay of technetium-95 produces molybdenum-95, where Z decreases by one due to neutron formation:

$$^{95}_{43}Tc \rightarrow {}^{95}_{42}Np + {}^0_{+1}e$$

Electron capture occurs when an unstable nucleus captures an electron from its inner orbital, which produces a neutron ^1_0n. Like positron emission, a proton transforms into a neutron. The conversion of potassium-40 to argon-40 is an example, where Z decreases by 1:

$$^{40}_{19}K + {}^0_{-1}e \rightarrow {}^{40}_{18}Ar$$

Radioactive decay can produce a nucleus that exists in an excited state; however, this nucleus will fall to a more stable energy state and release **gamma (γ) rays** (photons $^0_0\gamma$) with a wavelength ~10^{-12} m. The metastable technetium-99 exists in an excited state but decays to a lower energy state nucleus that has no change in A or Z:

$$^{99}_{43}Tc \rightarrow {}^{99}_{43}Tc + {}^0_0\gamma$$

Spontaneous fission is another decay process whereby a heavy nucleus ($Z > 89$) splits into lighter nuclei, resulting in neutron formation and large amounts of energy. Uranium-236 can undergo fission to produce neutrons:

$$^{236}_{92}U \rightarrow {}^{96}_{39}Y + {}^{136}_{53}I + 4\,{}^1_0n$$

The various radioactive decay processes are all examples of first order kinetics.

Elementary Reactions

Determining the Rate Law of Elementary Reactions

The reaction equation can be used to determine the rate law of elementary reactions. The stoichiometry of the molecules colliding—and thus reacting—provides insight into the rate law. If the coefficients of a hypothetical reaction have the following values of $a = 1$, $b = 2$, $c = 1$, and $d = 2$. Then the equation would take the following form:

$$A + 2B \rightarrow C + 2D$$

The terms A and B are the **reactants**, and C and D are the **products**. If the rate of reaction is expressed in terms of the reactants, the rate will have a negative sign.

$$Rate\ of\ reaction\ for\ reactants = -\frac{\Delta[A]}{\Delta t} = -\frac{1}{2}\frac{\Delta[B]}{\Delta t}$$

Elementary Reactions Involving the Simultaneous Collision of Three Particles

Termolecular reactions involve the combination of three reactant molecules,

$$O + O_2 + M \rightarrow O_3 + M^*$$

and can occur in the atmosphere where atoms and small radicals can combine. The reaction shown above is one elementary reaction that involves the formation and reaction of ozone. The term "M" typically represents an atom or another reactant such as O_2 or N_2, which can absorb energy during the reaction. Molecularities that involve over three reactant molecules do not generally occur because the probability of four reactants coming together is relatively small. Similarly, the rate equation for a termolecular reaction is

$$A + B + C \rightarrow D + E; \quad Rate = k[A][B][C]$$

and is proportional to the product reach reactant molecule concentration $[A][B][C]$. Consider the following termolecular reaction:

$$I + I + M \rightarrow I_2 + M \quad or \quad 2I + M \rightarrow I_2 + M$$

The rate law is predicted to be

$$Rate = k[I][I][M] \quad or \quad Rate = k[I]^2[M]$$

and has an overall order of three.

Collision Model

Collision Models

The **rate constant** is the product of the **collision frequency Z**, the fraction of collisions with an energy exceeding the activation energy denoted by f, and a **steric factor** or proper molecular orientation denoted p. The collision of reactants can affect the rate constant because increasing the temperature and concentrations of reactants can lead to more collisions with proper orientation and sufficient energy. Unimolecular or biomolecular elementary reactions are first or second order where the rate is proportional to the reactant product concentrations. For example, in a bimolecular reaction, the rate at which the product is produced is proportional to Z because there is a definite fraction of collisions that produce the product. The term **Z** is proportional to the number of A reactant molecules (e.g., n_A) multiplied by the number of B molecules (e.g., n_B). Therefore, the rate in a bimolecular reaction is proportional to [A] times [B] because the concentration of A and B is proportional to n_A and n_B.

Successful vs. Unsuccessful Collisions

Most collisions between reactant molecules do not result in a chemical reaction. The reactant molecules, atoms, or ions not only need enough kinetic energy to surmount E_a, but must also collide at the correct orientation. This observation would explain why nitrogen and oxygen, in the air, do not completely convert to nitric oxide (NO) at room temperature and one atmospheric pressure even though there are about 10^{30} collisions. Most reactant molecules hit each other at the wrong orientation and bounce off. The image in the following section indicates that at $T_2 = 500$ K, more reactant molecules will react because they have enough energy and hit each other at the right orientation.

The Maxwell-Boltzmann Distribution

At relatively lower temperatures, there are larger fractions of reactant molecules that do not react as shown by the distribution in the figure below. At $T_1 = 300$ K there is a smaller fraction of molecules that have the required energy to form an activated complex and subsequently a set of products than at higher temperatures (such as 500 K). The fraction of molecules with Energy $E > E_a$ is equal to the area under the curve and will depend on the temperature. At $T_2 = 500$ K, the kinetic energy distribution changes and the area under the curve, where $E > E_a$, increases. Because the rate constant is proportional to f ($k = Zfp$), more reactant molecules collide with enough kinetic energy to form an activation complex equal to or greater than E_a. The progress of the chemical reaction is illustrated in the image by a

potential energy surface where the reactant molecules combine to form an activated complex, located at the maximum, which subsequently forms the reaction product that is relatively more stable.

Maxwell-Boltzmann Kinetic Energy Distribution and a Potential Energy Surface

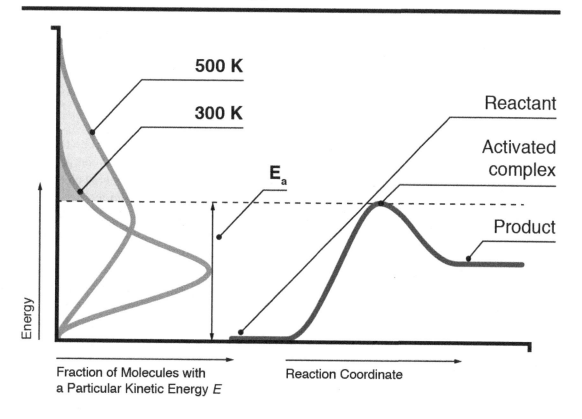

Reaction Energy Profile

Unimolecular Reactions

Unimolecular reactions occur through collisions in the gas or aqueous phase and are explained by the Maxwell-Boltzmann thermal distribution of a substance's energy. This distribution in terms of particle speeds will vary with a specific temperature but will have an average speed corresponding to the maximum in a distribution curve. The average kinetic energy associated with the motion of a particle is related to the number of particles by the equation:

$$Kinetic\ energy = \frac{1}{2}mu^2 = \frac{3}{2}\frac{RT}{N_A}$$

In the above equation, "u" is the average molecular speed, "m" is the mass of the particle, and "N_A" is Avogadro's number. In a chemical reaction that may involve more than one reactant, the collision of the reactant molecules, with sufficient kinetic energy greater than the activation energy barrier E_a, will allow the formation of products.

Elementary Reactions Follow a Single Reaction Coordinate

The following elementary chemical reaction,

$$NO + O_3 \rightarrow NO_2 + O_2$$

is a bimolecular reaction that involves the formation of a new N-O bond and the breaking of an O-O$_2$ bond. The **reaction mechanism** can show the step or steps in which these bonds break and form, which may be described by a specific reaction coordinate that may involve a specific set of motions within the molecule. For instance, a simple reaction coordinate may describe a changing bond distance between nitrogen and oxygen ON—O, or between the two oxygen atoms O—O$_2$. In practice, more than one coordinate is often used to better understand the reaction mechanism (e.g., 2-D coordinate for two bond distances). In either case, a relatively high energy or short-lived structure called an **activated complex** will form [ON—O—O$_2$]$^{2+}$, and subsequently stabilize to form a new set of products.

The Energy Profile

Suppose we are given the stable reactant structure O_3 where the bond distances between the two oxygen atoms in O$_2$-O are all fixed in space and not allowed to move. The interaction energy corresponding to one point can be calculated for this one geometry, which is the overall electrostatic energy due to the repulsion of electrons, the attraction of electrons to protons, and the repulsion of protons found in O_3. If the distance between the oxygen atoms is increased, O$_2$—O, the interaction energy would change. If the distance were increased again, O$_2$——O, then the interaction energy would be different from the previous value. If this process were repeated until the oxygen atom was completely dissociated, forming O$_2$, then there would be several energy points corresponding to the reactant, the activated complex, and product geometries. If each interaction energy point were plotted as a function of distance, then it would yield an energy profile generally called the **potential energy surface**. The image below illustrates an example of a potential energy surface (PES) for the reaction of NO and O_3, which shows the energy transition from the initial reactants to the activated complex, and the final products.

PES Corresponding to the Reaction Pathway of NO and O3

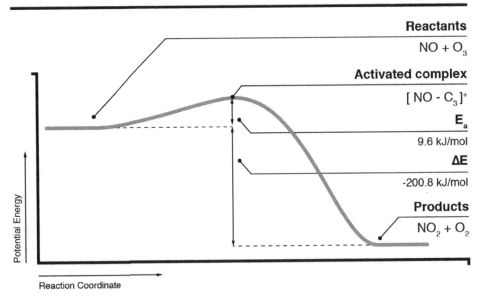

If the activated complex, $[ON-O-O_2]^{2+}$, corresponds to the highest energy structure, then it is called a **transition state**. Note that the reaction coordinate can represent several complex motions, which include the movement of the oxygen atom from reactant to another.

The Arrhenius Equation

Recall that the rate constant (k) is related to the activation energy barrier (E_a) by the following equation:

$$k = Zpf = Zpe^{-E_a/RT}$$

This rate constant equation can be simplified to the following form:

$$k = Ae^{-E_a/RT} \; ; \; A = Zp$$

The rate equation is called the **Arrhenius equation,** and it emphasizes the dependence of the rate constant, k, on the temperature and activation energy barrier. The frequency factor, A, is a constant and is related to the product of the frequency of collisions (Z) and the fraction of molecules having the proper orientation. The equation often takes the following form:

$$\underbrace{\ln(k)}_{y} = \underbrace{\left(-\frac{E_a}{R}\right)}_{m} \underbrace{\left(\frac{1}{T}\right)}_{x} + \underbrace{\ln(A)}_{b} \quad or \quad \ln\frac{k_2}{k_1} = \frac{E_a}{R}\left(\frac{1}{T_1} - \frac{1}{T_2}\right)$$

The first equation takes on a linear form ($y = mx + b$) whereby the slope is related to the activation energy. For a series of experiments that have different rate constants with respect to temperature, plotting $ln(k)$ as a function of the inverse temperature ($1/T$) would allow one to calculate the activation energy barrier using the second equation.

Introduction to Reaction Mechanisms

Using Collision Theory to Determine the Rate Law of Initial Steps

For a single unimolecular elementary reaction where there is one reactant, the rate law is the product of the reactant concentration and rate constant. These unimolecular reactions have an order of one, and as the number of reactants increases, so does the percentage of collisions that result in the formation of the products. Similarly, for a single bimolecular elementary reaction, the rate law will equal to the product of the rate constant and each reactant concentration, and will have an overall order of two because there are two reactants involved.

Collision theory is closely related to kinetic molecular theory. Based on the fact that as temperature increases, movement increases, collision theory's premise is that reactions happen faster as temperature increases due to an increased number of reactants' collisions with each other and with catalysts. Of note, from a biological perspective, there is a limit to this temperature increase because, if it is a reaction involving biological protein catalysts (enzymes), the enzyme will eventually denature, "cook," and lose functionality.

Balancing Multistep Reactions

A chemical reaction can occur in multiple elementary steps. The reaction of carbon monoxide and nitrogen dioxide occurs in two biomolecular elementary steps. The overall or net chemical reaction is

$$CO(g) + NO_2(g) \rightarrow CO_2(g) + NO(g)$$

and can be referred to as a **general reaction mechanism**. The first elementary reaction involves the collision between two nitrogen dioxide molecules.

$$NO_2(g) + NO_2(g) \rightarrow NO_3(g) + NO(g)$$

In the second step, nitrate and carbon monoxide collide to form carbon dioxide and nitrogen dioxide:

$$CO(g) + NO_3(g) \rightarrow CO_2(g) + NO_2(g)$$

The overall balanced chemical equation is determined by the number of molecules on each side of the equation, which is specified by the coefficients and is not related to the rate.

$$NO_2(g) + NO_2(g) \rightarrow NO_3(g) + NO(g)$$

$$\underline{CO(g) + NO_3(g) \rightarrow CO_2(g) + NO_2(g)}$$

$$\cancel{NO_2(g)} + NO_2(g) + CO(g) + \cancel{NO_3(g)} \rightarrow \cancel{NO_3(g)} + NO(g) + CO_2(g) + \cancel{NO_2(g)}$$

$$Net\ equation: \qquad NO_2(g) + CO(g) \rightarrow NO(g) + CO_2(g)$$

In the net equation above, the order of the reactants and products is not important, but the equation must be balanced in each elementary step to obtain the correct net equation. Nitrate is not shown in the net reaction; it is an intermediate reaction species that quickly reacts when it is produced.

A number of mechanisms may be postulated for most reactions, and experimentally determining the dominant pathway of such reactions is a central activity of chemistry.

Chemical reactions occur on timescales on the order of 10^{-15} s (femtoseconds), which makes it extremely difficult to observe the reaction mechanism. Chemists often propose a reaction mechanism based on experimental observation. For example, consider a solution containing a spiropyran chromophore or a molecule that changes colors when heated. A one-step reaction mechanism shown in the image below illustrates the C-O bond, which breaks and leads to a conjugated product structure that is colored. Reaction mechanisms tend to be provisional because there may be additional experiments that lead to a more accepted mechanism.

One-Step Reaction Mechanism:
Conversion of Spiropyran *(1)* to Merocyanine *(2)*

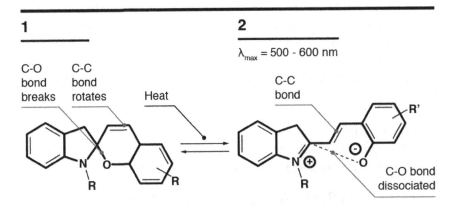

For example, the mechanism shown above is believed to occur in two steps, where the first step involves breakage of the C-O bond and the second step involves rotation about a C-C bond.

Once the mechanism is proposed, a rate law can be formulated, and if the rate law is consistent with experimental rate measurements, it indicates that the proposed mechanism is likely correct.

A Reaction Intermediate
A reaction intermediate is a species that is only present while a reaction is occurring. It is produced by an elementary step and consumed by another, so it does not appear in the net reaction. Consider the reaction of hydrogen gas with molecular iodine to produce hydrogen iodide gas:

$$H_2(g) + I_2(g) \rightarrow 2HI(g)$$

This reaction mechanism occurs in two elementary steps. In the first step, molecular iodine will break down into iodine, and in the second step, two iodine atoms react with hydrogen gas:

$$I_2 \leftrightarrow 2I \qquad Fast\ and\ reversible$$

$$I + I + H_2 \rightarrow 2HI \qquad Fast$$

Atomic iodine is produced in the first elementary step only to be consumed in the second step when it collides with hydrogen gas. Atomic iodine is a reaction intermediate that is only presently briefly when the reaction takes place. In practice, these reaction intermediates are short-lived and can be difficult to observe experimentally.

Using the Experimental Detection of a Reaction Intermediate to the Determine Reaction Mechanism
Because reaction intermediates are present for and at very short periods of time, they can be difficult to observe experimentally and may not be included in a reaction mechanism. Reaction intermediates are not the same as activated complexes or transition states, which are highly unstable and nearly impossible to observe. Reaction intermediates are relatively lower energy molecular structures, relative to activated complexes, and are at a minimum on the potential energy surface. By observing reaction intermediates, more information about the reaction mechanism can be obtained. For example, the conversion of spiropyran (SP) to merocyanine was believed to occur in one step. However, through absorbance measurements, a reaction intermediate (cis isomer) was found to absorb light at 400 nm, which provided support for a two-step reaction mechanism.

Reaction Mechanism and Rate Law

With Irreversible Elementary Steps, Reaction Rate is Determined by the Rate-Limiting Step
Suppose the following reaction $2A + B \rightarrow 2C$ occurs irreversibly in two steps, with the first step being slow and the second elementary reaction being fast.

$$A + B \overset{k_1}{\rightarrow} C + D \qquad Slow$$

$$D + A \overset{k_1}{\rightarrow} C \qquad Fast$$

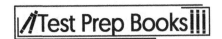

The rate of k_2 is greater than k_1. The net reaction, or overall equation, is just the sum of the first and second steps and can be written as

$$A + B \rightarrow C + D$$

$$D + A \rightarrow C$$

$$A + A + B + \cancel{D} \rightarrow C + C + \cancel{D}$$

$$Net\ equation: \quad 2A + B \rightarrow 2C$$

The "D" term is a reaction intermediate and does not show up in the net equation. Note that as soon as A and B react, D is produced and is quickly consumed in the second step to form C. The first slow step determines the rate of disappearance of B. The rate law will be determined by the rate-determining step, which is the slowest elementary reaction step in the two-part reaction mechanism shown above. Therefore, the predicted rate equation is:

$$Rate = k_1[A][B]$$

The experimental rate law should be equal to this rate.

Steady-State Approximation

Determining the Rate Law in Elementary Reactions

As just mentioned, the rate law can be determined in reactions when the first elementary reaction is the rate-limiting step. If the first elementary reaction is not the rate-limiting step however, the rate-law expression must be determined from other approximations, such as steady state. The overall reaction rate can also be determined from steady-state approximations when the rate-limiting step is unknown (be it actually the first step or otherwise); however, it is best reserved for consecutive reactions where it is known that the first step is not the rate-limiting step because that ensures there is limited buildup of the reaction intermediate. This is important because the assumption underlying steady-state approximations is that the concentration of a reaction intermediate remains constant over the duration of a reaction because the rate at which a reaction intermediate is consumed is equal to the rate at which it is generated. In other words, the rate of change of the reaction intermediate is assumed to be zero.

Multistep Reaction Energy Profile

Constructing Energy Profiles for Multistep Reactions

Reactions that occur in a series of collisions, in which intermediates are produced, are called **multistep reactions.** Each step of the process is consists of a single step elementary reaction. Consider the following real-world multistep reaction involved in the depletion of the ozone layer:

$$Reaction\ 1: NO(g) + O_3(g) \rightarrow NO_2(g) + O_2(g); slow$$

$$Reaction\ 2: NO_2(g) + O(g) \rightarrow NO(g) + O_2(g); fast$$

$$Overall\ Reaction: O_3(g) + O(g) \rightarrow 2O_2$$

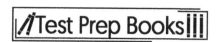

Constructing an energy profile for this multistep reaction is a bit more complicated than sketching one for an elementary reaction. Both steps include their own activation barrier, E_{A1} and E_{A2} for reaction 1 and reaction 2, respectively. The slower step, in this case reaction 1, has a higher activation barrier than the faster step. The result would look something like what is shown below:

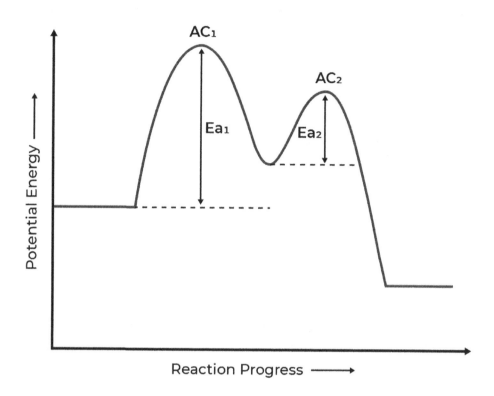

Catalysis

Catalysts Lower Activation Energy

A **catalyst** is a chemical compound or element that participates in a chemical reaction without being consumed in the process. The catalyst must take part in at least one reaction step and then regenerate in a later step. A catalyst increases the rate of a chemical reaction by decreasing the activation energy barrier, E_a, which increases the reaction rate because the reaction rate is proportional to the exponential of $-E_a/RT$. In addition, the activated complex or transition state lowers in energy and is stabilized along the reaction path. The image below shows a chlorine-catalyzed potential energy surface where chlorine reacts with ozone (O_3) in the first step, and then subsequently with the oxygen atom in a second step.

Potential energy surface for a catalyzed and uncatalyzed reaction

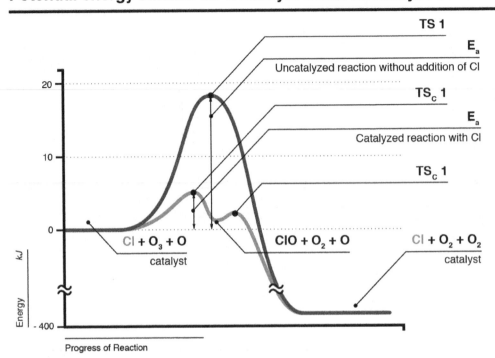

Without a Cl catalyst, the reaction of the ozone and oxygen atoms would take place in a single step, with a relatively larger activation barrier. One transition state (TS 1) is found along the uncatalyzed pathway, which connects the reactants to the products. When Cl is added, TS1 is stabilized to a lower energy structure (double-headed arrow) and is near a minimum on the potential energy surface. The catalyzed reaction now occurs in two steps with two transition states (TS_c 1 and TS_c 2) with a reaction intermediate that connects the reactants and products.

Catalysts Can Create New Reaction Pathways

For the catalyzed reaction, the two activation barriers (Ea_1 and Ea_2) are smaller relative to the uncatalyzed reaction barrier (E_a). Therefore, the overall rate of reaction with the addition of a catalyst will increase. The image above shows how Cl initially reacts with ozone in the first step and is later regenerated in the last step as a product. Addition of a catalyst changes the reaction mechanism and the shape of the potential energy surface. The two-step mechanism is

$$Cl(g) + O_3(g) \rightarrow ClO(g) + O_2(g) \quad Slow$$

$$\underline{ClO(g) + O(g) \rightarrow Cl(g) + O_2(g) \quad Fast}$$

$$O_3(g) + O(g) \rightarrow 2O_2(g) \; Net\; reaction$$

The species ClO is a reaction intermediate that would be found at a local minimum, between TS_c 1 and TS_c 2, along the pathway. The species O and O_2 react with the reaction intermediate and are found in the overall equation. No reaction intermediates are present along the uncatalyzed pathway.

Note that the rate of the second step is expected to be faster than the first step because $Ea_2 < Ea_1$.

Acid-Base Catalysis and Reaction Rate

Catalysis can be carried out with an acid (e.g., H^+) or base (e.g., OH^-), which accelerates the reaction through protonation or deprotonation. Esterification chemical reactions are examples of acid-base catalysis in organic chemistry. The image below shows the reaction of an acid and alcohol with a carboxylic acid. The acid catalyst can be an acid such as hydrofluoric or sulfuric acid, whereby the proton (H^+) from the acids binds or protonates the carboxylic acid. This acid-catalyzed process lowers the activation energy barrier, creating reaction intermediates (e.g., cations or anions) and speeds up the rate of reaction. The rate of reaction will be proportional to the concentration of the acid or base. Note that in each case, the catalyst reforms in the last step.

Acid Catalysis (Fischer Esterification Reaction) and Base Catalysis

Acid Catalysis: carboxylic acid to ester

Base Catalyst: alcohol to ketone

Surface Catalysis

In **heterogeneous surface catalysis**, a solid substrate catalyst (e.g., platinum) is used to facilitate the conversion of reactant to product gas molecules. This process often involves the lowering of the activation energy barrier and the formation of new intermediates along the reaction pathway. For example, the reaction of ethene and hydrogen gas with a solid platinum catalyst promotes the formation of intermediates and increases the likelihood of molecular or atomic collision because these

species stay near the catalyst surface. The image below shows a general three-step reaction mechanism of the catalytic process.

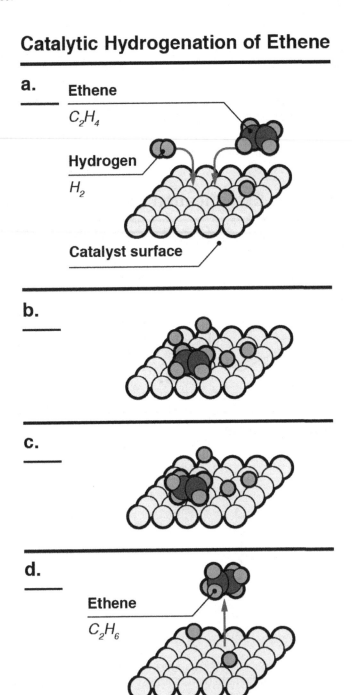

In a process called **chemisorption**, hydrogen gas and ethene first bind to the platinum surface via bonding forces (Step A), but hydrogen gas dissociates to a hydrogen atom intermediate while the ethene pi bonds (Step B) form bonds to the metal surface. Each hydrogen atom then migrates to one side of the ethene molecule (Step C), forming another intermediate followed by the formation of ethane molecule (Step D).

Enzymes in Reactions

Biological enzymes (E) are efficient catalysts and relatively large proteins (>1 million atomic mass units) that can convert thousands of reactant molecules to products in less than a second. Enzymes are specific to substrates (S)—the reactant molecule—and will form an **enzyme-substrate (ES) complex** through intermolecular forces (hydrogen bonding, dipole-dipole, and London dispersion forces) such that the activation barrier is lowered, thereby accelerating the conversion of reactants to products (P). Enzymes have the capability of stabilizing the activated complex or transition state and even forming reaction intermediates. The image below shows how a reactant molecule binds to the pocket or active site of the enzyme, forming a new intermediate or enzyme-substrate complex. It can be seen that the figure on the left is the uncatalyzed pathway with a relatively high activation energy. In the figure on the right, an enzyme create a new pathway by lowering the activation energy, thereby increasing the rate of reaction.

Comparing an Enzyme-Catalysed Reaction with an Uncatalyzed Pathway

Uncatalyzed

Activation energy

S

P

Energy

Progress of reaction

Catalyzed

E + S

Activation energy

E + P

ES

Energy

Progress of reaction

Thermodynamics

Endothermic and Exothermic Processes

Temperature Changes in a Sample

In a chemical system, **temperature** is a measure of the average kinetic energy for the particles in that system or object. In everyday objects (e.g., table, iced beverage), the temperature is also a measure of how hot or cold an object is. All atomic particles or molecules in an object are in motion because the particles' average kinetic energy is proportional to the object's temperature. Temperature changes in a system are indicative of energy changes.

Temperature of a System and Its Energy

When a chemical system such as water is heated, e.g., cold to hot, the kinetic energy of the system will increase because it is proportional to the temperature, T~KE. For example, one liter, or 1000 mg, of solid water at -10°C will have less energy compared to one liter of water at 10°C.

Measuring Temperature Changes to Determine if Reactions are Exothermic or Endothermic

Endothermic reactions are chemical reactions that absorb heat from the environment to make them proceed. When heat is added to a reaction, the surrounding environment feels cold. For example, if solid ammonium nitrate is dissolved in water in a flask, it requires the addition of heat, and the flask will feel cold. Chemical bonds are broken, generally, in endothermic reactions. Molecules prefer to stay intact, so when they are dissociated into smaller components, they require more energy—in the form of heat—to break the bonds between the atoms.

Exothermic reactions are chemical reactions that involve the release of heat energy as the reaction proceeds. These types of reactions cause an increase in temperature in the surrounding environment. When solid calcium chloride is dissolved in water, calcium hydroxide, hydrogen chloride gas, and heat are produced. The flask feels hot as the reaction occurs. When chemical bonds are formed, energy is released in the form of heat.

The process of cooking is endothermic because food absorbs heat from the pan it is in. Photosynthesis is another endothermic reaction where carbon dioxide and water mix with heat from the sun to form glucose energy and oxygen. Combustion reactions are exothermic, such as when fossil fuels are burned. The condensation of water vapor into water is also exothermic. As water goes from a gaseous state to a liquid state, heat energy is released.

The change in heat energy of a chemical reaction is known as the **enthalpy** of the reaction. It is written as the ΔH of the reaction. Endothermic reactions have a positive ΔH, whereas exothermic reactions have a negative ΔH. Enthalpy takes into account both the reactant and product side of the reaction. It is the difference in the amount of the heat absorbed by bonds that are broken in the reactant molecules and the amount of heat released by the bonds made to form the product molecules.

Potential Energy (PE) of Products and Reactants in Endothermic and Exothermic Reactions

If the products have a lower PE than the reactants, then the reaction is exothermic, and energy is released. If the products have higher PE than the reactants, then the reaction is endothermic, and

energy is absorbed. The image below shows the difference between an exothermic and endothermic reaction. The direction of the arrows only indicates if energy is released or absorbed. The arrows are not representations of actual PE.

Potential Energy Level Diagrams: Exothermic Reaction vs. an Endothermic Reaction

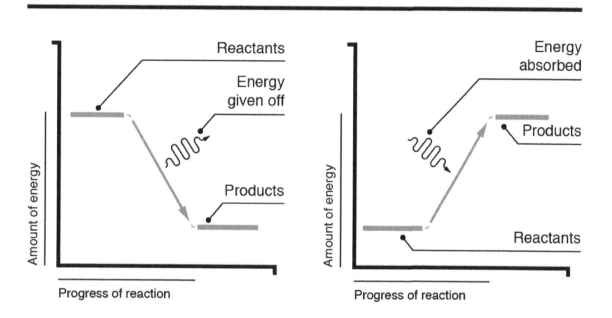

Kinetic Energy of Products and Reactants in Endothermic and Exothermic Reactions

The **total energy** of an isolated chemical system is conserved and is equal to the sum of its potential and kinetic energy (PE and KE). In an exothermic reaction, the products will have lower PE than the reactants, but the KE and temperature of the products will be greater (KE~T) compared to the reactants. Similarly, the products in an endothermic reaction will have higher PE than the reactants, but will have lower KE and temperature compared to the reactants.

Energy of Forming a Solution

There are three parts to the formation of a solution. The breaking of bonds between solute particles must occur as well as the breaking of bonds between the solvent particles. These are both endothermic processes. Then the solvent and solute particles must bond together, which is an exothermic process. Depending on the amount of heat energy needed for each of these three processes, the resulting sum of the heat energy added or released determines whether the solution is formed endothermically or exothermically, respectively.

Energy Diagrams

Energy Diagrams

Energy diagrams are graphs that depict the overall potential energy states of the reactants and products of a chemical reaction, as well as the energy level that is needed to make the reaction go forward. During the transition state, which is when bonds are partially broken and partially formed in between

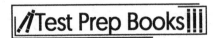

the reactants and the products, the energy level is higher than both the reactants and products. The **activation energy** is the amount of energy needed to start the reaction and reach the transition state. In endothermic reactions, the energy level of the products is greater than that of the reactants. In exothermic reactions, the opposite is true: the energy level of the reactants is greater than that of the products.

Comparison of the Potential Energy Graphs of Exothermic and Endothermic Reactions

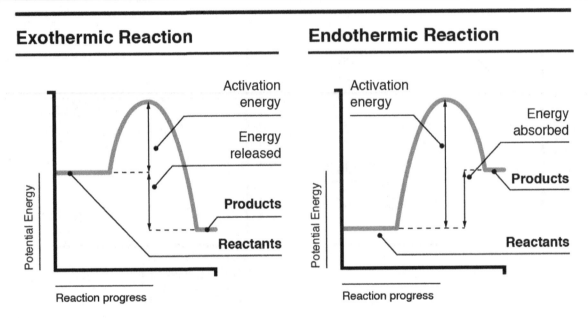

Heat Transfer and Thermal Equilibrium

Kinetic Energy of Molecules and Temperature
Because the average kinetic energy (KE) is proportional to the absolute temperature, warmer objects will contain molecules that have more kinetic energy compared to cooler objects.

Transferring Energy Through Collisions
When a set of molecules (m_1) containing more kinetic energy collides with another set of molecules (m_2) with less kinetic energy, some of that kinetic energy from m1 will be transferred to m2. The molecules that had been in m_2 now have more KE, and the temperature will also be greater because KE~T. When m_1 and m_2 collide, which may result in bond breaking, the bonds in m_2 will vibrate more or move to a greater extent because m_2 acquired more energy.

When m_1 and m_2 are in thermal or physical contact, heat can be transferred. **Heat** is the transfer of energy due to temperature differences that cause atoms or molecules to move around.

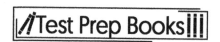

The transfer of thermal energy is due to temperature differences between the system and surroundings. In contrast, the exchange of work energy occurs when a force (*F*) is applied to object such that it moves across a certain distance (*d*):

$$W = F \times d$$

The engine motors found in automobiles contain several pistons and can be used to demonstrate work energy. The transfer of heat and work energy can be exemplified with the piston shown in the image below. Work done by the expansion of a gas is related to the previous work equation and is given by:

$$W = -P\Delta V$$

P is the result of pressure from the gas molecules that push against the piston or walls, and the change in volume (ΔV) is related to how far the piston moves. The negative sign applies if work is done by the gas such that it moves the piston upward.

Exchange of Work and Heat in a Piston

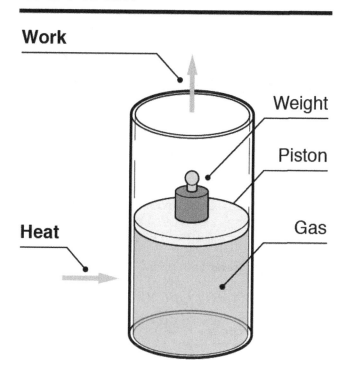

Suppose the system is defined as the gas, while the surroundings is everything other than the gas (e.g., piston and weight). If the temperature of the surroundings increases, heat will be transferred to the gas or colder body. Consequently, the average kinetic energy of the gas molecules will increase. More molecular collisions will occur against the container wall causing the pressure to increase. The gas will begin to expand. Because the gas applies a pressure (equal to Force/Area) against the piston and weight, work energy is transferred to the surroundings.

Thermal Equilibrium

Over time, as each set of molecules (m_1 and m_2) continues to collide with one another, these molecules will have the same average kinetic energy, thereby approaching a thermal equilibrium because KE~T.

Heat Capacity and Calorimetry

Heating a System and Calorimetry

When a chemical system such as water is heated, e.g., cold to hot, the kinetic energy of the system will increase because it is proportional to the temperature, T~KE. For example, one liter, or 1000 mg, of solid water at -10°C will have less energy compared to one liter of water at 10°C.

Constant pressure calorimetry experiments are conveniently carried out in labs because they allow for the calculation of heat lost or gained by a system. At constant pressure, the enthalpy of reaction, fusion, or vaporization of a substance is equal to its heat:

$$\Delta H = q_p$$

To find the amount of heat that is gained or lost, an insulator, e.g., Styrofoam® cup, is used to contain a heat bath or substance such as water, which has a known and experimentally-determined specific heat capacity of 4.184 J/g °C. The amount of heat (q) absorbed or lost by the water bath is related to the following equation:

$$q = mc\Delta T$$

The heat is equal to the product of its mass (m), the specific heat capacity (c), and the change in temperature. By placing a hot or cold object into the heat bath, a measurable temperature change can be recorded, which will allow for the quantification of heat (q).

When a chemical system or object is placed in thermal contact with the water bath, that object will either lose or gain heat. The heat lost or gained is proportional to the mass and temperature of the system and heat bath. If an unknown metal with a mass 50.0 g, at an initial temperature of 100.0°C, is thrown into a room temperature water bath (21.0°C) with a mass of 100.0 g, it will lose heat to water.

$$-q_{metal} = +q_{water}$$

If the final temperature of the metal and water bath is 25.0°C, then:

$$+q_{water} = (100.0g)(4.184 J / g°C)(25.0 - 21.0°C) = -q_{metal}$$

$$+q_{water} = 1.67 \times 10^3 J$$

$$-q_{metal} = -1.67 \times 10^3 J$$

Regardless of whether the chemical system undergoes cooling or a chemical/physical change, there will be a change in temperature whereby the magnitude of heat loss in the system is equal in magnitude to the heat gained in water. The amount of energy removed from the metal is equal to the amount of energy gained by the bath. The heat bath or surroundings gained 1.7×10^3 J of energy because its temperature increased, and the metal system lost 1.7×10^3 J energy, as its temperature decreased. In contrast, if a cold metal coin were dropped in the heat bath, the energy of the metal system would increase because its temperature went up. The heat bath would lose energy because its temperature would decrease.

The First Law of Thermodynamics

The **First Law of Thermodynamics** says that energy (E) is conserved and is described as the change in internal energy (ΔU) equal to the sum of heat (q) and work (W) energy:

$$\Delta U = q + W$$

The system and surroundings must be carefully defined to keep track of heat and work energy. The **system** is the chemical reaction, e.g., the reactants and products. The **surroundings** include everything else and not the chemical components.

Transferring Thermal Energy

In thermodynamics, as just mentioned, the terms **system** and **surroundings** are used to keep track of heat and work energy. For example, suppose heat was being transferred from a hot to a cold object. If both the cold and hot object were in direct contact with one another, then heat is transferred through conduction. The hot object would be assigned a value of -q because it is losing heat to the cool object, and the cold object would be assigned a value of +q because it is gaining heat from the hot object. The hot object can be described as the **system**, and the cold object is the **surroundings**. Therefore, the system loses heat to its surroundings. Placing a kettle containing boiling water over a cool kitchen counter would exemplify heat transfer. The kettle is heating the kitchen counter. The kettle is the system (hot body) and the kitchen counter the surroundings (cold body).

Heating a System

Heat is not matter. **Heat exchange** refers to a process where energy moves from a hot to a cold body that are in physical or thermal contact. Suppose a hot solid aluminum cube (1 in³) is placed on top of a cool solid iron cube (1 in³); heat will gradually be transferred from the hot substance (the aluminum cube) to the cooler substance (the iron cube). Heat is transferred from a hot to cold object and never the other way around. At the atomic level, the aluminum atoms vibrate, or oscillate, at a greater frequency compared to the iron atoms. The average kinetic energy (KE) of the aluminum atoms is greater than that of the iron molecules. Because the aluminum cube is in thermal contact with the iron cube, some kinetic energy is lost to the iron cube in the form of heat.

Specific Heat Capacity

Heat capacity, which is the ratio of heat change to temperature change, or Q/T, is another parameter involving gas behavior. Much like specific heat, substances that have large heat capacities don't change as much with temperature increases, but particles with small heat capacities will heat up much faster with the same temperature increase.

Heat capacity multiplied by temperature change (when volume is constant) is equal to a substance's **internal energy**. Heat capacity times a temperature change (when pressure is constant) is equal to **enthalpy**.

Water can moderate the temperature of air by absorbing or releasing stored heat into the air. Water has the distinctive capability of being able to absorb or release large quantities of stored heat while undergoing only a small change in temperature. This is because of the relatively high specific heat of water, where **specific heat** is the amount of heat it takes for one gram of a material to change its temperature by 1 degree Celsius. The specific heat of water is one calorie per gram per degree Celsius, meaning that for each gram of water, it takes one calorie of heat to raise the temperature of water by 1 degree Celsius. Coastal cities such as San Francisco tend to be cooler than inland cities because water by the coast absorbs a relatively large amount of heat from the sun.

Every chemical substance (e.g., metals, nonmetals) will not lose or absorb heat to the same capacity. Metals tend to lose or absorb heat more easily compared to nonmetal substances. The units of heat are **Joules (J)**. When a substance loses heat, a negative sign is used to indicate heat loss (e.g., -100 J). If a substance absorbs or gains heat, a positive sign is used to specify that heat is absorbed (e.g., +100 J). If +100 J was absorbed by a 20 g of copper and +100 J was absorbed by 20 g of water, the temperature change in each substance would be different. Heat capacity and specific heat capacity values are often assigned to metal and nonmetal substances. The **heat capacity** is defined as the amount of heat absorbed or lost per gram of that substance (J/g) and is an extensive property that changes with respect to the amount of substance. In contrast, the **specific heat capacity** (J/(g °C)), is an intensive property that doesn't depend on the amount of substance and is the amount of heat per gram of substance needed to raise the temperature by one degree Celsius (°C).

Changes in Energy in Chemical Reactions

If a chemical system undergoes an **exothermic reaction**, the energy of the reactants will decrease, and the produced products will have relatively lower energy. Energy lost by the chemical system (e.g., reactants and products) is transferred to the surroundings. For an **endothermic reaction**, the chemical system will absorb energy (e.g., heat or work) from the surroundings, and the resulting products will have relatively higher energy than the starting reactants. Energy lost by the surroundings is transferred to the chemical system.

Energy of Phase Changes

Energy and Phase Changes

For a substance to melt or boil, energy must be added to the chemical system. Solid to liquid and liquid to gas phase transitions will absorb energy and the chemical system will increase in energy. For instance, as the temperature of water increases from -10 to 0°C, energy will be transferred to ice, disrupting the crystal lattice structure such that individual water molecules begin to move freely. In contrast, condensation and freezing are processes that result in energy being released from the chemical system. The energy of the chemical system decreases.

Conservation of Energy as Seen in Transfers Between Systems

The amount of energy (e.g., internal energy, heat, or work) gained by the system is equal in magnitude to the amount of energy lost by the surroundings, but opposite in sign. The table below lists sign conventions that are used when keeping track of heat and work energy.

Sign Conventions (+ and -) for Heat (q), Work (w), and ΔU		
q	+ system "gains" thermal energy	System loses thermal energy
w	+ work done "on" the system'	Work done "by" the system
ΔU	+ Energy flows "into" the system	Energy flows "out of" the system

Consider a problem where the gas is the system, and everything else is the surroundings. If a gas inside a piston warms up and gains 650 J of heat, and then expands to perform 350 J on the surroundings, then the change in internal energy would be:

$$\Delta U = q + W = +650\,J + (-350\,J) = +300\,J$$

The amount of energy (e.g., internal energy, heat, or work) gained by the system is equal in magnitude to the amount of energy lost by the surroundings, but opposite in sign.

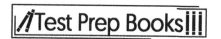

Introduction to Enthalpy of Reaction

The Enthalpy Change of a Reaction

At constant pressure, if the chemical system loses energy to its surroundings (exothermic reaction), the enthalpy change will be negative ($-\Delta H_{rxn}$). In contrast, if the chemical system absorbs energy from the surroundings (endothermic reaction), the enthalpy of reaction will be positive ($+\Delta H_{rxn}$).

Molar Enthalpy

The **molar enthalpy of fusion, ΔH_{fus},** is the amount of energy that is absorbed or released when one mole of a substance (e.g., water) changes from a solid to liquid (via melting). ΔH_{fus} is represented by the short flat line shown in the image below. At 0°C, the solid and liquid water phases will coexist and will be in equilibrium with one another. As heat is added, the temperature will remain constant.

When the substance transitions from a liquid to a solid, heat will be released. As the temperature increases from over 0°C to less than 100°C, more heat is added, and water will exist in the liquid phase. At 100°C, water is in equilibrium between the liquid and gas phase. As heat is transferred to the system, the temperature will remain constant. The **molar enthalpy of vaporization ΔH_{vap},** is the amount of energy needed to convert one mole of a liquid substance to a gaseous substance and is represented by the relatively long flat line. If heat is removed from water vapor, it will condense to liquid water. The energy when the liquid vaporizes or condenses will be equal in magnitude but opposite in sign. The image below shows the heating curve for water.

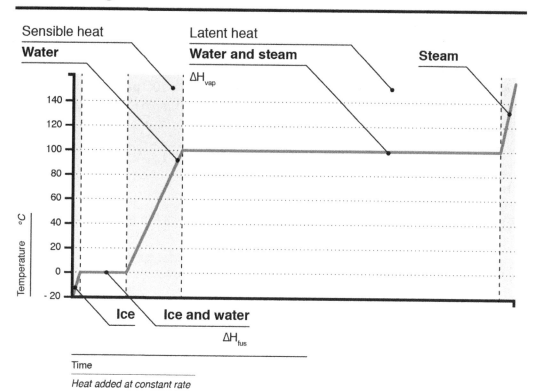

Bond Enthalpies

Changes in Potential Energy with Making and Breaking Bonds

The amount of energy needed to form or break a given bond is equal in magnitude. The **bond energy** is the energy needed to break a chemical bond. For example, the amount of bond energy needed to break the single H-H bond is about 432 kJ/mol.

Estimating the Average Energy Required to Break All of the Bonds

If the bond enthalpy or energies for every bond is known within the reactant molecule, the average energy for all bonds can be found by adding up each of the bond enthalpies. If the average energy for all bonds in the products is known, then the average amount of energy released can be determined. In an exothermic reaction, the amount of chemical energy that is released will be higher than the input energy. In an endothermic reaction, the amount of input energy will be higher than the chemical energy that is released.

Atomic Attraction and Repulsion

The potential energy of a chemical system, such as the hydrogen molecule (H_2), includes the electrostatic interaction energy between the negatively-charged electrons of one atom (H atom) with the positively-charged protons found on each hydrogen atom. The potential energy also accounts for the repulsion between the nuclei or protons on each hydrogen atom and keeps each nucleus separate. The chemical bond in the hydrogen molecule, H-H, is the result of the attractive and repulsive forces that keep each atom bound to one another. The equilibrium bond length corresponds to the energy minimum on the potential energy surface. The atoms in the bond will vibrate about this energy minimum.

Bond Energy

The **bond energy** can be described as the amount of energy required to break apart one mole of covalently-bonded gases. Bond energies are measured in kilojoules per mole of bonds (kJ/mol). Again, one mole of something is equal to Avogadro's number: $6.02214078 \times 10^{23}$ particles (or molecules, atoms, etc.). To calculate the number of moles of something, some simple formulas can be used:

$$Moles = \frac{Mass\ (g)}{Relative\ mass\ (g/mole)}$$

Example: How many moles are there in 30 grams of helium?

On the periodic table, helium's relative mass is approximately 2. Using this information and the formula yields this result:

$$Moles = \frac{30\ g}{2\ g/mole} = 15\ moles$$

During a chemical reaction, some bonds are broken, and some are formed. Bonds do not break or form spontaneously; they require energy to be added or released. The energy needed to break a bond is the **bond energy**. Generally, the shorter the bond length, the greater the bond energy.

When atoms combine to make bonds and form a compound, energy is always released, normally as heat. Certain types of bonds have similar bond energies, despite each molecule being different. For

example, all C-H bonds will have a value of roughly 413 kJ/mol. There are published lists of average bond energies for reference.

Using bond energies, it is possible to calculate the **enthalpy change** within a system. When a chemical reaction occurs, there will always be an accompanying change in energy. Energy is released to make bonds, so the enthalpy when breaking bonds is positive. Conversely, energy is also required to break bonds. Thus, the enthalpy change within a system is negative because energy is released when forming bonds.

- Some reactions are **exothermic**—where energy is released during the reaction, usually in the form of heat—because the energy of the products is lower than the energy of the reactants. In exothermic reactions, energy can be thought of as a **product.**

- Some reactions are **endothermic**—where energy is absorbed from the surroundings because the energy of the reactants is lower than the energy of the products. In endothermic reactions, energy can be thought of as a **reactant**.

It's possible to look at two sides of a chemical reaction and work out whether energy is gained or lost during the formation of the products, thus determining whether the reaction is exothermic or endothermic. Here is an example:

Two moles of water form two moles of hydrogen and one mole of oxygen:

$$2H_2O(g) \rightarrow 2H_2 + O_2(g)$$

The sum on the reactant's side (2 moles of water) is equal to four lots of H-O bonds, which is 4 x 460 kJ/mol = 1840 kJ/mol. This is the input.

The sum on the product's side is equal to 2 moles of H-H bonds and 1 mole of O=O bonds, which is 2 x 436.4 kJ/mol and 1 x 498.7 kJ/mol = 1371.5 kJ/mol. This is the output.

The total energy difference is 1840 − 1371.5 = +468.5 kJ/mol.

Because the energy difference is positive, it means that the reaction is endothermic and that the reaction will need energy to be carried out.

Enthalpy of Formation

Calculating the Standard Enthalpy of Reactions

Consider the following hypothetical chemical equation where the formation of Product C can be written in two steps with the enthalpy of reactions given for each:

$$2A + B \rightarrow 2C \quad Overall\ equation$$

$$Step\ 1: 2A + 2B \rightarrow 2D\ ; \Delta H_1 = -800\ kJ/mol$$

$$Step\ 2: 2C + B \rightarrow 2D; \Delta H_2 = -600\ kJ/mol$$

For a given reaction step (e.g., Step 1), the enthalpies of formation ΔH_f° for all reactants and products are often used to find the standard enthalpy of reaction ΔH_{rxn}°, which is given by:

$$\Delta H_{rxn}^\circ = \sum n\Delta H_f^\circ \, (products) - \sum m\Delta H_f^\circ \, (reactants)$$

To understand this equation, suppose that ΔH_f° for A, B, and D in Step 1 above are +200, 0.00, and -200 kJ/mol, respectively. Then:

$$\Delta H_{rxn}^\circ = \left[\overset{D}{\overbrace{\overset{n}{2} \times (-200)}} \right] - \left[\overset{A}{\overbrace{\overset{m}{2} \times (+200)}} + \overset{B}{\overbrace{\overset{m}{2} \times (0.00)}} \right] = -800 \; kJ/mol$$

$$\Delta H_{rxn}^\circ = (-400) - [(+400) + (0.00)]$$

$$\Delta H_{rxn}^\circ = -400 - [400] = -800 \; kJ/mol$$

There is only one product, D, which is multiplied by its coefficient (n = 2 from equation) and ΔH_f°. The procedure for the reactants is similar, except that ΔH_f° for A and B are summed (Σ). Tables of enthalpies of formation ΔH_f° are often given for different elements and compounds. For pure molecular elements and solid metals, ΔH_f° is set to a reference value of zero. In some cases, the ΔH_{rxn}° of a metal from its metal oxide may be given by $\Delta H_{rxn}^\circ = n\Delta H_f^\circ$, where ΔH_f° corresponds to the metal oxide. Because ΔH_f° for metals is zero, it is easy to compare ΔH_{rxn}° for the extraction of different metals.

Using Calorimetry to Determine the Heat Capacities, Enthalpies of Vaporization, Enthalpies of Fusion, and Enthalpies of Reactions

Constant pressure calorimetry will quantify the change in energy for the system and surroundings and can be used to find the specific heat or enthalpies of different substances (e.g., ΔH_{rxn}, ΔH_{vap}, ΔH_{fus}). For instance, from the previous section, because

$$-q_{metal} = -1.67 \times 10^3 \; J$$

$$-q_{metal} = (50.0 \; g)(c_{metal})(25.0 - 100.0°C)$$

$$c_{metal} = \frac{-q_{metal}}{(50.0g)(-75.0°C)} = \frac{-1.67 \times 10^3 J}{-3750.0°C} = 0.45 \; J/g \; °C$$

The dissolution of 1.00 g of calcium chloride ($CaCl_2$) into a 100.0 g water bath is an exothermic reaction that will transfer energy into the heat bath. The enthalpy of reaction for the salt is:

$$+q_{water} = -q_{salt}$$

$$\Delta H_{rxn} = \frac{-q_{salt}}{1.00 \; g \; salt} \; ; units \; of \; J/g$$

Hess's Law

Thermal Energy Transfer with Surroundings

Thermal energy will be transferred from the chemical system to the surroundings, or from the surroundings to the system, to reach thermal equilibrium because heat moves from hot to cold. In an exothermic reaction, thermal energy will be transferred from the chemical system, or hot products, to the surroundings. In an endothermic reaction, the cold products will absorb thermal energy or heat from the surroundings.

Hess's Law

Hess's law explains that an overall or net chemical equation can be written in two or more steps and is equal to the sum of enthalpy changes for each reaction step. Essentially, it says that the energy change of any reaction or transformation is the same whether it occurs in one step or many. This means, for example, the energy needed to melt a block of ice is the same, whether the ice is melted all at once versus in multiple steps.

Recall again the formation of Product C, which can be written in two steps with the enthalpy of reactions given for each:

$$2A + B \rightarrow 2C \quad Overall\ equation$$

$$Step\ 1: 2A + 2B \rightarrow 2D\ ;\ \Delta H_1 = -800\ \text{kJ/mol}$$

$$Step\ 2: 2C + B \rightarrow 2D;\ \Delta H_2 = -600\ \text{kJ/mol}$$

To find the enthalpy change, ΔH, in the overall equation, Steps 1 and 2 must be combined such that the correct reactants and products, with the proper stoichiometric coefficients, are equal in the overall equation. Step 1 shows that A and B are on the reactant side just like in the overall equation. However, Step 2 shows C on the reactant side and not the product side. The reaction in Step 2 must be reversed to obtain C on the product side, and the sign of ΔH_2 must be reversed according to Hess's law ($600\ kJ/mol$). Once Step 2 is reversed, the overall equation is:

$$2A + 2B \rightarrow 2D\ ;\ \ \Delta H_1 = -800\ \text{kJ/mol}$$

$$2D \rightarrow 2C + B;\ \ \Delta H_2 = +600\ \text{kJ/mol}$$

$$-----------------------$$

$$2A + B \rightarrow 2C$$

$$\Delta H = \Delta H_1 + \Delta H_2 = -800 + 600 = -200\ \text{kJ/mol}$$

Note that the two reactions are summed to give the final overall reaction, and that the enthalpies of the reaction for both steps are summed to give the enthalpy change for the net equation.

Equilibrium

Introduction to Equilibrium

Reversible Reactions in the Lab and Real World

Many chemical reactions are **reversible**, which means that the reactants can form the products, or the products can react to form the reactants. Depending on the environmental and experimental conditions, the reaction can be driven in either direction. Substances can move between different states of matter by changing the conditions of the reaction. When heat is added to water, it will become gaseous water vapor. If the temperature is lowered, i.e., heat is removed from the environment, the water vapor will condense into liquid water. Heat can also be used to dissolve a salt in water to make an aqueous solution. The conditions can be changed to then precipitate the salt out of the aqueous solution. The transfers of protons and electrons in acid-base and redox reactions, respectively, are also reversible reactions. The conditions of the solutions can cause acid and base ions to either disassociate or reattach to each other. Similarly, electrons can be transferred back and forth in chemical equations that involve both reduction and oxidation reactions.

There are many relevant, biological examples of reversible reactions that we experience every day. Hemoglobin is a protein that is used to transport oxygen in the blood throughout the human body. Oxygen molecules bind to the hemoglobin and when hemoglobin reaches the cells and tissues that need oxygen, the reaction is reversed, and the oxygen molecules are released from the hemoglobin molecule. As another example, olfactory molecules bind to the receptors in the nose to trigger a reaction from the brain. Once the reaction is triggered, the molecules are released.

The environment also has many reversible reactions that occur every day. Carbon is constantly transferred between the biosphere and the atmosphere. When living organisms exhale, they release carbon dioxide from their bodies into the atmosphere. Plants and other anaerobic elements of the biosphere then reabsorb carbon dioxide from the atmosphere and use it to produce energy. The **hydrosphere** is the system of the Earth that contains all of its water, such as oceans, seas, and lakes. Many substances—such as carbon dioxide, sulfur, and nitrogen—dissolve easily in water. When the water heats up, such substances can enter the atmosphere as a gas. The cycle continues when the substances condense again with drops in temperature and reenter the hydrosphere.

Equilibrium Conditions

Chemical reactions reach a state of equilibrium when the rate of the forward reaction is equivalent to the rate of the reverse reaction. The concentrations of the reactants and products are proportional to each other with regard to the stoichiometry of the equation and are no longer changing at equilibrium. The reaction quotient, Q, represents the proportional concentrations of reactants and products in the equation and at equilibrium, it is equal to the equilibrium constant, K. The equilibrium constant can be calculated from the equilibrium concentrations of the products and reactants of a chemical reaction. In the reaction $2H_2 + S_2 \rightarrow 2H_2S$, the equilibrium concentrations of the components are 0.25 moles $2H_2$, 0.625 moles S_2, and 0.75 moles $2H_2S$. K can be calculated using the following generic equation for the chemical reaction aA + bB $\rightarrow$ cC + dD, using the equilibrium concentrations of the reactants and products:

$$K = \frac{[C]^c[D]^d}{[A]^a[B]^b}$$

For hydrogen and sulfur reacting, K can be calculated as:

$$K = \frac{[H_2S]^2}{[H_2]^2[S_2]}$$

Given the equilibrium concentrations:

$$K = \frac{(0.75)^2}{(0.25)^2(0.625)} = 14.4$$

Using Concentration Over Time Graphs to Understand the Establishment of Chemical Equilibrium

Simple graphs can be used to see the changes in the rate of a reaction and in the concentrations of the reactants and products as a reaction progresses and reaches equilibrium. The rates of the forward and reverse reactions equalize at equilibrium. Depending on the dynamics of the reaction, there may be

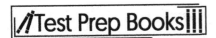

more products or more reactants at the equilibrium, but both concentrations should change and then become steady as equilibrium is reached.

Reaction Rate vs. Concentration Graphs

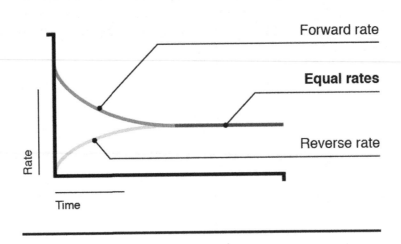

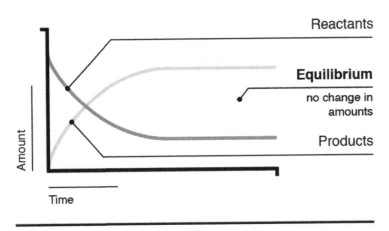

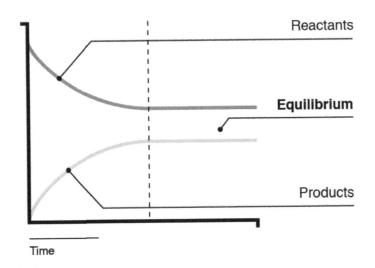

Direction of Reversible Reactions

Net Conversion Between Reactions and Products Based on Rate vs. Equilibrium State

In reversible reactions, if the forward reaction proceeds at a faster rate than the reverse reaction, there is a net conversion of the reaction's reactants to the products. On the other hand, if the reverse reaction proceeds at a faster rate than the forward reaction, there is a net conversion of the reaction's products to the reactants.

Reaction Quotient and Equilibrium Constant

The Reaction Quotient, Q

The reactant quotient, Q, measures the proportions of the reactants and products at any given point in the reaction, not just at equilibrium. In the equation aA + bB → cC + dD, the reaction quotient can be calculated as:

$$Q = \frac{[C]^c[D]^d}{[A]^a[B]^b}$$

The value of Q gives a picture of the progression of the reaction. As its value approaches that of the equilibrium constant K, the reaction gets closer to an equilibrium state. When Q = K, equilibrium is reached.

Q and K in Reversible Reactions

When Q < K, product formation is favored, and if the reaction is reversible and Q > K, reactant formation is favored. At the start of a reaction that has only reactants, the concentrations of A and B are high, and C and D are zero, so Q would be zero. As the reaction progresses to produce more products, the value of Q can become infinitely high. The measure of Q is only dependent on concentrations of reactants and products that change as the reaction progresses. It is not affected by substances whose concentrations are independent of the reaction, such as with solids that are in contact with an aqueous solution or gas or pure solids or liquids that are in contact with a gas. When two reactions occur consecutively due to the formation of an intermediary substance, the Q and K values of the resulting reaction are calculated by multiplying the individual Q values and individual K values for each equation.

Important Reversible Reactions

Examples of common and important reversible reactions include the transfer of electrons between species in oxidation-reduction reactions, the transfer of protons in acid-base reactions, and the dissolution of a solid in a solution.

Calculating the Equilibrium Constant

Determining Equilibrium Constants in Experiments or the Concentrations at Equilibrium

If the value of K is known, as well as the initial concentrations, the concentration of the products or reactants at equilibrium can also be calculated. Take the conversion of nitrogen tetroxide, N_2O_4, to nitrogen dioxide, NO_2, for example. The chemical reaction is written as $N_2O_4 \rightarrow 2NO_2$. The equilibrium constant, K, for the reaction is 4.6×10^{-3}. If the initial concentrations are 1 mole of N_2O_4 and 0 moles of NO_2, an ICE table can be used to calculate the equilibrium concentrations of the reactant and product.

Equation	N_2O_4	$2NO_2$
Initial Moles	1	0
Relative Change in Moles	-X	2X
Equilibrium Moles	1 - X	0 + 2X

$K = 4.6 \times 10^{-3} = (2X)^2 / (1-X)$

$X = 0.033$.

Thus, the following concentrations exist at equilibrium:

Equation	N_2O_4	$2NO_2$
Initial Moles	1	0
Relative Change in Moles	-X	2X
Equilibrium Moles	0.967	0.066

Magnitude of the Equilibrium Constant

Reasoning Qualitatively About Equilibrium Systems

Every chemical reaction has one specific value of K at a given temperature. The value of K is different for the same reaction at different temperatures. Large values of K represent more products than reactants at equilibrium. Small K values represent more reactants than products at equilibrium. Aqueous reactions often have either very large or very small values of K because the reactants either dissolve readily in water or they do not. There are few substances that will reach equilibrium when only half of the ions have been dissociated. Looking at the relative magnitude of K can give a qualitative insight into the drive of the chemical reaction and how equilibrium is reached.

Properties of the Equilibrium Constant

Reversing Reactions

When a reaction runs in the reverse direction, such that the original products become the reactants, the value of K is inverted. Written as K', the equilibrium constant for the reverse reaction is the reciprocal of that for the forward reaction, or $K' = \frac{1}{K}$.

K and the Coefficients of a Reaction

For any given reaction, if the stoichiometric coefficients are multiplied by a common factor, the equilibrium constant, K, is raised to that common factor: $K' = (K)^n$, where K' is the equilibrium

constant for the multiplied reaction, K is the equilibrium constant for the original reaction, and n is the factor by which the stoichiometric coefficients are multiplied. Consider our previous reaction: $N_2O_4 \rightarrow 2NO_2$. The equilibrium constant, K, for the reaction is 4.6×10^{-3}. If we were to multiply the entire reaction by 2, our K' value would be K squared:

$$2\ N_2O_4 \rightarrow 4NO_2,\ K' = (4.6 \times 10^{-3})^2 = 2.116 \times 10^{-5}$$

K and Adding Reactions

When multiple reactions are added together, the equilibrium constant, K, for the overall reaction is simply the product of the K's for the added reactions: $K' = K_1 \times K_2 \times K_3$ ..., where K' is the equilibrium constant for the desired overall reaction, K_1 is the equilibrium constant for the first reaction, K_2 is the equilibrium constant for the second reaction added, and so on. Consider trying to calculate the equilibrium constant for the reaction: $2\ NO(g) + Br_2(g) \rightleftarrows 2\ NOBr(g)$. To do so, the following two reactions and their respective equilibrium constants are provided:

$$2NO(g) \rightleftarrows N_2(g) + O_2(g);\ K_1 = 1 \times 10^{30}$$

$$N_2(g) + Br_2(g) + O_2 = (g) \rightleftarrows 2NOBr(g); K_2 = 2 \times 10^{-27}$$

Thus, $K' = K_1 \times K_2 = (1 \times 10^{30}) \times (2 \times 10^{-27}) = 2 \times 10^3$.

Algebraically Manipulating Q and K

Any of the aforementioned manipulations of K can be combined together. Additionally, because K and Q have identical mathematical forms, it is possible to conduct these same manipulations discussed for K on Q.

Calculating Equilibrium Concentrations

ICE Tables and Concentrations of Reactants and Products

An ICE table is a useful tool for calculating the changes in concentrations of reactants and products to reach equilibrium from different stages of a reaction. In an **ICE table**, there are three rows of information: (I) initial concentration, (C) change in concentration, and (E) equilibrium concentration. The concentrations of the reactants and products must be proportional to the stoichiometry of the equation. Take the example of the reaction of hydrogen and sulfur to form hydrogen sulfide: $2H_2 + S_2 \rightarrow 2H_2S$. Let's say the initial concentrations are known to be 1 mole each of hydrogen and of sulfur, and when equilibrium is reached, there are 0.75 moles of H_2S. An ICE table can be used to determine the changes in the concentrations and the equilibrium concentrations of all of the substances in the chemical reaction.

Equation	$2H_2$	S_2	$2H_2S$
Initial Moles	1	1	0
Relative Change in Moles	-2X	-X	2X
Equilibrium Moles	?	?	0.75

From the table, you can see that the 2X change in moles of hydrogen sulfide is equal to 0.75 moles. The coefficient "2" comes from the stoichiometry of the chemical reaction. Once X is calculated, the rest of

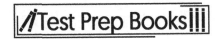

the table can be filled in and the equilibrium concentrations for all of the components in the reaction can be calculated.

Equation	$2H_2$	S_2	$2H_2S$
Initial Moles	1	1	0
Relative Change in Moles	-2X	-X	2X
Equilibrium Moles	0.25	0.625	0.75

Representations of Equilibrium

Particulate Representations of Equilibrium Concentrations

Recall that **particulate representations** are visual representations of different molecules and how they change within a chemical reaction. They can be useful for looking at the relative amounts of each element in a reaction and the stoichiometry of the equation. They can also illustrate the relationship of reactants and products at equilibrium. When looking at a particulate representation, it's important to understand what the white space represents. In an aqueous solution, white space, unless otherwise indicated, represents water because it would be too difficult and crowded to show all of the water molecules in the particulate representation. Contrary to a common misconception, in a gas, the white space does not represent air, unless it specifically is noted as such. Instead, it represents empty space or a vacuum.

Introduction to Le Chatelier's Principle

Using Le Chatelier's Principle to Predict the Response of a System to Different Stresses

Le Chatelier's principle is a law of equilibrium that states that if a system at equilibrium is subjected to a change in the environment, the system will react accordingly to counter the change and restore its state of equilibrium. Disruptions to the system can include changes to the temperature, volume, or pressure of a gas, or the concentration of any of the reactants or products. The equilibrium constant, K, only changes with changes to the temperature of the system. With all other changes, the value of K remains the same. For exothermic reactions, energy, or heat, is released from the system. If the temperature is decreased, or energy is taken out of the system, formation of the products would be favored because more energy, or heat, would need to be produced to restore equilibrium and vice versa for energy being added to an exothermic reaction.

For an endothermic reaction, in the same situation of energy being taken away from the system, the formation of the reactants would be favored because energy is needed to form the products. If energy was added, the reactants would use the energy to react and form more products. When the volume of a gas increases, the pressure decreases and vice versa. If volume increases and pressure decreases in a system, the reaction will run toward the side of the equation that has the largest number of gas moles to increase the pressure of the system and restore equilibrium. With a decrease in volume and an increase in pressure, the reaction would run toward the direction of the fewest number of gas moles.

If the concentration of the reactants was increased or that of the products was decreased, the reaction would run in the forward direction to form more product/s. If the concentration of the reactants was decreased or that of the products was increased, the reaction would run in the reverse direction to form more reactants. Diluting the reactants with water is one way in which the concentration of the reactants could be decreased.

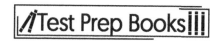

Some reactants and products have different colors. Looking at the mix of colors of the solution at equilibrium can elucidate the proportion of reactants and products in the solution. For example, if the reactant is blue, the product is yellow, and at equilibrium the solution is perfectly green, a green color is evidence that the reactant and product may be present in equal amounts. The pH of a solution can also be checked to monitor the dissociation of acids and bases in a solution at equilibrium.

Qualitative Reasoning with Le Chatelier's Principle

Le Chatelier's principle states that when a property of a chemical reaction changes and the reaction becomes unbalanced, the system will compensate for the property that changed to restore equilibrium to the system. Equilibrium concentrations can be calculated quantitatively using ICE tables that show the initial concentration of the substances in a reaction and the change that takes place. The value of K—the equilibrium constant—indicates the proportion of reactants and products that are present at equilibrium. At equilibrium, both reactants and products are present; most reactions do not completely favor one side of the reaction. The rates of the forward and reverse reactions are equal at equilibrium so that the concentrations of the reactants and products are constant and do not change. Q—the reaction quotient—becomes equal to K at equilibrium. If a change in the system created more products, the reaction would run in the reverse direction to create more reactants in order to restore equilibrium and bring the products and reactants back to their equilibrium proportions. If the change created more reactants, the reaction would run in the forward direction to create more products until the equilibrium proportions were reached again.

Reaction Quotient and Le Chatelier's Principle

Stresses that Shift Q or K

The reaction quotient, Q, can be measured at any given point of the chemical reaction. Once the reaction has reached equilibrium, Q becomes equal to K, the equilibrium constant. K describes the state of the reactants and products when the reaction has reached equilibrium and the concentrations are no longer changing with time. The forward and reverse reactions occur at equal rates to keep the reactant and product concentrations steady. In reality, stresses can occur to change the equilibrium status of the chemical reaction. When a stress occurs, the value of Q changes immediately and is no longer equivalent to the value of K.

Changes in volume, pressure, concentration, and temperature all cause the value of Q to change. The reaction then shifts to counteract the stressor and bring the values of Q and K back to equivalency. If there was an increase in any of these properties, the reaction would run in the opposite direction to decrease the increases. As mentioned, temperature is the only stressor that also changes the value of K. The value of K is dependent on steady state conditions, which includes a certain temperature. If the temperature changes, the value of K changes, and the concentrations of the reactants and products at equilibrium changes.

Introduction to Solubility Equilibria

Solubility Products and the Solubility of Salts

When a salt substance is dissolved in a solvent, it is considered a **reversible reaction** because the substance can be precipitated out of the solvent again. The reaction quotient, Q, of the reaction describes the concentration of the reactants and products at any time of the reaction. The equilibrium constant for solubility reactions is K_{sp}. The value of K_{sp} is dependent on the solubility of the salt in the solvent. Large K_{sp} values indicate greater dissociation of the salt molecule and greater solubility.

The Free Energy Change for Dissolution of a Substance

The **thermodynamic favorability**, or **free energy change ΔG**, of solubility of a salt in a solvent is dependent on the sum of three factors. The first is the separation of the salt ions from each other in the molecule. This requires breaking bonds and is an endothermic process. The second is the separation of the solvent ions from each other, which is another endothermic process due to the breaking of molecular bonds. The third is the creation of new molecules through the formation of new bonds, which is an exothermic process. Another factor to consider is the entropy of the system. When a salt dissolves in a solvent, entropy increases, which is thermodynamically favorable. If the sum of the endothermic and exothermic reactions is favorable and the entropy increases, the salt will be favorably soluble in the solvent.

Common-Ion Effect

The Common-Ion Effect

Salts with the greatest solubility in water are sodium, potassium, and nitrate salts, due to their polar nature. Polar solutes dissolve readily in polar solvents, which includes water because it is a polar molecule. If the solvent solution of the reaction has an ion in common with the salt molecule, the solubility of the salt will be affected in a negative manner. From the understanding of Le Chatelier's principle, the addition of a solvent with the same ions as the salt disrupts the system as if more salt was added to the reaction. The reaction would not run in the regular direction to achieve equilibrium. It also explains why salts do not readily dissolve in sea water and other natural bodies of water.

pH and Solubility

The Effect of pH on the Solubility of a Salt

Just as the common-ion effect influences the solubility of a salt per Le Châtelier's principle, so too does pH. If one of the ions comes from an acid or a base, the solubility of the salt is affected by the pH of the solution. One example is acid mine drainage and the production of ferrous hydroxide. Ferrous disulfide is commonly found in the waste rock of mines. Through many oxidation-reduction reactions, it becomes hydrogen ions, sulfate ions, and ferrous hydroxide. When acid mines drain into a stream, the acidity of the stream is determined by its pH value and free hydrogen ions. However, the total acidity would be a better gauge of the drainage because of the high level of dissolved ferrous hydroxide in the stream. Another example is the increased solubility of carbonates in acid rain compared to pure rain water because calcium bicarbonate can be formed between the carbonates and carbon dioxide.

Free Energy of Dissolution

Free Energy Change, ΔG_0

Reactions must use energy to do work, and that available energy is called **Gibbs free energy (G)**. Gibbs free energy, or G, represents the thermodynamic work potential for a system that is at a constant temperature and pressure (it may help to know that Gibbs free energy used to be called "available energy"). Free energy can be found by identifying the changes in enthalpy and entropy of a system.

$$\Delta H_{reaction} = \Delta H_{products} - \Delta H_{reactants}$$

$$\Delta S_{reaction} = \Delta S_{products} - \Delta S_{reactants}$$

A reaction can only be **spontaneous,** or occur without any influence of an outside force, if G is negative, and G depends on both entropy and enthalpy:

$$G = \Delta H - T\Delta S,$$ where T is in Kelvin and can never be negative.

Gibbs Free Energy (G)

$G = \Delta H - T \Delta S$

Summary Spontaneous and Non-Spontaneous Reactions

	$\Delta H > 0$	$\Delta H < 0$
$\Delta S > 0$	**Spontaneity depends on T** spontaneous at higher temperatures	**Spontaneous at all temperatures**
$\Delta S < 0$	**Nonspontaneous** proceeds only with a continous input of energy	**Spontaneity depends on T** spontaneous at lower temperatures

Process	Products	Reactants	Sign	Meaning
Enthalpy	Lower #	Higher #	-	*Exothermic (energy released)*
Enthalpy	Higher #	Lower #	+	Endothermic (energy absorbed)
Entropy	Lower #	Higher #	+	*More disorder*
Entropy	Higher #	Lower #	-	Less disorder
BOLDED reactions are favorable				

In terms of **collision theory**:

- *H > 0 and S > 0:* If an endothermic reaction has high entropy and temperatures are low with slow particle movement, free energy (energy not invested into other chemical reactions) will be used up. However, if *T* is high enough, the extreme kinetic energy will make G negative and the reaction will be spontaneous.

- *H > 0 and S < 0:* If an endothermic reaction has low entropy, the reaction will require additional energy to proceed. This reaction has two unfavorable properties and will never be spontaneous.

- *H < 0 and S > 0:* If an exothermic reaction has high entropy, the reaction will always be spontaneous. It will never require additional energy since it has two favorable properties.

- *H < 0 and S < 0:* If an exothermic reaction has little random particle movement due to low temperatures, the reaction may not be spontaneous. Movement might be too low and slow for the reaction to proceed (collision theory).

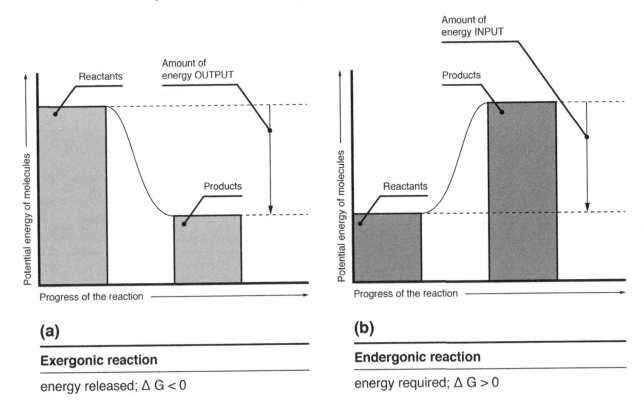

(a)

Exergonic reaction

energy released; Δ G < 0

(b)

Endergonic reaction

energy required; Δ G > 0

Acids and Bases

Introduction to Acids and Bases

PH and POH

The strength of acids and bases is determined by their ability to dissociate and form H_3O^+ and OH^-, respectively. Acids can be measured using a **pH scale**, which measures concentration of H_3O^+, and bases can be measured using a **pOH scale**, which measures concentration of OH^-.

Water Autoionization

Water can act as either an acid or a base. When mixed with an acid, water can accept a proton and become an H_3O^+ ion. When mixed with a base, water can donate a proton and become an OH^- ion. Sometimes water molecules donate and accept protons from each other; this process is called **autoionization**. The chemical equation is written as follows: $H_2O + H_2O \rightarrow OH^- + H_3O^+$.

Pure Water

Water autoionizes and produces some H_3O^+ and OH^- ions in a reversible reaction from water molecules interacting with each other. Pure water is a neutral solution. This means that there is always an equal number of H_3O^+ and OH^- molecules, and the pH = pOH, which equals 7.0 at 25 °C. The equilibrium constant of water is written as K_w and is equal to 10^{-14} at 25 °C.

Temperature and K_w

It is important to note that the K_w of water is temperature dependent. As such, at temperatures other than 25 °C, the pH of pure, neutral water will deviate from 7.0.

Ions in Liquid Solution

Generally, when a molecule becomes part of a solution, its ionic components are attracted to other ions to form neutrally charged entities. For example, when NaCl is added to water, the Na^+ ions are attracted to the OH^- ions and the Cl^- ions are attracted to the H^+ ions. Ions that are left in the solution can be determined by measuring the conductivity of the solution. Solute molecules that dissociate easily and exist almost completely as ions are called **strong electrolytes**. Solute molecules that remain mostly as molecules with few ions dissociated are called **weak electrolytes**. Strong electrolytes can conduct electricity through the solution by transferring electrons.

pH and pOH of Strong Acids and Bases

Strong Acids

Strong acids completely dissociate in solution to produce H_3O^+ ions and acid anions. Because no acid molecules remain whole, the concentration of the acid is equivalent to the concentration of the H_3O^+ ions and the pH of the acid can be easily calculated, as pH = $-\log[H_3O^+]$. Examples of strong acids include HCl, HBr, HI, $HClO_4$, H_2SO_4, and HNO_3.

Strong Bases

Similarly, strong bases dissociate completely to produce OH^- ions in solution. Groups I and II from the periodic table combine with hydroxides to form strong bases.

Weak Acid and Base Equilibria

Weak Acids

Weak acids only partially ionize in water, so the concentration of H_3O^+ ions does not equal the initial concentration of the acid. Most of the acid molecules remain unionized. Equilibrium is reached for weak acids when the unionized acid molecules are equivalent to the conjugate base. The pH of a weak acid can be calculated from the initial acid concentration and the pK_a of the acid, which is the inverse log of the acid dissociation constant, K_a. The magnitude of K_a is dependent on structural factors of the acid, including bond strength, solvation, and electronegativity of the atoms. Larger K_a values indicate a greater propensity for the acid to dissociate. Examples of weak acids include carboxylic acids, formic acid, hydrofluoric acid, and hydrogen sulfide.

Weak Bases

Ammonia, amines and pyridines, nitrogenous bases, and conjugate bases are all examples of weak bases. They do not react completely with water, and many unionized molecules remain in the solution. The concentration of hydroxide ions does not equal the initial concentration of the base as it does with strong bases. Equilibrium is reached when there is a mix of unionized weak base molecules and conjugate acid in the solution. The equilibrium constant, pK_b, is equivalent to the inverse log of the base dissociation constant, K_b. Using pK_b and the initial concentration of the base, the pH of the solution can be calculated.

Neutralization Reactions

When acids and bases react together, they create a neutralization reaction because the hydrogen ion from the acid and the hydroxide ion from the base produce a water molecule. They also generally form a salt molecule from the cation and anion of the acid and base. **Neutralization reactions** usually have an equilibrium constant, K, with a value greater than one, so the reaction goes to completion until all of the hydrogen and hydroxide ions have been used to form water molecules.

Percent Ionization of a Weak Acid or Base

Weak acids and bases undergo only partially dissociate at equilibrium. The percent ionization of a week acid or base is a measure of the relative number of acid or base molecules that dissociate. We can use the fact that the ionization of weak acid occurs at equilibrium to calculate the percent ionization of a weak acid. Consider the following chemical equation:

$$HA + H_2O \leftrightarrows H_3O^+ + A^-$$

For this system:

$$K_a = \frac{[H_3O^+] + [A^-]}{[HA]}$$

Remember that to convert pH to $[H^+]$, pH = -log$[H^+]$, where $[H^+]$ refers to the molarity of protons in the solution.

Acid-Base Reactions and Buffers

Reactions Between a Strong Acid and Strong Base

When a strong acid and strong base mix together, the chemical reaction that occurs is $H_3O^+ + OH^- = H_2O$. The equilibrium constant, K, for the formation of water from ions is 10^{-14} at 25 °C, so the reaction goes

to completion, where all of the acid and base molecules are dissociated and form salt and water molecules. The pH of the final solution is determined by the limiting reactant, which is the lower concentration reactant, either acid or base. Once all of the limiting reactant is dissociated and used to make water and salt molecules, the remaining dissociated ions of the excess reactant contribute to the pH of the solution.

Reactions Between a Weak Acid and Strong Base

Neutralization reactions also occur between strong bases and weak acids and between strong acids and weak bases. At the same pH, a greater quantity of strong base is required to neutralize a weak acid than a strong acid because the initial concentration of the weak acid is higher. The weak acid resists neutralization by the strong base for a longer time because of the number of unionized molecules present.

Reactions Between a Weak Base and Strong Acid

The same situation that occurs between a strong base and weak acid occurs between a strong acid and a weak base. This situation can be represented by the following chemical equation:

$$B(aq) + H_3O^+(aq) \leftrightarrows HB^+(aq) + H_2O(l)$$

A buffer solution is formed in cases where the weak base is in excess. In such cases, the Henderson-Hasselbalch equation can be used to determine the pH of the solution. The pH must be determined from the moles of excess hydronium ion and the total volume of the solution in cases where the strong acid is in excess. If the strong acid and weak base and equimolar, the solution will be slightly acidic and the pH can be determined via the following equation:

$$HB^+(aq) + H_2O(l) \leftrightarrows B(aq) + H_3O^+(aq)$$

Reactions Between a Weak Acid and a Weak Base

The equilibrium state reached in reactions between a weak acid and weak base can be represented by the following equation:

$$HA(aq) + B(aq) \leftrightarrows A^-(aq) + HB^+(aq)$$

Acid-Base Titrations

Titration for Neutralization Reactions

Titration experiments can aid in quantifying the changes in reactants that are occurring in neutralization reactions. When a base is added to a strong or weak acid, the pH does not change much until the reaction comes close to the equivalence point. Then, once the **equivalence point** (the point when the moles of the titrant and the moles of the titrate are in stoichiometric proportion) is neared, the pH changes rapidly. A color-changing chemical can be added to the experiment to visually determine when the equivalence point is drawing near. Once the equivalence point is determined, it can be used to determine the concentration of the titrant. The pH of the buffer can be calculated at any time using the **Henderson-Hasselbalch equation**:

$$pH = pK_a + log \frac{[A^-]}{[HA]}$$

At the halfway point to the equivalence point, the ratio of anion to acid concentration is 1:1, making the log of the ratio equal to zero. Therefore, the pK_a of the acid is equal to the pH of the solution at that point. The contents of strong acid solutions and weak acid solutions differ at the halfway point to the equivalence point. For a strong acid, the main species are hydroxide ions, anions from the acid, and cations from the base. For a weak acid, the main species are hydronium ions, anions from the acid, cations from the base, and undissociated acid molecules. For both types of acids, at that point, the total positive charge of the solution is equivalent to the total negative charge.

Using Titration Curves to Evaluate the Number of Labile Protons for Polyprotic Acids
Polyprotic acids are acids that have more than one hydrogen ion to donate. The concentration of these acids can still be determined using a titration curve, but the titration curve will have multiple equivalence points. The number of equivalence points is equal to the number of labile protons available to donate. From the molecular composition, the concentrations of the different ionic species present at any region of the curve can be determined.

pH and pK_a

pH vs PK Relationship
The **protonation state** of a conjugate acid-base pair describes where the proton of the molecule is at that given time or that given pH. If the proton is attached to the molecule, the acid is protonated. If the proton is not attached to the molecule, the conjugate base is deprotonated. Comparing the pH and the pK_a of a solution can elucidate the ratio of the acid and base forms, and the protonation state, of the solution at that pH. If the pH is lower than the pK_a, there is more acid present than deprotonated conjugate base. If the pH is higher than the pK_a, there is more conjugate base present than acid molecules. This information is useful for determining the required pH of a solution needed for acid-catalyzed chemical reactions and when using acid-base indicators in an experiment. Identification of the protonation state also helps with determining labile protons in acids and protein side chains of different molecules.

Acid-Base Indicators
pH can be measured using a pH meter, test paper, or indicator sticks. Acid-base indicators, in general, are special substances (usually weak acids or bases) that have certain properties that respond to the pH of a solution because these certain properties express themselves differently in their protonated versus deprotonated state. The most common variable property of acid-base indicators is color. Phenolphthalein is a common acid-base indicator. It is colorless, weak acid that dissociates when added to water, forming pink ions when the solution is basic enough that the pH is high. Under acidic conditions, the anion concentration is too low for the pink color to be detected.

Properties of Buffers

Buffer Solutions
Buffer solutions resist changes in pH when small amounts of acid or base are added. Weak acids and bases make good buffers because they require a large amount of the opposite type of solution to effect a change in pH due to the large quantity of unionized molecules that are present.

Henderson-Hasselbalch Equation

pH Changes in Buffered vs. Unbuffered Solutions

The pH of a solution can be determined using the Henderson-Hasselbalch equation, which is written as:

$$pH = pK_a + log\frac{[A^-]}{[HA]}$$

The acid dissociation constant, K_a, is constant for a solution. The pK_a value is the inverse log of the K_a, so it is also constant for a solution. The ratio of anion to acid changes as a chemical reaction continues to completion. For the log of the anion to acid ratio to increase by one unit, thus increasing the pH of the solution by one, the ratio must change tenfold. Therefore, when small amounts of acid or base are added to a weak acid or base, which make good buffers, the pH of the buffer will not change very much. Weak acids and bases have a lot of unionized molecules in solution that resist change when bases or acids, respectively, are added to them. The conjugate bases and conjugate acids can absorb the hydrogen ions and hydroxide ions easily. Strong acids and bases, on the other hand, become completely dissociated, and their pH changes rapidly as bases and acids, respectively, are added to them, and their ions are neutralized.

Buffer Capacity

Conjugate Acid-Base Pairs

Acids and bases can be distinguished by their ability to lose or gain a proton. An **acid** has a proton that it can lose, turning it into a base. On the other hand, a **base** can gain a proton. When an acid molecule reacts with water and loses a proton to form a hydronium ion, the resulting ion is its conjugate base. The acid molecule and the base ion are called a **conjugate acid-base pair**. The water molecule and the hydronium ion that are formed are also a conjugate acid-base pair. Weak bases have strong conjugate acids, weak acids have strong conjugate bases, and vice versa for both scenarios. The conjugate base of a strong acid is a weaker acid than the water molecule acting as a base, so it does not react as a base in the solution and does not readily gain a proton.

The acid and base ionization constants, K_a and K_b, respectively, can be calculated by multiplying the concentration of each dissociated ion and dividing the product by the concentration of the unionized acid or base molecules at equilibrium. For example, for the chemical reaction $HCl + H_2O \rightarrow H_3O^+ + Cl^-$, $K_a = \frac{[H_3O^+][Cl^-]}{[HCl]} = 1.3 \times 10^6$. The large value of K_a indicates that most of the molecules are dissociated at equilibrium and HCl is a strong acid. When K_a and K_b are multiplied, they equal the **ionization constant of water**, K_w. Taking the inverse log of the acid and base ionization constants, $pK_a + pK_b = 14$ because strong acids and bases are indicated by lower pK_a and pK_b values, respectively, and vice versa.

Solution pH and Acid Strength and Concentration

The strength of an acid depends on two things: how easily the acid can dissociate and its concentration. At the same pH, a strong acid and a weak acid have the same concentration of H_3O^+ because pH is a measure of the concentration of H_3O^+ ions. However, because strong acids completely dissociate, it takes a lower concentration of strong acid molecules to reach the same concentration of hydronium ions compared to weak acids that do not fully dissociate. With weak acids, there are many more unionized acid molecules in solution. Therefore, the weak acid must have a much higher concentration than the strong acid to achieve the same pH. At the same initial concentration, a strong acid has a lower pH compared with a weak acid at the same concentration because it has fully dissociated and created a greater concentration of hydronium ions than the weak acid.

Factors that Make a Good Buffer

Buffers comprise a mix of a weak acid and its conjugate base or a mix of a weak base and its conjugate acid. The conjugate base or conjugate acid absorbs the hydrogen ion or hydroxide ion, respectively, from the acid or base that is added to the solution. As just mentioned, weak acids and bases have a lot of unionized molecules in solution that resist change when bases or acids, respectively, are added to them, so they make good buffers. The conjugate bases and conjugate acids can absorb the hydrogen ions and hydroxide ions easily. Strong acids and bases become completely dissociated, and their pH changes rapidly as bases and acids, respectively, are added to them, and their ions are neutralized, so they make poor buffers. The higher the concentration of the buffer solution, the larger the capacity of the buffer to absorb the ions of the added acid or base.

When choosing a buffer for an experiment, the capacity of the buffer should be within one or two pH units of the endpoint desired pH value. If a base is being titrated with an acid, the titration buffer should have a pK_a value that is a little lower than the pH at the halfway point to the equivalence point. If an acid is being titrated with a base, the titration buffer should have a pK_a value that is slightly higher than the pH at the halfway point to the equivalence point.

Applications of Thermodynamics

Introduction to Entropy

Entropy Increases When Matter is Dispersed

When transitioning from the solid to liquid or liquid to gas phase, the entropy of the system will increase. In each case, the particles are spread out and further apart from one another, such that there are more ways they can move and occupy a given space. For example, water vapor has higher entropy than liquid water because the gaseous water molecules have more kinetic energy, and consequently take up more volume because the molecules are further apart. These molecules can occupy more points in space, where each point will have a different geometrical configuration. Liquid water will have higher entropy than solid water because these molecules have enough kinetic energy to slide past one another, thereby occupying another point in space within the liquid. The water molecules in ice may be further apart compared to liquid water, but these water molecules are arranged to form a solid lattice where hydrogen bonding is maximized. Consequently, solid water has the least entropy because it is more ordered compared to other phases.

$$H_2O(s) < H_2O(l) < H_2O(g)$$

$$\rightarrow Increasing\ Entropy\ S\ \rightarrow$$

When transitioning from a solid to liquid or from high to low order, the change in entropy, ΔS, will be positive: $H_2O(s) \rightarrow H_2O(l); +\Delta S$. If the phase transition goes from less to more order, the change in entropy is negative: $H_2O(l) \rightarrow H_2O(s); -\Delta S$.

In a chemical reaction, if the number of particles in the products is greater than that of the reactants, as represented by the stoichiometric coefficients in the balanced chemical equation, the entropy increases. Conversely, if the number of products is less than that of the reactants, the entropy decreases.

$$A + B \rightarrow 3C; +\Delta S$$

$$C + D \rightarrow E; -\Delta S$$

For gaseous particles contained in a piston at a fixed temperature, if the volume of the gas increases, then the entropy increases because the particles have more space to move around.

Entropy Increases When Energy is Dispersed

As the temperature of a collection of molecules or particles increases, the entropy increases, and the distribution of kinetic energy within those particles becomes more dispersed. For instance, the Maxwell-Boltzmann distribution in the image below shows that a hot gas will have more entropy compared to a cold gas because the kinetic energy has a greater spread among the fraction of gaseous particles.

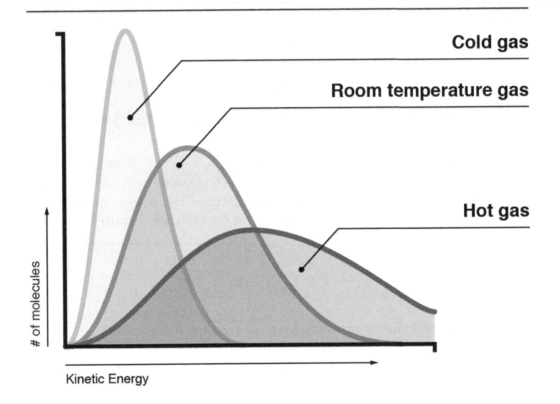

As the Temperature of the Gas Increases, the Kinetic Energy Is More Dispersed, Which Increases Entropy

S (hot gas) > S (room temperature gas) > S (cold gas)

Cold gas

Room temperature gas

Hot gas

# of molecules

Kinetic Energy

Absolute Entropy and Entropy Change

Qualitative Reasoning with Entropy

Entropy, *S*, is a measure of disorder in a chemical system. In physical processes, the entropy can change when a substance transitions between different phases: $Solid \leftrightarrow Liquid \leftrightarrow Gas$. In a chemical process, when reactants are converted to products, the entropy or disorder of the system can increase or decrease.

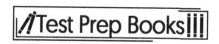

It is possible to calculate the entropy change for a process from the absolute entropies of the constituent species involved before and after the chemical process occurs:

$$\Delta S^\circ_{reaction} = \sum \Delta S^\circ_{products} - \sum \Delta S^\circ_{reactants}$$

Exothermic Reactions that Involve Decreases in Entropy

In nature, chemical reactions are favored toward an increase in entropy, $+\Delta S$, and a decrease in enthalpy $(-\Delta H)$ in a system. Whether a reaction occurs or not will depend on the balance of the factors that make up the Gibbs Free energy, ΔG. Endothermic and exothermic reactions can result in entropy changes to both the chemical system and its surroundings. In an exothermic reaction, where energy is lost to the surroundings, there is a decrease in enthalpy $(-\Delta H_{system})$. In some exothermic reactions (e.g., ammonia production), there may be a decrease in entropy $(-\Delta S_{system})$ because heat is transferred to the surroundings, thereby increasing the disorder $(+\Delta S_{surr})$. However, in an exothermic reaction, because heat is released to the surroundings, the overall entropy from both the surroundings and system can still be positive: $+\Delta S_{total} = \Delta S_{system} + \Delta S_{surroundings}$.

The equation is the **Second Law of Thermodynamics**, which says that for a process to be spontaneous, the total entropy must always increase. The second law also is written as:

$$\Delta S_{surroundings} = \frac{-\Delta H_{system}}{T}$$

and is connected to Gibbs free energy, ΔG. The value of ΔH_{system} for an exothermic reaction is negative, making the entropy of the surroundings positive $-(-\Delta H_{system}) \sim + \Delta S_{surroundings}$. Substituting $\Delta S_{surroundings}$ and rearranging gives: Gibbs Free energy ΔG: $+\Delta S_{total} = \Delta S_{system} + \frac{-\Delta H_{system}}{T}$.

$$-T\Delta S_{total} = \Delta G = \Delta H_{system} - T\Delta S_{system}$$

Note that $-T\Delta S_{total}$ is called ΔG.

Gibbs Free Energy and Thermodynamic Favorability

Free Energy Change, ΔG_0

As mentioned, reactions must use energy to do work, and that available energy is called **Gibbs free energy (G)**. Gibbs free energy, or G, represents the thermodynamic work potential for a system that is at a constant temperature and pressure (it may help to know that Gibbs free energy used to be called "available energy"). Free energy can be found by identifying the changes in enthalpy and entropy of a system.

$$\Delta H_{reaction} = \Delta H_{products} - \Delta H_{reactants}$$

$$\Delta S_{reaction} = \Delta S_{products} - \Delta S_{reactants}$$

Thermodynamically Favorable Meaning

The **Gibbs free energy, G,** is often used to determine whether a chemical reaction is **thermodynamically favored**, which means that the reaction will proceed favorably toward the products at equilibrium.

$$A \rightleftharpoons B; K > 1$$

The double-headed arrow indicates that the chemical reaction is at equilibrium, whereby the reactants are converted to products and the products to reactants. However, the reaction is favored toward the products as shown by the longer arrow, and the equilibrium constant, K, will be greater than one.

If a chemical reaction is **spontaneous**, meaning that it is thermodynamically favorable, the direction of the reaction will proceed toward the products (e.g., B). For a spontaneous reaction, the standard change in Gibbs free energy between the reactants and products is negative $-\Delta G° = G_{products} - G_{reactants}$.

Determining the Signs of ΔH^0 and ΔS^0 for a Physical or Chemical Process

The standard (298 K and 1 atm) Gibbs free energy change $\Delta G°$ is given by:

$$\Delta G° = \Delta H° - T\Delta S°$$

and is equal to the standard change in enthalpy ($\Delta H°$) and the negative product of temperature and the standard change in entropy ($\Delta S°$). The table below lists the signs of $\Delta G°$ concerning the various possible signs of $\Delta H°$ and $\Delta S°$.

Scenario	Spontaneity		K	$\Delta G°$	$\Delta H°$	$-T\Delta S°$	$\Delta S°$
	$\Delta G°$ **Sign Conventions for Different Signs of** $\Delta H°$ **and** $\Delta S°$						
1	Spontaneous		> 1	$-$	$-$	$-$	$+$
2	Nonspontaneous		< 1	$+$	$+$	$+$	$-$
3	Spontaneous	(low T)	> 1	$-$	$-$	$+$	$-$
	Nonspontaneous	(high T)	< 1	$+$			
4	Nonspontaneous	(low T)	< 1	$+$	$+$	$-$	$+$
	Spontaneous	(high T)	> 1	$-$			

A calculation to determine if the reaction is spontaneous is not necessary if $\Delta H° < 0$ and $\Delta S° > 0$ because the sum of $\Delta H°$ and $-T\Delta S°$ will give a negative $\Delta G°$ as shown in Scenario 1 on the table.

Situations When Both Enthalpy and Entropy are Needed to Determine Favorability

In some cases, the signs and magnitudes of entropy and enthalpy must be quantified to determine the temperature conditions at which the reaction is thermodynamically favorable. In Scenario 3 from the previous table, exothermic reactions can be favorable at relatively low temperatures. However, increasing the temperature can make the reaction unfavorable, pushing it toward the reactants because $-T\Delta S > -\Delta H$ if $\Delta S < 0$. For instance, the production of ammonia (Haber process) is an exothermic reaction whereby the entropy of the system decreases. As the temperature is increased, the forward reaction becomes unfavorable:

$$N_2(g) + 3H_2(g) \overset{400°C}{\rightleftharpoons} 2NH_3(g); -\Delta H_{system} \text{ and} -\Delta S_{system}$$

The melting of ice and the dissolution of sodium nitrate in water are examples of endothermic reactions. Sodium nitrate is a common ingredient found within instant cold packs.

$$NaNO_3(s) \overset{H_2O(l)}{\rightleftharpoons} Na^+(aq) + NO_3^-(aq); +\Delta H_{system} \text{ and} + \Delta S_{system}$$

In an endothermic reaction, as the chemical system absorbs energy from the surroundings, the entropy of the system will increase ($+\Delta S_{system}$), but there will be a decrease in entropy to the surroundings ($-\Delta S_{surr}$). In Scenario 4 of the table above, endothermic reactions are not spontaneous at low temperatures given $+\Delta H_{sys}$ and $+\Delta S_{system}$. However, at higher temperatures, the magnitude of $-T\Delta S > \Delta H$ $if \Delta S > 0$ and the combined enthalpy and entropy terms give a negative standard Gibbs free energy, $-\Delta G°$. The chemical process is thermodynamically favorable.

The Signs of ΔH^0 and ΔS^0 and the Direction of a Reaction

In Scenario 2 in the table above, $\Delta G° > 0$ and the reaction is nonspontaneous, given that $\Delta H° > 0$ and $\Delta S° < 0$ where $-T\Delta S° > 0$. Scenarios 3 and 4 are slightly more complicated and will generally require a quantitative calculation. Given Scenario 3, if the temperature is relatively small and if the $\Delta H°$ term is greater in magnitude than $-T\Delta S°$, the reaction is spontaneous ($\Delta G° < 0$). At equilibrium, the reaction will favor the products ($K > 1$). As the temperature increases, the magnitude of $-T\Delta S°$ becomes more significant and more positive such that it will overtake the negative $\Delta H°$ term. As a result, the overall reaction will be nonspontaneous, where $\Delta G° > 0$, and at equilibrium, the chemical process will favor the reactants ($K < 1$).

Thermodynamic and Kinetic Control

Thermodynamically Favorable Reactions Do Not Necessarily Proceed at a Measurable Rate

The standard free Gibbs energy, $\Delta G°$, only indicates whether the products will be favorable or not at equilibrium, but provides no insight into the reaction rate, or how fast or slow the reaction process will occur. A thermodynamically-favorable, or spontaneous process ($-\Delta G°$), will not necessarily occur at a measurable or observable rate. Some spontaneous processes will not happen within a person's lifetime, and some will occur in less than a second for a given set of reaction conditions (e.g., high or low temperatures). For example, the conversion of diamond (tetrahedral sp^3 carbons) to graphite (trigonal planar sp^2 carbons) is spontaneous ($\Delta G° = -2.87 \ kJ/mol$) at room temperature and one atmospheric pressure. The reaction is also exothermic and increases in entropy, but the rate of reaction is extremely slow. The conversion of diamond to graphite would not occur to a great extent nor would it be noticeable in our lifetime.

Processes Under "Kinetic Control"

Chemical processes that are spontaneous but do not proceed to the products at a considerable rate will be under **kinetic control**. The conversion of diamond to graphite is under kinetic control due to the high activation barrier at moderate temperatures and pressures (e.g., 1 atm and 298K). Even though the formation of graphite is favorable ($-\Delta H$ and $+\Delta S$) and extremely slow, it does not indicate that the chemical system will be in equilibrium. For example, suppose in one case there is one reaction mixture that contains 85 percent diamond and 15 percent graphite at one point in time. In a second case, at a much later time, let's assume that the equilibrium mixture is 0.5 percent diamond and 99.5 percent graphite.

However, in the first scenario, even though the reaction is spontaneous, the chemical system is not at equilibrium because it is under kinetic control. In the second scenario, the reaction will be closer to or at chemical equilibrium because the lowest value of $\Delta G°$ has been reached (as shown in the image below). At low temperatures, the reaction of diamond to graphite is under kinetic control; however, at higher temperatures, the reaction will be under thermodynamic control, where equilibrium can occur. For graphite formation, increasing the temperature would increase the thermodynamic favorability of the reaction (as described in Scenario 1 of the previous table titled $\Delta G°$ *Sign Conventions for Different Signs of $\Delta H°$ and $\Delta S°$*). As seen in the image below, ΔG° decreases as the reaction proceeds to graphite. At equilibrium, ΔG° reaches a minimum, and mostly contains graphite.

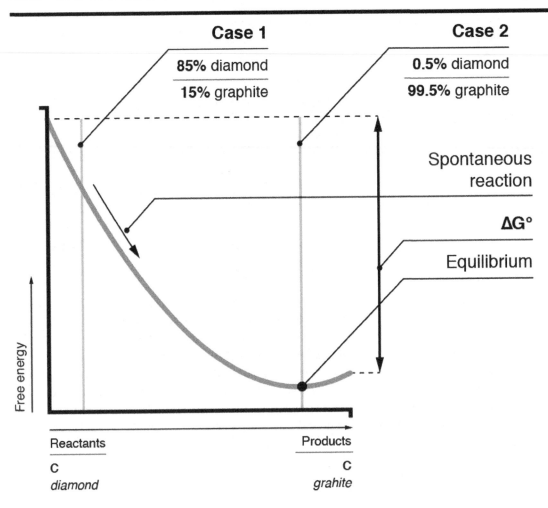

ΔG° Decreases as the Reaction Proceeds to Graphite

Free Energy and Equilibrium

Thermodynamically-Favorable Reactions and ΔG₀

The standard Gibbs free energy, $\Delta°G$, determines if a chemical process is thermodynamically favorable. If ΔG° < 0, the reaction is favorable (spontaneous) and proceeds toward the products. If ΔG° > 0, the reaction is not favorable (nonspontaneous) and proceeds toward the reactants.

Using the Relationship Between K and Q to Predict the Direction of a Reaction

Standard Gibbs Free Energy Change, ΔG°, and the equilibrium constant, K, of a chemical reaction are related by the following equation:

$$\Delta G° = -RT \ln K$$

In this equation, R is the universal gas constant and T is the temperature in Kelvin. This equation assumes that the reaction is happening at standard conditions of 1 M solutions and 1 atm pressure of the system. The free energy change reaction can also be rearranged to written in terms of K:

$$K = e^{-\Delta G^{\circ}/RT}$$

RT, which represents the thermal energy of the system, is equal to 2.4 kJ/mol at room temperature of 298 K. When ΔG° is greater than zero, the value of K is less than one and becomes very small. The reactants are favored and are in much greater proportion to the products at equilibrium. When ΔG° is less than zero, the value of K is greater than one and the products are favored in the reaction. The change in the magnitude of K increases greatly as the value of ΔG° gets farther away from the value of RT.

Determining Whether a Reaction is Exothermic or Endothermic with Free Energy and K

Looking at the relationship between K and ΔG° elucidates information about the qualitative relationship between the thermodynamics of the system and the equilibrium of the system. Enthalpic changes increase the value of ΔG°, whereas entropic changes decrease the value of ΔG°. The free energy of a reaction can also qualitatively be determined by the positive or negative value of ΔG°. If ΔG° is positive, the reaction is **endergonic**, which means that energy is being absorbed from the environment to run the reaction. K has a value less than one. These reactions are nonspontaneous. If ΔG° has a negative value, the reaction is **exergonic**, and the bonds that are being formed are stronger than the bonds that are being broken, so energy is released into the environment. K has a value greater than one. These reactions are mostly spontaneous, although some may require a small input of energy to start the reaction.

The Relation Between the Equilibrium Constant and Free Energy

$$N_2 + 3H_2 \leftrightarrow 2NH_3$$

The driving force of reactions is G, and if it is a very large negative number, the reaction may go almost to completion. Equilibrium, as indicated by the double arrows in equations, does not mean that the reaction has stopped, as particles are moving and dynamic. It also does not mean that there are equal amounts of products and reactants. For example, spontaneous reactions will have more products since they are so favorable. Many reactions are reversible, so learning the conditions and factors that will cause a system to reach equilibrium can be useful.

Altered conditions affect equilibrium. For example, adding more reactants yields more products. Increasing the external pressure will affect the reaction depending on where the most gas particles are (as determined by coefficients). Greater pressures will cause the side with the greatest gas particles to react faster since they will be moving faster, pushing the direction of the reaction to the other side. Increasing temperature will always change the equilibrium constant (K) because K is dependent on temperature as shown in this equation (valid with reversible reactions that are at equilibrium):

$$\Delta G = -RT \ln K$$

As previously discussed, the **equilibrium constant (K)** is a value that expresses the ratio of the products over the reactants. If K is greater than one, there are more products at equilibrium (forward reaction is favored), while a value less than one indicates that the reverse reaction is favored. The equilibrium constant can be calculated using the molarity of each product (to the power of its coefficient) divided by the molarity of each reactant (to the power of its coefficient) as shown in the equation below:

$$aA + bB \leftrightarrow cC + dD$$

$$Kc = \frac{[C]^c[D]^d}{[A]^a[B]^b}$$

While K shows the equilibrium values, the reaction quotient Q shows how far away the reaction is from equilibrium. It is the same equation but shows the ratio of products over reactants at any given time. Eventually, once a reversible reaction reaches equilibrium, Q will be equal to K.

Gas equilibrium constants are measured in terms of **partial pressure**, rather than concentration, and the equation is:

$$Kp = \frac{[PC]^c[PD]^d}{[PA]^a[PB]^b}$$

The equilibrium constant for heterogeneous solutions is the same except no substances other than gases are included.

Catalysis describes the increase in the rate of a reaction caused by the addition of a catalyst. The addition of a catalyst causes reactions to speed up and require less activation energy. Only small amounts of catalysts are required, and they are not consumed by the reaction. Catalytic activity is indicated using the symbol z, and it is measured in mol/s. Catalysts usually react with one—or sometimes more than one—reactant, forming intermediates. Catalysts provide alternative mechanisms for chemical reactions that involve different transition states and lower activation energy.

Coupled Reactions

Using Electricity to Drive Unfavorable Reactions

An external electrical current (e^-) can be used to promote a thermodynamically-unfavorable process. For example, an external charger can be connected to the positive and negative terminal of a car battery to jump start or charge the battery. For a conventional lead-acid battery, the reactions that lead to the charging of the battery are:

$$2e^- + H^+(aq) + PbSO_4(s) \rightarrow Pb(s) + HSO_4^-(aq) \ \ Reduction$$

$$PbSO_4(s) + 2H_2O(l) \rightarrow 2e^- + HSO_4^-(aq) + 3H^+(aq) + PbO_2(aq) \ Oxidation$$

If these half-reactions are reversed, the cell potential E°_{cell} is positive such that the reaction is spontaneous $\Delta^{\circ}G = -nFE^{\circ}_{cell} < 0$ ($n = -2 \ mole \ e^-$ and $F = 9.6485 \times 10^4 \ C \cdot mole \ e^{-1}$).

Similarly, in the **Downs cell**, battery terminals are connected to electrodes dipped in molten sodium chloride, which supply an electrical current causing globules of sodium metal to form.

Using Light to Drive Unfavorable Reactions

The Photoionization of an Atom

When an atom absorbs a specific frequency of light energy, the negatively-charged electron can be separated from the atom. The photoionization of an atom is not spontaneous and is highly endothermic

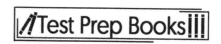

because it requires a relatively large and specific amount of light energy. The threshold frequency of light needed for the ionization of electrons from an atom has been reported by Albert Einstein, who generally called such phenomena the **photoelectric effect**.

$$Metal \overset{Light}{\rightleftharpoons} Metal^+ + e^-$$

The Conversion of Carbon Dioxide to Glucose Through Photosynthesis

In photosynthesis, a biological chemical reaction, oxygen gas and sugar are produced from carbon dioxide. The reaction is not thermodynamically favorable.

$$6CO_2(g) + 6H_2O(l) \overset{light}{\rightleftharpoons} C_6H_{12}O_6(aq) + 6O_2(g); \Delta G° = +2870 \, kJ/mol$$

But through the absorption of photons or light in the ultraviolet-visible spectrum (400–700 nm), the reaction can occur favorably in multiple steps.

Coupling Reactions

Many nonspontaneous processes, $\Delta G° > 0$, can be coupled with a spontaneous process, $\Delta G° < 0$, to become thermodynamically favorable. The overall reaction of glucose and adenosine triphosphate (ATP) to adenosine diphosphate (ADP) is a favorable reaction and contains two separate steps where a spontaneous and nonspontaneous reaction are coupled:

$$Glucose + Phosphate \, ion \rightarrow Glucose - 6 - Phosphate + H_2O(l); \quad \Delta G° = +12.7 \, kJ/mol$$

$$\underline{ATP + H_2O(l) \rightarrow ADP + \, Phosphate \, ion \, ; \hspace{3cm} \Delta G° = -30.5 \, kJ/mol}$$

$$ATP + Glucose \rightarrow ADP + \, Glucose - 6 - Phosphate; \hspace{1.5cm} \Delta G° = -17.8 \, kJ/mol$$

Water and phosphate are reaction intermediates not shown in the final net equation. The overall value of $\Delta G°$ is the sum of the first and second steps, which is negative and thermodynamically favorable.

Galvanic (Voltaic) and Electrolytic Cells

Components of Electrochemical Cells

An example of an oxidation-reduction reaction is the electrochemical cell, which is the basis for batteries. The electrochemical cell comprises two different cells, each equipped with a separate conductor, and a salt bridge. The **salt bridge** isolates reactants, but maintains the electric current. The **cathode** is the electrode that is reduced, and the **anode** is the electrode that is oxidized. Anions and cations carry electrical current within the cell and electrons carry current within the electrodes.

Galvanic Versus Electrolytic Cells

In the galvanic cell, the anode and cathode are surrounded by different electrolyte solutions, and there is a semi-permeable membrane between the two solutions. The anode metal is oxidized, so it releases electrons. The electrons travel across the electron bridge to the cathode. The cathode metal is reduced and incorporates the electrons in its metallic molecule. The reaction occurs spontaneously, and electric neutrality is maintained because the ions that are generated can flow between the compartments through the membrane. In the electrolytic cell, an external source of energy is needed to create a potential difference between the two electrodes. Once that is created, the electrons flow from the anode to the cathode. Both electrodes are placed in the same compartment of electrolyte solution for electrolytic cells. In both types of electrochemical cells, one half-reaction occurs at each electrode. The oxidation half-reaction occurs at the anode, and the reduction half-reaction occurs at the cathode.

Oxidation and Reduction of Electrochemical Cells

As mentioned, an electrochemical cell consists of two **electrodes**, which conduct the electricity, and an **ionic conductor**, which is often an electrolyte solution. Of the electrodes, one is an **anode**, where oxidation occurs and electrons are leaving, and the other is a **cathode**, where reduction occurs and electrons are flowing to it.

Galvanic vs. Electrolytic Cells

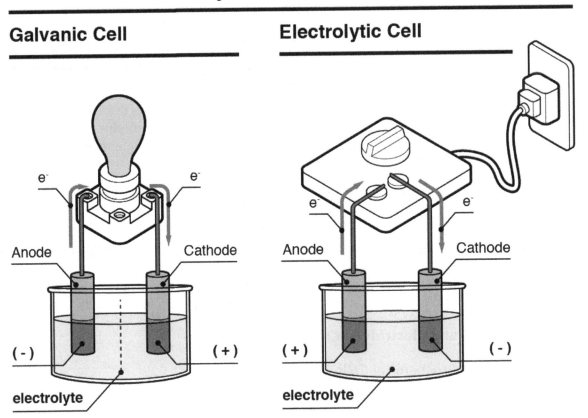

Cell Potential and Free Energy

Electrochemistry

Electrochemistry is the subfield of chemistry that involves chemical reactions in which electrons move. The movement of electrons is called **electricity**. **Oxidation-reduction reactions**, or **redox reactions**, fall under the field of electrochemistry because they involve the movement of electrons from one molecule to another. There are two types of electrochemical reactions that can occur: electrolytic or galvanic. In **electrolytic reactions**, an electrical current causes a chemical change to occur. In a **galvanic reaction**, the chemical reaction causes the production of electricity.

Electrical Potential of Galvanic Cells

The electric potential of galvanic cells can be calculated by adding the electric potentials of each half-reaction. The electric potential of reduction half-reactions can be looked up in a table of standard reduction potentials. This table gives the energy potential for reactions at their standard states, which is a concentration of 1 mol/L, pressure of 1 atm, and temperature of 25 °C. The oxidation energy potential is equivalent to the negative of the reduction energy potential. At standard state, the energy potential of a galvanic cell is:

$$E^0_{cell} = E^0_{red} + E^0_{ox}$$

In the case of a galvanic cell with a zinc anode and a copper cathode, the energy potential of copper being reduced is 0.339 V. The energy potential of zinc being reduced is -0.762 V. However, because zinc is being oxidized, the energy potential for the oxidation reaction in this galvanic cell is 0.762 V. Adding the reduction and oxidation energies together, the energy potential for the whole galvanic cell is 1.101 V. When the galvanic cell is not subject to standard conditions, the **Nernst equation** must be used to calculate the energy potential of the cell:

$$E_{cell} = E^0_{cell} - (RT/nF)*\ln Q$$

where E^0_{cell} is the cell potential at standard state, R is the universal gas constant, T is temperature in Kelvin, F is the Faraday constant, n is the number of moles of electrons that are transferred, and Q is the reaction quotient. Q, the reaction quotient, is a measure of the relative amounts of the reactants and products at any given time in the reaction. When Q = 1, the reaction has reached equilibrium at standard state conditions and will not run anymore. K is an equilibrium constant that represents the value of Q when equilibrium has been reached. When Q > K, there are more products present than reactants compared to the equilibrium state. The reaction should run in reverse, if possible. If Q < K, there are more reactants than products present compared to the equilibrium state, so the reaction should run in the forward direction. The farther the conditions of the cell from standard state, the greater the magnitude of the cell potential relative to E^0. As the conditions get closer to standard state, the magnitude of the cell potential decreases relative to E^0. Once equilibrium is reached, the cell potential equals zero.

The Free Energy of an Electrochemical Cell

A measure of the free energy of a cell can determine whether the chemical reaction of a galvanic cell would run spontaneously. The term ΔG measures the **Gibb's free energy** of a system and is determined by the equation $ΔG = -nFE^0$, where n equals the number of moles of electrons transferred, F equals the Faraday constant, and E^0 equals the standard cell potential. If ΔG is negative, the reaction is favored will run spontaneously. If it is positive, the reaction is not favored and will not run spontaneously. Because the Faraday constant and number of moles are always positive numbers, the term "-nF" will always be a negative number because of the negative sign in front of the term. Therefore, E^0 must have a positive value for the reaction to run spontaneously and ΔG to have a negative value.

Cell Potential Under Nonstandard Conditions

The Cell Potential In Real Systems

Most systems in the real world exist under nonstandard conditions. In such systems, the concentration of the active species involved in the reaction dictate the cell potential. The magnitude of the cell potential increases the further the reaction is from equilibrium. At equilibrium, the cell potential is zero. This is because the cell potential is a driving force for the system to reach equilibrium.

Electrochemical Systems and Equilibrium

It is important to remember that Le Châtelier's principle, and general tenets pertaining to equilibrium, are not applicable to electrochemical systems, such as galvanic cells, because such systems are not in equilibrium.

The Magnitude of the Standard Cell Potential Relative to the Equilibrium State of the System

As mentioned, the absolute value of the standard cell potential increases the further a system is from equilibrium. As a system approaches equilibrium, the standard cell potential approaches zero, and reaches zero at equilibrium. Thus, E^0 has a greater magnitude when a system is far from equilibrium.

Electrolysis and Faraday's Laws

Faraday's Laws

The English scientist Michael Faraday came up with quantitative laws to express the magnitude of changes that occur in electrochemical reactions. First, the quantity of chemical changes that occur due to the current of the system are proportional to the amount of electricity that is used. Electricity is mostly measured in **Coulombs**, which is the quantity of charge transferred in one second across a conductor. The **current** of the system is the number of electrons that move per period of time. In other words, the changes in the chemical elements are proportional to the quantity of electrons that are transferred in the cell. Second, when the same amount of electricity passes through different electrolytes, or charged ionic species, the amount of chemical change is dependent on the mass of the substances. The amount of substance removed from the oxidized electrode may not be the same as the amount of substance deposited on the reduced electrode because the formula weights required to gain or lose an electron may be different for the two substances.

Organic Chemistry

Organic chemistry is a branch of chemistry that involves the study of the structures, properties, and reactions of organic compounds. *Organic compounds* are molecules that have carbon as the backbone molecule. Organic chemistry studies range from understanding the nature of various molecules to describing chemical reactions in laboratories and living organisms.

Structures and Properties

Organic compounds are carbon-based compounds, meaning the carbon (C) atoms form the skeleton or backbone of the molecule. In organic compounds, the C atoms bind to other atoms, mostly hydrogen (H), oxygen (O), and nitrogen (N). Organic molecules can also contain other atoms, including sulfur (S), phosphorus (P), and halogens (X) such as fluorine (F), chlorine (Cl), bromine (Br), and iodine (I). *Functional groups* are the different atoms and how they are bonded together. Functional groups give the organic compound its molecular properties and reactivity.

In organic chemistry, functional groups help predict the chemical behavior of an organic compound because molecules possessing the same function groups undergo similar patterns of chemical reactions. Adjacent functional groups may also influence this reactivity. Functional groups are used to classify organic compounds into different chemical classes. An example of this is *alcohols*. The molecular and structural formula of an *alcohol* is shown below. The functional group is the –OH group.

The Molecular and Structural Formula of an Alcohol

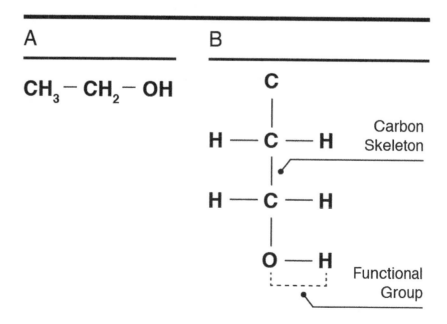

131

Structural Formulas and Bonding

Hydrocarbons are organic compounds that consist of only H and C atoms. They are classified into two major categories: aliphatic hydrocarbons (open chain) or cyclic hydrocarbons (closed chain). These groups are further divided. The overall classification of hydrocarbons is illustrated in the chart below:

The Classification of Hydrocarbons

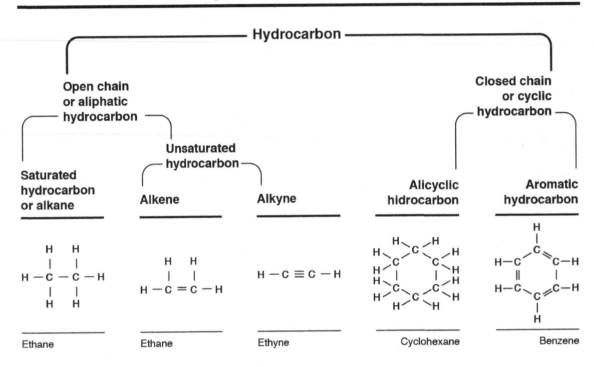

Aliphatic Hydrocarbons (Open Chain)

Aliphatic hydrocarbons have two open ends, meaning the C atoms at two terminals of the chain do not connect together. Because an aliphatic hydrocarbon chain has an open structure, it is often called an open chain hydrocarbon. Aliphatic hydrocarbons can contain single, double and/or triple C-C bonds. These different types of bonds classify aliphatic hydrocarbons into saturated hydrocarbons and unsaturated hydrocarbons.

Saturated hydrocarbons are the simplest form of hydrocarbons, and a type of aliphatic hydrocarbons. They contain only single bonds and are called alkanes. The single bonds are σ-bonds. Therefore, saturated hydrocarbons are more stable (or less susceptible to chemical reactions) than unsaturated hydrocarbons, and there is full rotation between all C-C bonds. In these molecules, the carbon bonds have the maximum number of H atoms possible, and are fully saturated with H atoms. Saturated hydrocarbons can have a linear or branched structure. In a linear structure, all of the carbon atoms form one single line. In a branched structure, the carbon chain splits off, creating multiple "branches" of

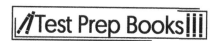

carbon chains. The general molecular formula of saturated hydrocarbons (or alkanes) is C_nH_{2n+2}, where n is the number of C atoms in the chain. Below are examples of short-chain saturated hydrocarbons.

Examples of Short-Chain Saturated Hydrocarbons

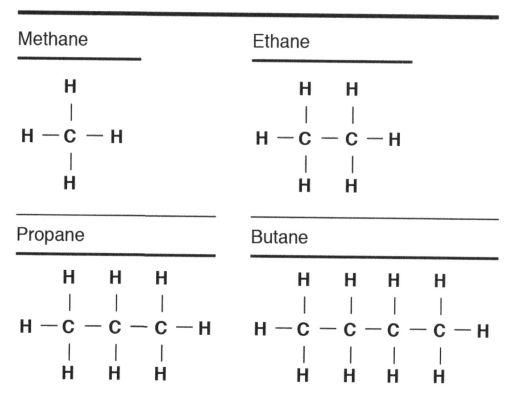

Unsaturated hydrocarbons are hydrocarbons that contain one or more double or triple bonds between C atoms in the hydrocarbon chain. They are further categorized into two groups: alkenes and alkynes.

Alkenes are hydrocarbons with double bonds. They have the general molecular formula of C_nH_{2n}. In their double bonds, one bond is a σ-bond and the other is a π-bond. Because of the presence of π-bonds, alkenes are more chemically reactive than saturated hydrocarbons. Additionally, there is no rotation

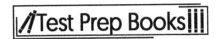

between the two C atoms in a double bond. Below are examples of the short-chain alkenes ethene and propene:

Examples of Short-Chain Alkenes

Ethene	Propene

$$\begin{array}{ccc} H & H & \\ | & | & \\ H-C & = C-H & \end{array} \qquad \begin{array}{ccc} H & H & H \\ | & | & | \\ H-C & -C & = C-H \\ | & & \\ H & & \end{array}$$

Alkynes are hydrocarbons with triple bonds. They have the general molecular formula C_nH_{2n-2}. The triple bonds in alkynes consist of one σ-bond and two π-bonds. Alkynes are quite unstable and are more reactive than alkanes and alkenes. Like alkenes, no rotation exists between the two C atoms in a triple bond. Below are examples of the alkynes ethyne and propyne:

Examples of Short-Chain Alkynes

Ethyne	Propyne

$$H-C \equiv C-H \qquad \begin{array}{ccc} & H & \\ & | & \\ H-C & -C & \equiv C-H \\ & | & \\ & H & \end{array}$$

Cyclic Hydrocarbons (Closed Chain)
Cyclic hydrocarbons have the terminal C atoms connected to each other, forming a closed, cyclic structure. Therefore, they are closed chain hydrocarbons. These hydrocarbons might have single, double, or triple bonds. Cyclic hydrocarbons are classified into alicyclic hydrocarbons and aromatic hydrocarbons.

Alicyclic hydrocarbons have one or more carbon rings and can contain either saturated or unsaturated bonds. They have similar properties to aliphatic hydrocarbons. Below are examples of alicyclic hydrocarbons:

Examples of Alicyclic Hydrocarbons

Cyclopropane

Cyclobutane

Cyclohexane

Cyclohexene

Aromatic hydrocarbons are cyclic hydrocarbons with σ-bonds and delocalized π-electrons between the carbon atoms that form the ring. Delocalized π-electrons are the result of a structure with resonance. This structure is often written as a molecule with a double bond between every other carbon atom, but this description is not entirely accurate. The double bonds are "shared" between all of the carbon atoms in the ring. This creates a very stable, unreactive structure. The geometry of these rings is flat and rigid. These compounds are also called arene, or aryl, hydrocarbons. Benzene is the simplest aromatic hydrocarbon. Historically, individuals thought that aromatic hydrocarbons contained a pleasant odor, resulting in the descriptive name. Aromatic hydrocarbons constitute a major part of crude oil and are highly flammable in nature.

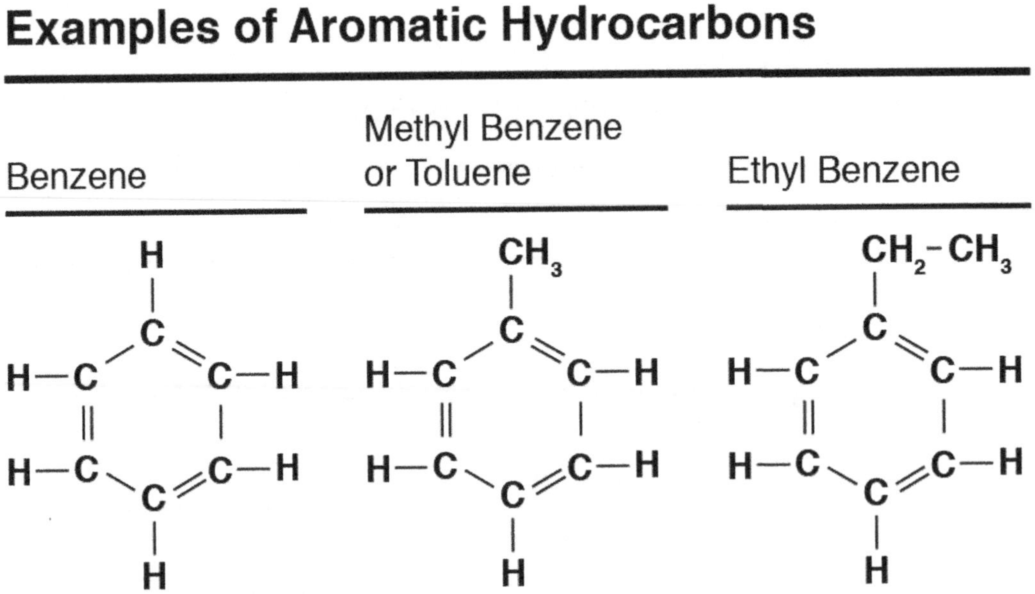

Examples of Aromatic Hydrocarbons

Benzene

Methyl Benzene or Toluene

Ethyl Benzene

Naming of Hydrocarbons

The naming of hydrocarbons is performed by the IUPAC (International Union of Pure and Applied Chemistry). According to this system, saturated hydrocarbons (alkane) are named by adding "ane" to the end of the prefix that corresponds to the number of C atoms present in the chain. An "alkyl" group is the name for a hydrocarbon with one H atom removed. These often serve as functional groups, and their name corresponds to the name of the alkane with the same number of C atoms. The table below lists alkanes and corresponding alkyl groups based on the number of C atoms in the chain.

C atoms	Prefix	Structure	Name	Alkyl Group	
C_1	Meth-	CH_4	Methane	Methyl	$-CH_3$
C_2	Eth-	CH_3-CH_3	Ethane	Ethyl	CH_3-CH_2-
C_3	Prop-	$CH_3-CH_2-CH_3$	Propane	Propyl	$CH_3-CH_2-CH_2-$
C_4	But-	$CH_3-(CH_2)_2-CH_3$	Butane	Butyl	$CH_3-(CH_2)_2-CH_2-$
C_5	Pent-	$CH_3-(CH_2)_3-CH_3$	Pentane	Pentyl	$CH_3-(CH_2)_3-CH_2-$
C_6	Hex-	$CH_3-(CH_2)_4-CH_3$	Hexane	Hexyl	$CH_3-(CH_2)_4-CH_2-$
C_7	Hept-	$CH_3-(CH_2)_5-CH_3$	Heptane	Heptyl	$CH_3-(CH_2)_5-CH_2-$
C_8	Oct-	$CH_3-(CH_2)_6-CH_3$	Octane	Octyl	$CH_3-(CH_2)_6-CH_2-$
C_9	Non-	$CH_3-(CH_2)_7-CH_3$	Nonane	Nonyl	$CH_3-(CH_2)_7-CH_2-$
C_{10}	Dec-	$CH_3-(CH_2)_8-CH_3$	Decane	Decyl	$CH_3-(CH_2)_8-CH_2-$

When alkanes are branched chains, the following rules are followed:

1. Find the longest chain of C atoms. This is the parent hydrocarbon. Name the alkane by using the rules listed above.

2. Identify all of the alkyl groups attached to the parent chain.

3. Number the parent chain so the first alkyl group is located closest to the number 1. If there are different alkyl groups equidistant from each terminal C atom, assign the lowest number to the one with the most substitutions or the one that that will come first in the name.

4. Name each substitution, starting with the number of the C atom, denoting its location and the name of the alkyl group. If more than one of the same alkyl group exists, list them all with commas separating the C atom number. In some cases, they may be on the same C atom.

Please note this is a simplified version of the rules. More rules exist for side chain priority.

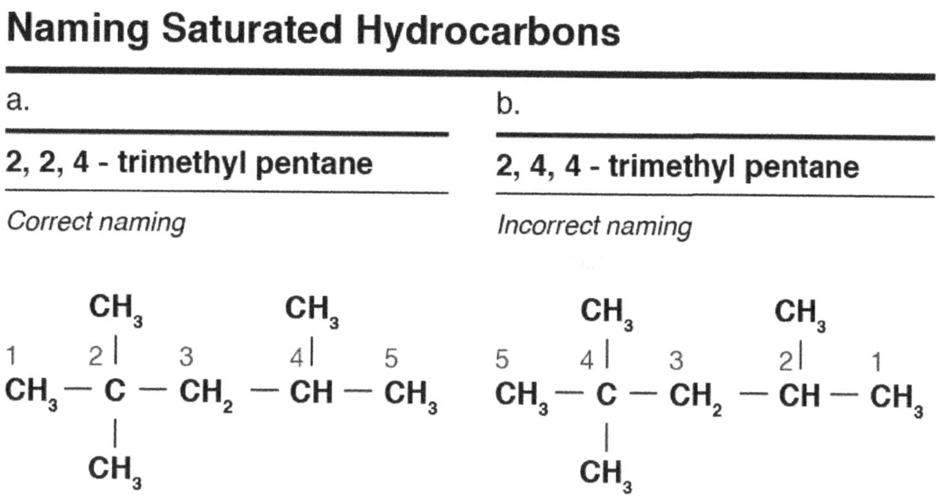

Naming Saturated Hydrocarbons

a.

2, 2, 4 - trimethyl pentane

Correct naming

$$\underset{\displaystyle \underset{1}{CH_3} - \underset{2}{\overset{\displaystyle \overset{CH_3}{|}}{\underset{\displaystyle \underset{|}{CH_3}}{C}}} - \underset{3}{CH_2} - \underset{4}{\overset{\displaystyle \overset{CH_3}{|}}{CH}} - \underset{5}{CH_3}}{}$$

b.

2, 4, 4 - trimethyl pentane

Incorrect naming

$$\underset{\displaystyle \underset{5}{CH_3} - \underset{4}{\overset{\displaystyle \overset{CH_3}{|}}{\underset{\displaystyle \underset{|}{CH_3}}{C}}} - \underset{3}{CH_2} - \underset{2}{\overset{\displaystyle \overset{CH_3}{|}}{CH}} - \underset{1}{CH_3}}{}$$

Similar to saturated hydrocarbons, unsaturated hydrocarbons are named using the longest C chain, but with adding "ene" (double bond) and "yne" (triple bond) to the end. The number of the C atom containing the double bond is added to the end of the name. The terminal C atom closest to the unsaturated bond is numbered 1. Below are examples of unsaturated hydrocarbons:

Naming Unsaturated Hydrocarbons

a.

4 - methyl pentene - 2

$$\underset{5}{CH_3} - \underset{4}{\overset{\displaystyle \overset{CH_3}{|}}{CH}} - \underset{3}{CH} = \underset{2}{CH} - \underset{1}{CH_3}$$

b.

5 - methyl hexyne - 1

$$\underset{6}{CH_3} - \underset{5}{\overset{\displaystyle \overset{CH_3}{|}}{CH}} - \underset{4}{CH_2} - \underset{3}{CH_2} - \underset{2}{C} \equiv \underset{1}{CH}$$

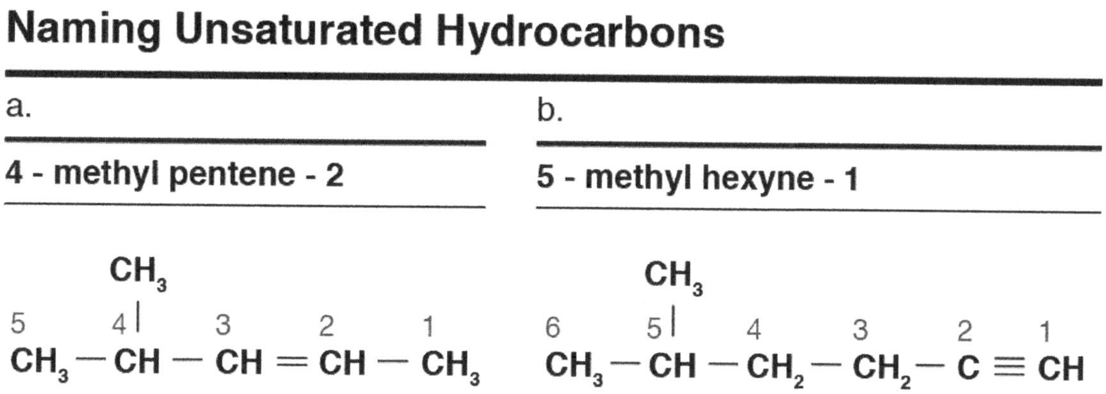

If the hydrocarbon chain contains two or three double bonds, then they are called "alka-di-ene" and "alka-tri-ene," respectively. Similarly, the unsaturated chains containing two or three triple bonds are called "alka-di-yne" and "alka-tri-yne," respectively.

Naming Unsaturated Hydrocarbons that Contain Two or Three Double Bonds

a.

2 - methyl - 1, penta-di-ene

$$CH_3$$
$$\underset{1}{CH_2} = \underset{2}{C} - \underset{3}{CH} = \underset{4}{CH} - \underset{5}{CH_3}$$

b.

5 - methyl - 1, 3 - hexa-di-yne

$$CH_3$$
$$\underset{6}{CH_3} - \underset{5}{CH} - \underset{4}{C} \equiv \underset{3}{C} - \underset{2}{C} \equiv \underset{1}{CH}$$

The C chain might contain various functional group(s), and they are named accordingly by adding the name of the functional group to that of the parent carbon chain. Common functional groups found in organic compounds are listed in the table below:

Name	Functional Group	Example
Alkyl	$-(CH_2)_n-CH_3$ (n=0, 1, 2,...)	$CH_3CH_2CH_3$ (propane)
Alkenyl	$-CH=CH-$	$CH_3-CH=CH_2$ (propene)
Amine	$-NH_2$	CH_3NH_2 (methyl amine)
Ether	$-O-$	CH_3OCH_3 (dimethyl ether)
Aldehyde	$-CHO$	CH_3CHO (acetaldehyde)
Ketone	$-CO-$	CH_3COCH_3 (acetone)
Carboxylic Acid	$-COOH$	CH_3COOH (acetic acid)
Ester	$-CO_2R$ (R=alkyl group)	$CH_3CO_2CH_3$ (methyl acetate)
Acyl Chloride	$-COCl$	CH_3COCl (acetyl chloride)
Amide	$-CONH_2$	CH_3CONH_2 (acetamide)
Hydroxyl	$-OH$	CH_3CH_2OH (ethanol)
Halide	$-F$, $-Cl$, $-Br$ and $-I$	CH_3Br (methyl bromide)

Isomers and their Classifications

Isomers are two or more molecules with the same molecular formula, but with different arrangements of atoms in the molecule, which can give them different chemical and physical properties. For example, ethanol (CH_3-CH_2-OH) and dimethyl ether (CH_3-O-CH_3) have the same molecular formula, C_2H_6O. However, they have different physical and chemical properties, due to the variation in their structural formula.

Isomerism is divided into two major categories: structural (constitutional) isomerism and stereoisomerism (spatial isomerism).

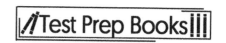

Structural isomerism is a type of isomerism in which the isomers have identical molecular formula, but the atoms in the molecules have different arrangements. There are four types of structural isomerisms: chain isomerism, position isomerism, functional group isomerism, and tautomerism.

1. *Chain isomerism* is also called skeletal isomerism. *Chain isomerism* occurs when the backbone atoms (usually C atoms) have different arrangements, such as straight or branch structures. Using the figure below as an example, butane and 2-methyl propane (isobutene) have the same molecular formula, C_4H_{10}, but different chain structures. This results in different physical and chemical properties.

The Same Molecular Formula (C_4H_{10}) can Yield Different Chain Structures

Butane	2 - methyl propane
$CH_3— CH_2—CH_2— CH_3$	$CH_3—CH—CH_3$ with CH_3 above the CH

2. *Position isomerism*, also termed regioisomerism, occurs when unsaturated (double or triple) bonds or functional groups have different locations in the chain. For example, 1-butene and 2-butene are position isomers because they have the same molecular formula (C_4H_8), but the double bonds are located at different positions. The same is true for 1-propanol and 2-propanol (isopropyl alcohol), where the functional group (-OH) is located on different C atoms. Position isomerism is also possible in aromatic compounds, based on the relative positions of the substituent atoms/groups in the benzene-ring structure.

Examples of Position Isomerism

1- Butene	2 - Butene	1 - Propanol	2 - Propanol
$CH_2 = CH — CH_2 — CH_3$	$CH_3 — CH \equiv CH — CH_3$	$CH_3 — CH_2 — CH_2 — OH$	$CH_3 — CH — CH_3$ with OH above the CH

1, 2 - dichlorobenzene	1, 3 - dichlorobenzene	1, 4 - dichlorobenzene
ortho - dichlorobenzene	meta - dichlorobenzene	para - dichlorobenzene

3. *Functional group isomerism* refers to isomers that have the same molecular formula but different functional groups. For example, ethanol (CH_3-CH_2-OH) and dimethyl ether (CH_3-O-CH_3) are functional isomers because they have same molecular formula (C_2H_6O), but have different functional groups—hydroxyl (-OH) and ether (-O-) groups.

A Functional Group Isomerism

Ethanol	Dimethyl ether

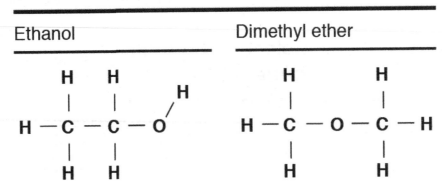

4. *Tautomerism* refers to a special type of structural isomerism resulting from spontaneous inter-conversion of functional groups to form two different compounds. This process primarily results in migration of a proton and switching between single and double bonds. Tautomerization is a chemical reaction that reaches equilibrium between the two tautomers (isomers). One example of tautomerism is "keto-enol tautomerism." It causes conversion of keto (-C=O) group to from a double bond ("ene" from alk<u>ene</u>) and an alcohol group (-OH) ("ol" from alcoho<u>l</u>). Therefore, it is called keto-enol tautomerism. See the example below in which a ketone, acetone (propanone), undergoes tautomerization to form an alcohol, 2-propenol:

The Tautomerization of a Ketone to an Alcohol

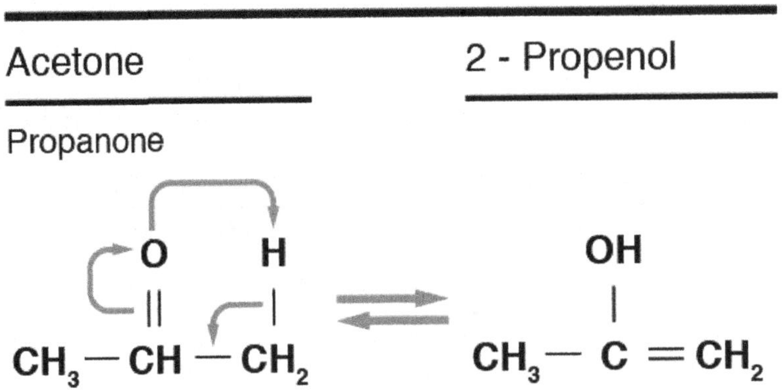

Acetone	2 - Propenol
Propanone	

Stereoisomerism is the type of isomerism in which molecules have similar molecular and structural formula, but they differ in their three-dimensional configurations and orientation of atoms in space. There are two types of stereoisomers: diastereomers and enantiomers.

1. Also called configurational or cis-trans isomers, the *diastereomers* have different orientations of atoms/groups along the C-C bond axis. Diastereomers generally contain a double bond or cyclic structure that has restricted bond rotation. The different isomers are designated "cis" and "trans." "Cis" isomers have similar atoms or groups on *the same side* of the C-C bond axis. "Trans" isomers have similar atoms or groups positioning on *the other side or across* the C-C bond axis. Examples of cis and trans isomers in double bond and cyclic structures are presented below:

Examples of cis and trans isomers in double bond and cyclic structures are presented below

Cis - 1, 2 - dichloro ethene	Cis - butene - 2	Cis - 1, 2 - dimethylcyclopropane
$\begin{array}{c} H \\ Cl \end{array} C = C \begin{array}{c} H \\ Cl \end{array}$	$\begin{array}{c} H \\ CH_3 \end{array} C = C \begin{array}{c} H \\ CH_3 \end{array}$	$H,\ CH_3 - C - C\ H,\ CH_3$ with CH_2 bridge
Trans - 1, 2 - dichloro ethene	Trans - butene - 2	Trans - 1, 2 - dimethylcyclopropane
$\begin{array}{c} Cl \\ H \end{array} C = C \begin{array}{c} H \\ Cl \end{array}$	$\begin{array}{c} CH_3 \\ H \end{array} C = C \begin{array}{c} H \\ CH_3 \end{array}$	$CH_3,\ H - C - C\ H,\ CH_3$ with CH_2 bridge

2. *Enantiomers* are stereoisomers that have a different arrangement around a chiral (asymmetric) center (usually a C atom). Enantiomers are mirror images of each other that cannot be superimposed on top of each other. A non-chemistry example this concept is your hands. If you hold them facing each other, you can see they are mirror images of each other. However, it is impossible to superimpose your left hand on top of your right hand.

When these isomers are in a symmetric environment, they have the same physical and chemical properties, but they behave differently in symmetric environments. Also, enantiomers can rotate plane-polarized light.

Most enantiomers are formed from a *chiral carbon*, which is a C atom that has four different atoms or functional groups attached to it. They are also called an asymmetric carbon. The four different atoms/groups can be arranged in two different ways around the chiral carbon, creating left-handed and right-handed configurations. These configurations are mirror images and do not superimpose. The number of optical isomers that can be formed from an organic compound is 2^n, where n equals the number of chiral C atoms in the molecule.

There are two naming conventions for enantiomers. One stems from their ability to rotate plane-polarized light. The enantiomer that rotates the light to the right is called the *dextrorotatory* or (+) or *d*-isomer. The enantiomer that rotates plane-polarized light left is called the *levorotatory* or (-) or *l*-isomer. The other naming method uses the Cahn-Ingold-Prelog priority rules, which ranks the functional groups based on their atomic number (a higher number has a higher priority). If the molecule is rotated so that the lowest priority group (4) faces you, the remaining three groups will either increase priority in a clockwise direction (the R enantiomer), or will increase priority in a counterclockwise direction (the S enantiomer). There is no correlation between d/l and R/S naming conventions.

Naming Enantiomers Using the Cahn-Ingold-Prelog Priority Rules

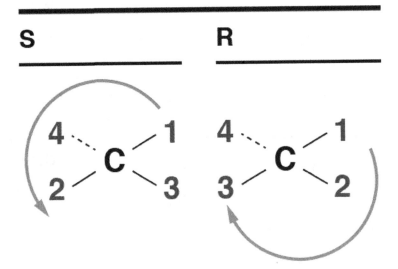

The example below shows a chiral carbon-containing compound, lactic acid [CH_3-CH(OH)-COOH]:

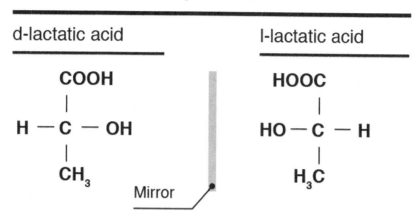

Normal visible light is a mixture of electromagnetic waves. The waves oscillate (up and down motion of the light wave) perpendicular to the direction the wave moves. In normal visible light, this oscillation can occur in all planes (directions) around the wave. *Plane polarized light* is light that is monochromatic (has only one wavelength), and all of the waves oscillate on the same plane. Plane polarized light is created by using a lamp that only produces light with one wavelength and then passing that light through an optical filter called a polarizer.

Creating Plane Polarized Light

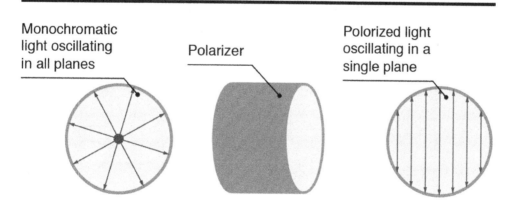

Optically-active compounds have the ability to rotate the plane polarized light. The instrument used to measure the angle of rotation of plane polarized light, while passing through an optically-active

compound, is called *polarimeter*. The overall process of studying the properties of optically-active compounds is presented in the figure below:

The Process of Studying Optically-Active Compounds

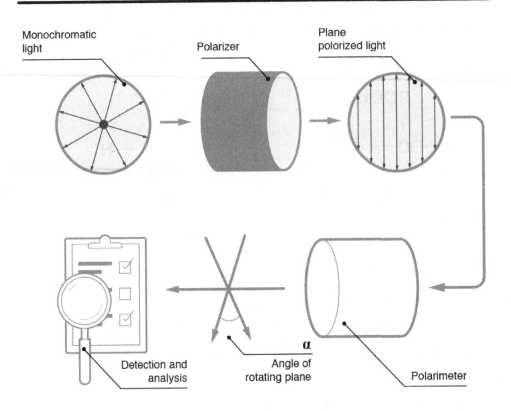

A *racemic mixture* contains an equal amount of *d*-(also called *R)* and *l*-(also called *S)* isomers, which do not rotate plane polarized light. For example, the mixture of equal ratio (1:1) of *d*-lactic acid and *l*-lactic acid is a racemic mixture. A racemic mixture is presented by *dl* or ± or *RS*. If the mixture is not a 1:1 ratio, a slash is used instead (i.e. *d/l* or *(+)/(-)* or *R/S)*.

The Classification of Isomers

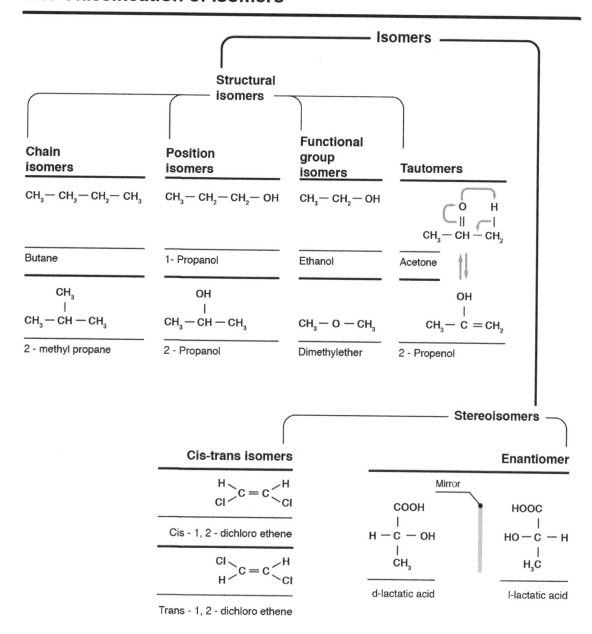

Catenation

Carbon atoms can undergo a unique process called catenation, which make them able to form a vast number of organic compounds found in nature. *Catenation* refers to the linkage of the same kind of atom to form longer chains and structures. Carbon catenation helps build numerous types of structures, containing open chain, branch chain, or cyclic configurations in organic compounds. C atoms can form single bonds (a, in the figure below), double bonds (b), or triple bonds (c). C atoms can also form ring structures (d).

The Catenation of Carbon

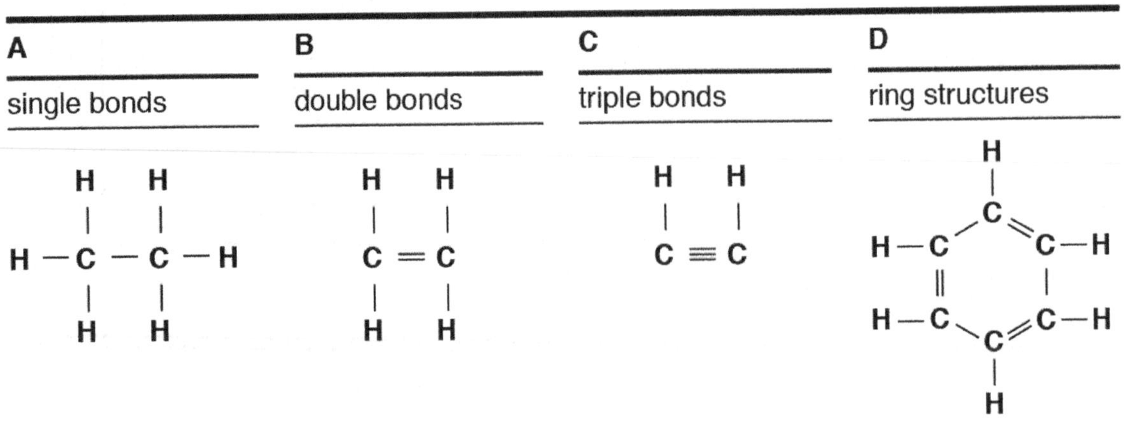

A	B	C	D
single bonds	double bonds	triple bonds	ring structures

The success of C atoms to undergo catenation is because of high stability of the covalent C-C bonds. This high bond stability is a result of the relative electron affinity of C atoms and the small atomic radius of the atoms to face across the intra-atomic orbitals and form C-C bonds (σ (sigma) and π (pi) bonds).

Covalent Bonds

A *covalent bond* is a type of chemical bond that is formed through sharing electron pairs between atoms. Contrary to ionic bonds (found mostly in inorganic molecules, e.g. Na^+Cl^-) that involve electron *transfer* between atoms, covalent bonds involve electrons *shared* between atoms. A covalent bond formation, through the sharing of electrons, helps the corresponding atoms attain a stable electronic configuration at the outer valence shell.

The C atom has four valence electrons in the outer electron orbital. The octet rule states that most atoms combine to result in eight valence electrons in the outer electron orbital. In order to get a stable atomic configuration, carbon can share those four electrons to form four covalent bonds. Below is an example of covalent bond formation in methanol (methyl alcohol). In this example, a C atom is attached to H and O through the sharing of electrons.

The covalent bond formation in methanol *(methyl alcohol)*

The C atom is attached to H and O through the sharing of electrons

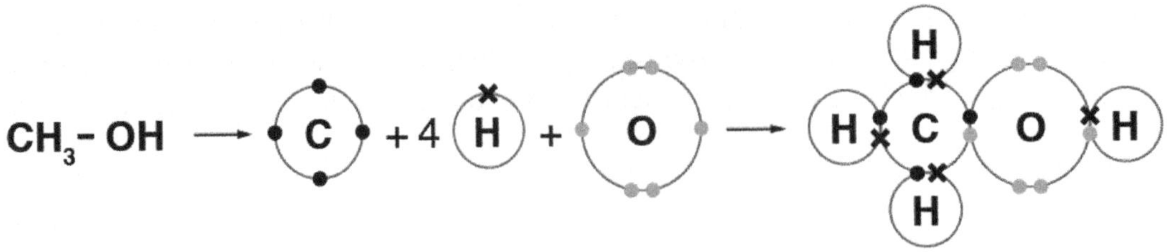

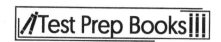

Atomic orbitals are the space, relative to the atom's nucleus, where one or more electrons persist. Covalent bonds are formed through an overlapping of the atomic orbitals. There are two types of covalent bonds: *sigma* (σ) and *pi* (π). Typically, a single bond is a σ-bond, and a multiple bond has both σ-bonds and π-bonds. For example, a double bond has one σ-bond and one π-bond, while a triple bond has one σ-bond and two π-bonds.

Sigma (σ) Bonds
A *sigma (σ) bond* is the strongest type of covalent bond. It forms from the direct and linear overlapping of atomic orbitals.

1. Formation of σ-bond by overlapping s-orbitals:

Sigma *(σ)* Bonds

1. Formation of σ-bond by overlapping s-orbitals

2. Formation of σ-bond by overlapping s-orbitals and p-orbitals

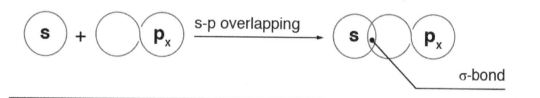

3. Formation of σ-bond by overlapping two p-orbitals

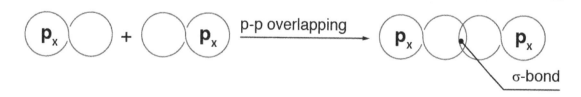

The nature of a σ-bond allows for rotation around the bond.

Pi (π) Bonds
Pi (π) bonds are formed as a result of parallel overlapping between two p-orbitals, which is a weaker and more diffuse bond than the linear σ-bond. π-bonds are weaker than σ-bonds and more susceptible to

fission in chemical reactions. The parallel nature of a π-bond prevents rotation around the bond. Any rotation would cause the π-bond to break.

Pi (π) Bonds Do Not Allow Rotation

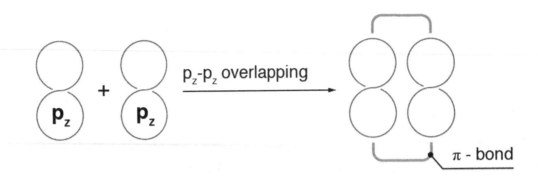

p_z-p_z overlapping

π - bond

Properties of Organic Compounds

Alkanes

Alkanes are saturated hydrocarbons with a general molecular formula of C_nH_{2n+2}. Some examples of alkanes are methane (CH_4), ethane (C_2H_6), propane (C_3H_8), and butane (C_4H_8).

The physical properties of alkanes are as follows:

1. *Physical State:* Lower molecular weight alkanes (C_1 to C_4) are odorless gases at room temperature (i.e., methane, ethane, propane, and butane). Higher alkanes with C numbers from C_5 to C_{17} are colorless liquids with a petroleum odor. The alkanes with a C number equal or greater than C_{18} are colorless, odorless, and a wax-like solid at room temperature.

2. *Melting and Boiling Points:* The melting and boiling temperatures of straight chain alkanes increase as the number of C atoms increases. Branched chain alkanes have lower melting and boiling points than the straight chain alkanes with the same number of carbons. For example, pentane, iso-pentane, and neo-pentane have the boiling points of 36°C, 28°C, and 10°C, respectively. This happens because the branched chain alkanes have a less surface area that contacts each other (the molecules cannot pack in as tightly), resulting in weaker intra-molecular attraction (van der Waals force). Therefore, this causes lower melting/boiling points.

3. *Solubility:* Alkanes are insoluble in water and other polar liquids. Instead, they are soluble in organic liquids. Alkanes are non-polar, so they are soluble in other non-polar liquids (e.g. benzene, ethanol, ether etc.).

4. *Relative Density:* The relative density, or specific gravity, of alkanes is less than 1. Therefore, alkanes are not as dense as water. With the increase in the chain length of alkanes, the relative density is gradually increased, but always remains less than 1.

5. *Combustion:* Alkanes are flammable, catch fire easily, and burn with a blue flame. The complete combustion of an alkane leads to the formation of carbon dioxide and water. During combustion, the

supply of oxygen has to be sufficient. Insufficient oxygen leads to the production of carbon monoxide, and the heat generated is less if sufficient oxygen is unavailable.

Alkanes only have sigma (σ) bonds, which are relatively stable in chemical reactions. Therefore, alkanes generally do not react with acids, alkalis, and oxidizing/reducing agents. However, in the presence of high temperature and pressure, σ-bonds undergo fission to produce free radicals. Free radicals are highly reactive, since they carry unpaired valence electrons. Once alkyl-free radicals are formed, they interact with the attacking reagents to form new compounds. The chemical reactions of alkanes are categorized three ways: substitution reactions, thermal decomposition reactions or pyrolysis, and isomerizations.

Alkenes

Alkenes are unsaturated hydrocarbons with one or more double bonds. Alkenes are represented by the general formula C_nH_{2n}. Examples of alkenes are ethene ($CH_2=CH_2$) and propene ($CH_3-CH=CH_2$).

The physical properties of alkenes are as follows:

1. *Physical State:* Lower molecular weight alkenes (C_2 to C_4) are gases at room temperature (i.e., ethene, propene and butene). Higher alkenes with the carbon number of C_5 to C_{15} are liquids. C_{16} and higher alkenes are solid. Ethene has a sweet odor, but other alkenes are colorless and odorless.

2. *Melting and Boiling Points:* The melting and boiling points of alkenes increase with the molecular weight. However, branched chain alkenes have lower melting and boiling points than those of the corresponding straight chain alkenes.

3. *Solubility:* Alkenes are insoluble in water, but soluble in organic solvents like benzene, ethanol, ether, etc.

4. *Relative Density:* The relative density of alkenes is less than 1. Therefore, they are less dense than water. With the increase in the chain length of alkanes, the relative density increases, but always remains less than 1.

5. *Combustion:* Alkenes are flammable, and they burn with a yellow flame. Complete combustion of an alkene in the presence of adequate oxygen produces carbon dioxide and water. When enough oxygen is not available, partial combustion of alkenes primarily produces carbon monoxide.

Alkenes are unsaturated hydrocarbons, which have double bonds. The double bond consists of one sigma (σ) bond and one pi (π) bond. The π-bond is prone to electrophilic addition reactions. Alkenes undergo three types of reactions: addition reactions, oxidation reactions, and polymerizations.

Alkynes

Alkynes are unsaturated hydrocarbons with one or more triple bonds. They are represented by the general formula C_nH_{2n-2}, where *n* equals 1, 2, 3 . . .n. Examples of alkynes are ethyne and 2-butyne.

The physical properties of alkynes are as follows:

1. *Physical State:* Lower molecular weight alkynes (C_2 to C_4) are gases at room temperature (i.e., ethyne, propyne and butyne). Alkynes with the carbon number of C_5 to C_{12} are liquids. Higher alkynes are colorless and odorless solids.

2. *Melting and Boiling Points:* The melting and boiling points of alkynes are higher than that of alkanes and alkenes of similar C-chain length. With the increase in molecular weight of the alkynes, the melting and boiling points increase accordingly.

3. *Solubility:* Alkynes are insoluble in water, but soluble in organic solvents.

4. *Relative Density:* The relative density of alkynes is higher than that of alkanes and alkenes, but still less than water.

5. *Combustion:* Alkynes are flammable, and they burn with a yellow flame. Like alkanes and alkenes, complete combustion of an alkyne in the presence of adequate oxygen produces carbon dioxide and water. When oxygen is inadequate, combustion of alkynes produces carbon monoxide.

A triple bond in an alkyne contains one σ-bond and two π-bonds. The presence of π-electrons in alkynes makes them suitable to interact with electrophiles. Alkynes undergo the following types of chemical reactions: addition reactions, oxidation reactions, and polymerizations.

Aromatic Compounds

Aromatic compounds are also called arenes. The term "aromatic" is a historical term related to the belief that aromatic compounds have a pleasant aroma. We now know this is not true. These molecules all contain a planar ring structure with conjugated, delocalized π-electrons. This configuration renders the aromatic structure unusually stable. Therefore, they are relatively more resistant to chemical fission or participation in chemical reactions. Aromatic compounds include benzene, benzene derivatives, and compounds that behave like benzene.

Aromatic Compounds are Stable Because of Their Structure

Benzenoid structure with delocalized π-electrons in the planar molecular orbital

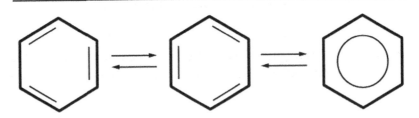

Aromatic compounds are classified into three major groups:

1. Benzene and substituted benzene compounds

2. Polycyclic or fused aromatic compounds

3. Heterocyclic aromatic compounds

Benzene and substituted benzene compounds contain a benzene ring with/without substituted groups or atoms in the ring. Examples are benzene, toluene (methyl benzene), ethyl benzene, and phenol.

Benzene and Examples of Substituted Benzene Compounds

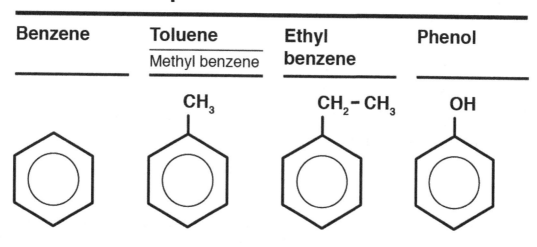

Polycyclic aromatic compounds contain two or more benzene rings fused together while sharing two adjacent C atoms. Examples of this type of compounds are naphthalene, anthracene, and phenanthrene.

Polycyclic Aromatic Compounds

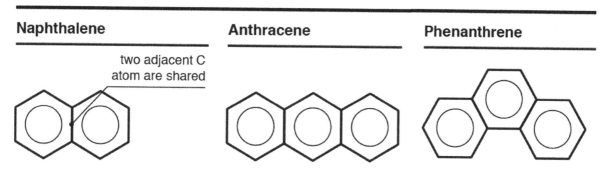

Heterocyclic aromatic compounds contain one or more heteroatoms (other than C atoms), like O, N, S, as a part of the aromatic ring. This causes a decrease in the ring's aromaticity and stability, and therefore increases its reactivity. Some examples of heteroaromatic compounds are furan, pyridine, imidazole, oxazole, and thiophene (or thiofuran).

Examples of Heterocyclic Aromatic Compounds

Furan	Pyridine	Imidazole	Oxazole	Thiophene
				Thiofuran

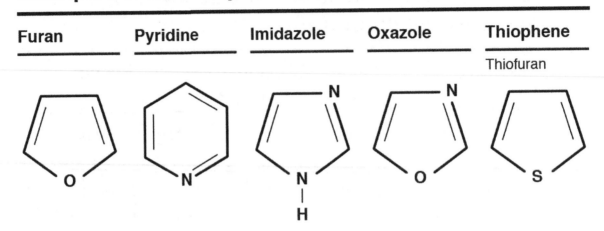

Benzene is a colorless liquid with a sweet smell. Benzene's melting and boiling points are 5.5°C and 80.1°C, respectively. Its specific gravity is less than water, and therefore it is less dense than water. It is a constituent of crude oil and it is flammable. It burns with a black flame. Because of its high stability, it is a good solvent for organic compounds.

Benzene has a hexagonal planar structure. Six H atoms remain attached to C atoms by σ-bonds. The functional group of an aromatic hydrocarbon formed by removing one H atom is called aryl functional group. For example, the function groups of benzene (C_6H_6) is called phenyl group (C_6H_5-).

The high stability of the benzene ring makes the molecule resistant to any reaction that would destroy the aromaticity of the molecule. Benzene can undergo two types of reactions: addition reactions and substitution reactions.

Alcohols

Alcohols are organic compounds that contain one or more hydroxyl (-OH) group in the aliphatic chain. Examples of alcohols are methanol

(CH_3—OH) and ethanol (CH_3—CH_2—OH). Alcohols can exist on an aliphatic side-chain of an aromatic ring. These alcohols are called aromatic alcohols or aryl alcohols. Examples are phenyl methanol (C_6H_5—CH_2—OH) and phenyl ethanol (C_6H_5—C_2H_4—OH).

Most alcohols contain one hydroxyl group. For example, methyl alcohol is CH_3-OH, ethyl alcohol is CH_3—CH_2—OH, and propyl alcohol is CH_3—CH_2—CH_2—OH.

Depending on the position of –OH on the C chain, monohydric alcohols are further classified as:

1. Primary or 1° alcohols have an –OH group attached to a C atom bound to two or three H atoms. See the classification examples below:

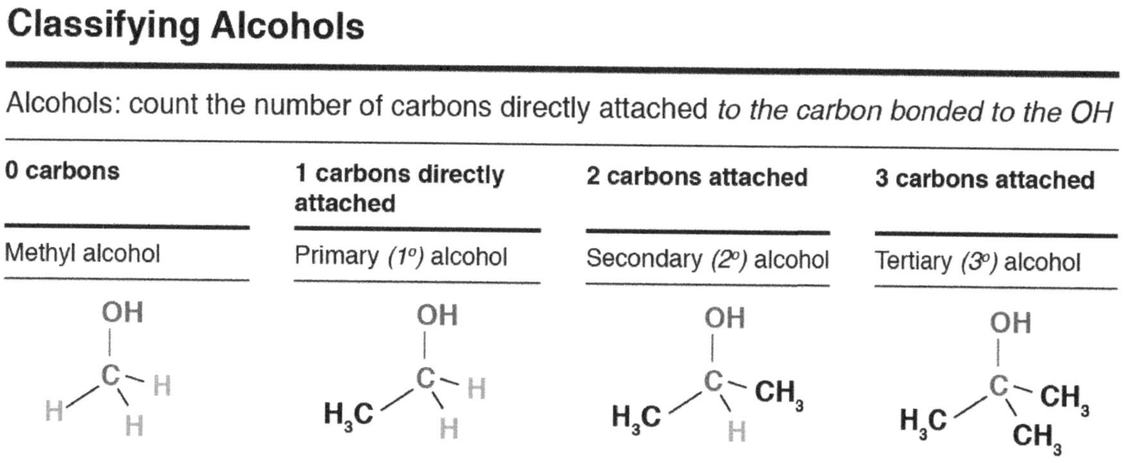

Classifying Alcohols

Alcohols: count the number of carbons directly attached *to the carbon bonded to the OH*

0 carbons	1 carbons directly attached	2 carbons attached	3 carbons attached
Methyl alcohol	Primary (1°) alcohol	Secondary (2°) alcohol	Tertiary (3°) alcohol

2. Secondary or 2° alcohols have an –OH group attached to a C atom bound to only one H atom. See the examples below:

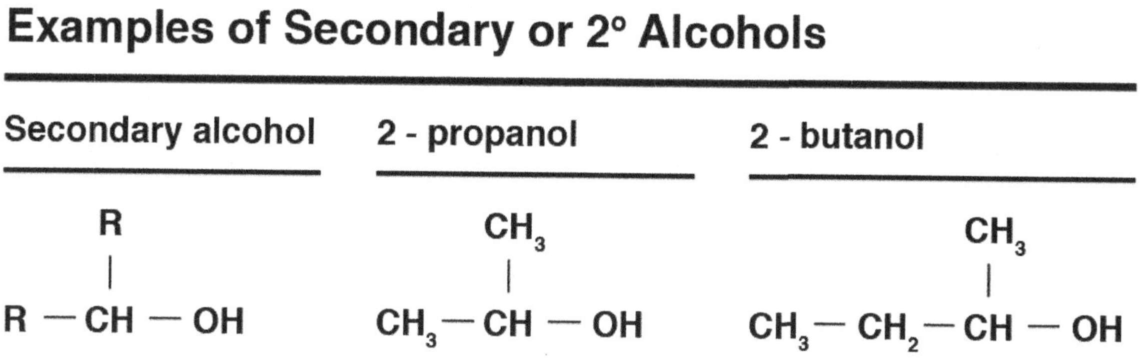

Examples of Secondary or 2° Alcohols

Secondary alcohol	2 - propanol	2 - butanol

3. Tertiary or 3° alcohols have an –OH group attached to a C atom not bound to any H atoms. All of the side chains have been substituted. Examples are illustrated below:

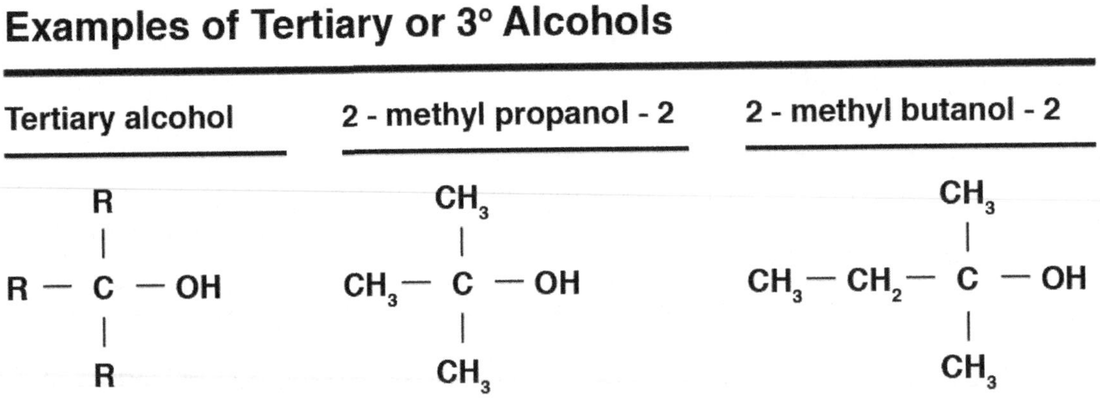

Examples of Tertiary or 3° Alcohols

Tertiary alcohol	2 - methyl propanol - 2	2 - methyl butanol - 2

$$
\begin{array}{ccc}
R & CH_3 & CH_3 \\
| & | & | \\
R - C - OH & CH_3 - C - OH & CH_3 - CH_2 - C - OH \\
| & | & | \\
R & CH_3 & CH_3
\end{array}
$$

Some alcohols contain more than one -OH group. Alcohols with two –OH groups are called "diols," three –OH groups are called "triols" and those with four –OH groups are called "tetraols." Examples of alcohols with more than one –OH group are shown below.

$OH - CH_2 - CH_2 - OH$ $OH - CH_2 - CH_2 - CH_2 - OH$ $OH - CH_2 - CH(OH) - CH_2 - OH$

Ethane-1,2-diol   Propane-1,3-diol   Propane-1,2,3-triol (glycerol)

The physical properties of alcohols are as follows:

1. *Physical State:* The lower molecular weight alcohols (C_1 to C_{12}) are colorless liquids, and higher molecular weight alcohols are wax-like solids. Lower molecular weight alcohols have pleasant odors.

2. *Melting and Boiling Points:* The higher electronegativity of the O in the –OH group makes alcohols polar. The polar molecules produce intra-molecular attractions, resulting in increases in melting and boiling points. In alkanes, this type of attractive force is absent. For example, the boing points of methane and ethane are -161.5°C, and -89°C, whereas the boiling points of methanol and ethanol are 64.7°C and 78.4°C, respectively.

3. *Solubility:* The –OH group makes alcohols polar. Methanol, ethanol and propanol are readily-soluble in water. However, as the size of the C chain increases, the solubility in water decreases because the C chain is non-polar.

4. *Relative Density:* The relative densities of alcohols are less than water.

5. *Combustion:* Alcohols are flammable, and they burn in the presence of oxygen to produce carbon dioxide and water.

Alcohols can act as a weak acid by donating a proton (H atom) in aqueous solutions.

An Alcohol Can Act as a Weak Acid

$$R-OH + H_2O \longrightarrow R-O + H_3O^+$$

Alcohol   Hydronium ion

Therefore, alcohols can interact with different chemical species including alkali metals, carboxylic acids, acyl halides, and oxidizing agents.

In reactions with alkali metals, the –OH group releases its proton to alkali metals to form alkoxide and hydrogen.

The Reaction between Alcohols and Alkali Metals

$$2R-O-H + 2Na \longrightarrow 2R-ONa + H_2$$

Alcohol   Sodium alkoxide

$$2CH_3-O-H + 2Na \longrightarrow 2CH_3-ONa + H_2$$

Methanol   Sodium methoxide

Alkoxides can react with water to reproduce an alcohol.

The Reaction between Alkoxides and Water can Produce an Alcohol

$$R-O-Na + H_2O \longrightarrow 2R-OH + NaOH$$

Alkoxide   Alcohol

Alcohols can react with organic acids (carboxylic acids) to produce esters and water. For example, the reaction between ethanol and acetic acid produces an ester, ethyl acetate.

An Example of an Alcohol Reacting with an Organic Acid to Produce an Ester and Water

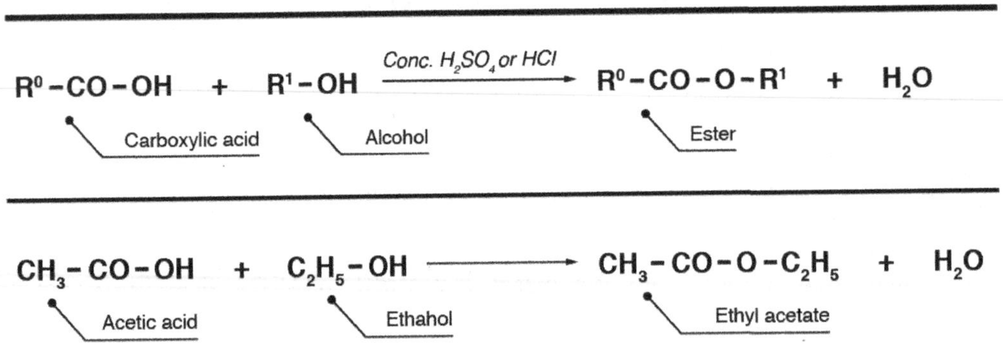
Conc. H_2SO_4 or HCl
R^0-CO-OH + R^1-OH → R^0-CO-O-R^1 + H_2O
Carboxylic acid
Alcohol
Ester
CH_3-CO-OH + C_2H_5-OH → CH_3-CO-O-C_2H_5 + H_2O
Acetic acid
Ethahol
Ethyl acetate

Alcohols react with acyl halide to form esters. For example, the reaction between ethanol and acetyl chloride forms ethyl acetate ester.

An Example of an Alcohol Reacting with an Acyl Halide

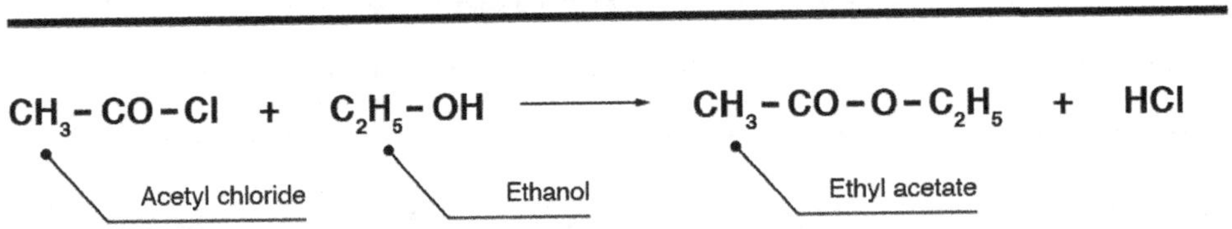
CH_3-CO-Cl + C_2H_5-OH → CH_3-CO-O-C_2H_5 + HCl
Acetyl chloride
Ethanol
Ethyl acetate

Phenols

Aromatic hydrocarbons that have one or more H atoms replaced by hydroxyl (-OH) groups are collectively called *phenols*. When a single H atom is replaced by an –OH group, it is called a carbolic acid (generally termed as phenol). Phenols are generally slightly acidic in nature. See examples below of certain phenolic compounds with different substituents in the ring:

Examples of Phenols

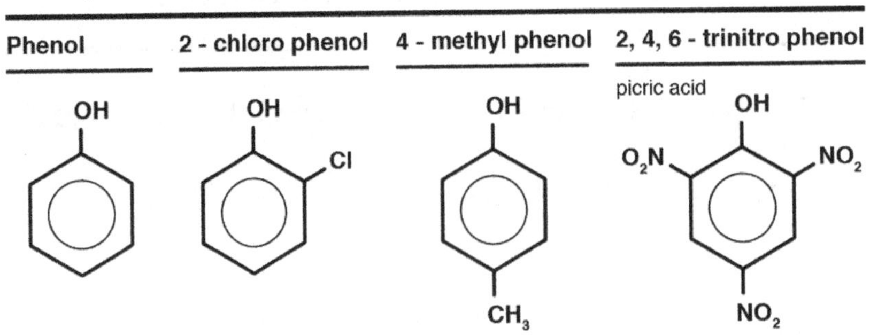
Phenol
2 - chloro phenol
4 - methyl phenol
2, 4, 6 - trinitro phenol
picric acid
OH
Cl
CH_3
O_2N
NO_2

Phenol, carbolic acid, specifically, is colorless crystalline solid. The –OH group in phenol is polar and forms intra-molecular hydrogen bonds. Therefore, phenol's molecular weight (94.1 g/mol), melting point (40.5°C), and boiling (181.7°C) points are higher than that of other organic compounds, such as toluene, with a similar molecular weight (92.1 g/mol), melting point, (-95°C), and boiling point (111°C).

As a class of organic compounds, phenols do not dissolve in water at room temperature. At higher temperatures hydrogen bonds break down, and phenols dissolve in water. Similarly, phenols are insoluble in other polar solvents, like alcohol and ether, at room temperature but dissolve at higher temperatures.

The chemical properties of phenols are as follows:

Phenols are more reactive than benzene due to the presence of the –OH group attached to the ring. However, the –OH group of a phenol is less reactive than that of an alcohol. This is because the delocalized π-electrons stabilize the O atom and the –OH group remains strongly attached to the benzene ring. Therefore, unlike alcohols, phenols do not react with the Lucas reagent. The bond between the –OH molecules in phenols is more polarized than that in alcohols and thus, they can readily donate a proton. Because of this, phenols are more acidic than alcohols.

Phenols Can Act as Acids

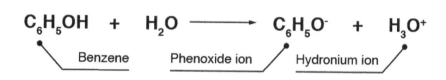

Phenols react with different chemicals. The major reactions of phenols are discussed below.

1. *Reduction of Phenol by Zinc:* When a phenol is passed through zinc crystals at high temperatures, it is reduced to form benzene, with the formation of zinc oxide.

Reduction of Phenol by Zinc

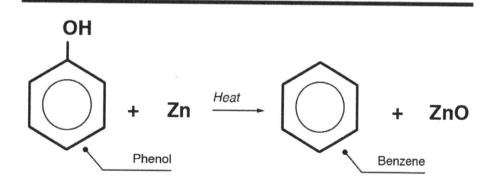

2. *Reduction of Phenol by Hydrogen:* When a phenol is treated with H_2 at 150°–175°C in the presence of a nickel catalyst, three molecules of H_2 are attached to form cyclohexanol.

Reduction of Phenol by Hydrogen

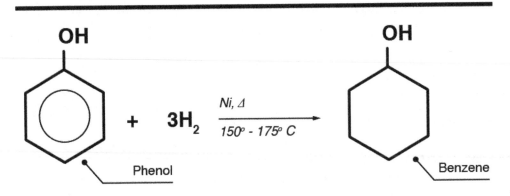

3. *Reaction with Ammonia*: At a high temperature and pressure, phenol reacts with ammonia to produce aniline (also called phenylamine or aminobenzene). The reaction uses zinc chloride as the reaction catalyst.

The Reaction of Phenol with Ammonia

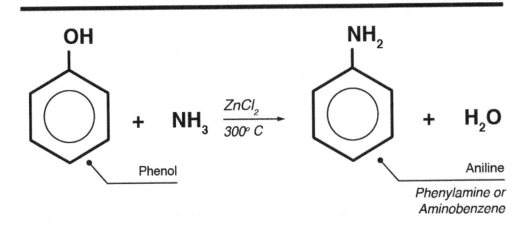

4. *Formation of Esters*: Like alcohols, phenols can also form esters. However, the reaction progresses slowly. Therefore, a phenol is first converted to a phenoxide, which further reacts with acyl halide or acid anhydride to form an ester. For example, sodium phenoxide reacts with acetyl chloride (CH_3COCl) or acetic anhydride (ethanoic anhydride) [$(CH_3CO)_2O$] to form phenyl acetate ester.

An Example of the Formation of an Ester from a Phenol

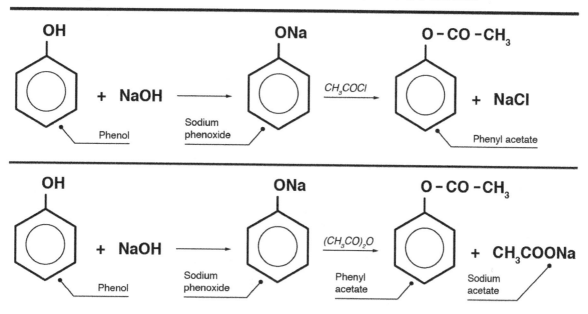

The above reactions can be used to commercially-synthesize aspirin (acetylsalicylic acid, ASA) and acetaminophen, which are widely used to treat fevers and pain.

The syntheses of aspirin and acetaminophen are shown below:

The syntheses of Aspirin and Acetaminophen

Sodium phenoxide — ONa, Pressure, CO_2, 125° C+ → Sodium salicylate — OH, COONa, H_2O, HCl → Salicylic acid — OH, COOH + NaCl

$(CH_3CO)_2O$ →

Acetylsalicylic acid / Asprin — OCOCH$_3$, COOH + CH$_3$COOH (Acetic acid)

4-aminophenol — OH, NH$_2$ + (CH$_3$CO)$_2$O (Acetic anhydride) → Acetaminophen — OH, NHCOCH$_3$ + CH$_3$COOH (Acetic acid)

5. *Williamson Ether Synthesis:* This refers to the formation of an ether from an organohalide and an alcohol. For example, phenol reacts with sodium hydroxide to form sodium phenoxide, which further reacts with methyl bromide to form an ether, methoxybenzene.

Williamson Ether Synthesis

Phenol — OH + NaOH → Sodium phenoxide — ONa, CH_3Br → Methoxybenzene — O–CH$_3$ + NaBr (Sodium bromide)

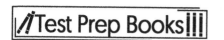

6. *Friedel-Crafts Alkylation:* In the presence of anhydrous aluminum chloride, phenols react with alkyl halides to form ortho- and para-alkyl phenol. For example, phenol reacts with methyl chloride to form ortho-methyl phenol and para-methyl phenol.

An Example of Friedel-Crafts Alkylation

2 **Phenol** $+ 2CH_3Cl$ **Methychloride** $\xrightarrow[AlCl_3]{\text{Anhydrous}}$ **o-methyl phenol** $+$ **p-methyl phenol** $+ 2HCl$

7. *Friedel-Crafts Acylation:* In the presence of anhydrous aluminum chloride, phenol reacts with acetyl chloride to form ortho- and para-acetyl phenol.

An Example of Friedel-Crafts Acylation

2 **Phenol** $+ 2CH_3COCl$ **Acetylchloride** $\xrightarrow[AlCl_3]{\text{Anhydrous}}$ **o-acetyl phenol** $+$ **p-acetyl phenol** $+ 2HCl$

8. *Reimer-Tiemann Reaction:* When phenol is mixed with chloroform and sodium hydroxide and heated to around 60°C, then phenol is converted into salicylaldehyde or 2-hydroxybenzaldehyde.

An Example of a Reimer-Tiemann Reaction

Phenol $+ CHCl_3 +$ **3NaOH** (Chloroform) (Sodium hydroxide) $\xrightarrow{60°}$ **Salicylaldehyde** $+ 3NaCl + 2H_2O$

Ethers

Ethers are organic compounds with two alkyl or aryl groups connected by an oxygen atom. They can be represented by R—O—R^1, where R and R^1 are alkyl or aryl groups.

Ethers can be classified into the following categories:

1. *Simple ethers* have the same alkyl group on each side of the O atom. Examples are dimethyl ether (CH_3—O—CH_3) and diethyl ether (C_2H_5—O—C_2H_5).

2. *Mixed ethers* have different alkyl groups on each side of the O atom. Examples are methyl ethyl ether (CH_3—O—C_2H_5) and methyl propyl ether (CH_3—O—C_3H_7).

3. *Cyclic ethers* form a ring structure. Three-membered ether rings (pictured below) are called epoxide compounds and are highly reactive (e.g. ethylene oxide).

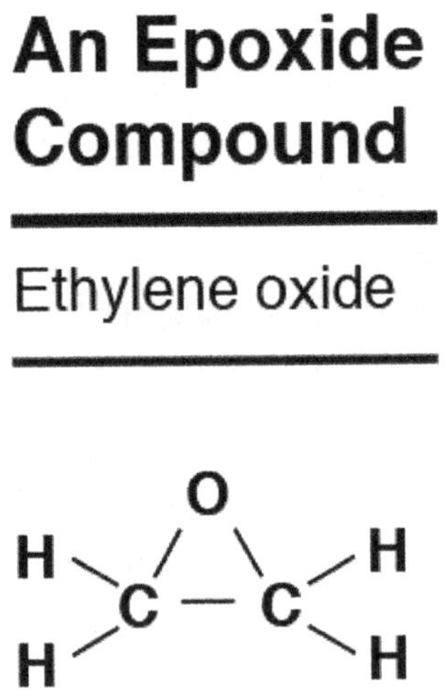

An Epoxide Compound

Ethylene oxide

---

4. *Aromatic ethers* contain at least one aryl group connected to the O atom (the other group can be an aryl or an alkyl group). Examples are methyl phenyl ether or methoxy benzene (C_6H_5—O—CH_3) and diphenyl ether (C_6H_5—O—C_6H_5).

The physical properties of ethers are as follows:

Low molecular weight ethers are gases. However, higher molecular weight ethers are colorless liquids. Ethers have slight solubility in water because of the formation of H bonds between the H atom of water and O atom of the ether. The solubility decreases with the increase in molecular weight.

Ethers have lower melting and boiling points than alcohols of similar molecular weight. This is because, unlike alcohol, ether molecules do not readily form intra-molecular hydrogen bonds. For example, dimethyl ether (CH_3—O—CH_3) and ethanol (CH_3—CH_2—OH) have the same molecular formula, C_2H_6O, but the boiling points of ethanol and dimethyl ether are 78.4°C and −24°C, respectively. At room temperature, dimethyl ether is a gas, but ethanol is a liquid.

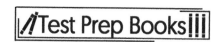

Ethers are less dense than water because they have a lower specific gravity. With an increase in molecular weight, their specific gravity increases. However, it remains less dense than water.

Ethers are one of the least reactive classes of organic compounds. They are less reactive than alcohols and phenols. When an ether is treated with a hydrogen halide, it undergoes fission to form alkyl halide and alcohol. In the case of a mixed ether, the halogen atom joins with the smaller alkyl group. For example, reaction between methyl ethyl ether and hydrogen iodide forms methyl iodide and ethanol.

An Example of a Reaction of an Ether with a Hydrogen Halide

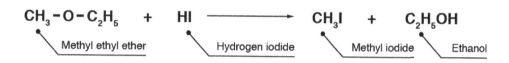

When treated with an excess of HI at a higher temperature, both alkyl groups are converted into alkyl halides.

An Example of a Reaction of an Ether with an Excess of a Hydrogen Halide at a High Temperature

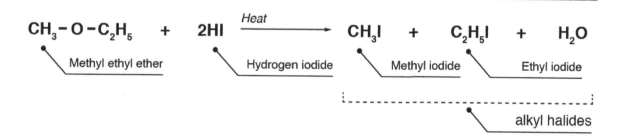

Carbonyl Compounds

Aldehydes and ketones are collectively termed *carbonyl compounds* because they contain the carbonyl (=C=O) functional group. The carbonyl group for aldehydes is on a terminal C atom. Some examples of aldehydes are formaldehyde or methanal (H—CHO), and acetaldehyde or ethanal (CH_3–CHO). The carbonyl group of ketones is on a central C atom. Examples of ketones are acetone or propanone (CH_3—CO—CH_3), and benzophenone or diphenylketone (C_6H_5—CO—C_6H_5).

Carbonyl Compounds

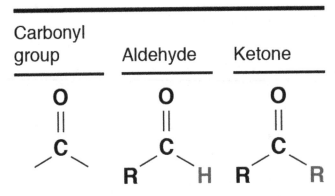

The physical properties of carbonyl compounds are as follows:

Formaldehyde (H—CHO) is a colorless gas, while acetaldehyde has a boiling point of 20.2°C. Aldehydes with C_3 to C_9 carbon numbers are colorless liquids, and higher carbon aldehydes are colorless solids. Ketones up to C_{11} are liquids, and C_{12} and higher ketones are solids. Lower C number aldehydes have an unpleasant odor, whereas higher C number aldehydes have a pleasant smell. Ketones have a pleasant smell as well.

Aldehydes and ketones with short C chains are fairly soluble in water through the formation of H bonds with water. Solubility decreases with the increase in molecular weight, and larger carbonyl compounds are insoluble in water. Carbonyl compounds are soluble in organic solvents.

Unlike alcohols, carbonyl compounds cannot readily form intra-molecular hydrogen bonds, resulting in relatively lower melting and boiling points.

The chemical properties of carbonyl compounds are as follows:

The bond between the C and O atoms in the carbonyl group is polarized ($-C^{\delta+}=O^{\delta-}$), making the carbonyl compounds highly reactive. The relatively positive-charged C atoms in carbonyl compounds are susceptible to nucleophile (electron-rich substrate) attack. Various chemical reactions of carbonyl compounds are summarized below.

1. *Grignard Reactions:* This reaction is used to produce a new C—C bond. The Grignard reagent reacts with the carbonyl group and is an organometallic compound with a general formula of R—MgX, in which R represents the alkyl or aryl groups, and X represents a halogen.

The Grignard reagent reacts with formaldehyde to form primary (1°) alcohols. For example, the reaction between formaldehyde and methyl magnesium bromide produces an unstable intermediate, which is further hydrolyzed to form ethanol.

A Grignard Reaction of a Carbonyl Compound

$$H-\overset{\overset{\displaystyle O}{\|}}{C}-H \ + \ CH_3-MgBr \longrightarrow H-\overset{\overset{\displaystyle OMgBr}{|}}{\underset{\underset{\displaystyle CH_3}{|}}{C}}-H \xrightarrow{H^+/H_2O} CH_3-CH_2-OH \ + \ BrMgOH$$

Formaldehyde   Grignard reagent   Intermediate compound   Ethanol   1° alcohol

The Grignard reagent reacts with other aldehydes (except formaldehyde) to form secondary (2°) alcohols. For example, the reaction between acetaldehyde and methyl magnesium bromide forms a secondary alcohol, 2-propanol.

The Grignard Reagent Reacts with Aldehydes to Form Secondary (2°) Alcohols

$$CH_3-\overset{\overset{\displaystyle O}{\|}}{C}-H \ + \ CH_3-MgBr \longrightarrow CH_3-\overset{\overset{\displaystyle OMgBr}{|}}{\underset{\underset{\displaystyle CH_3}{|}}{C}}-H \xrightarrow{H^+/H_2O} CH_3-\overset{}{\underset{\underset{\displaystyle CH_3}{|}}{C}H}-OH \ + \ BrMgOH$$

Acetaldehyde   Grignard reagent   Intermediate compound   2-propanol   2° alcohol

The Grignard reagent reacts with ketones to form tertiary (3°) alcohols. For example, the reaction between acetone and methyl magnesium bromide forms a tertiary alcohol, 2-methyl-2-propanol.

The Grignard Reagent Reacts with Ketones to Form Tertiary (3°) Alcohols

$$CH_3-\overset{\overset{\displaystyle O}{\|}}{C}-CH_3 \ + \ CH_3-MgBr \longrightarrow CH_3-\overset{\overset{\displaystyle OMgBr}{|}}{\underset{\underset{\displaystyle CH_3}{|}}{C}}-CH_3 \xrightarrow{H^+/H_2O} CH_3-\overset{\overset{\displaystyle CH_3}{|}}{\underset{\underset{\displaystyle CH_3}{|}}{C}}-OH \ + \ BrMgOH$$

Acetone   Grignard reagent   Intermediate compound   2-methyl-2-propanol   3° alcohol

2. *Oxidation Reactions:* Aliphatic aldehydes are readily oxidized by strong oxidizing agents, like potassium dichromate ($K_2Cr_2O_7$) and sulfuric acid (H_2SO_4), to form carboxylic acids. However, aromatic aldehydes are not easily oxidized.

An Oxidation Reaction of an Aliphatic Aldehyde Forms a Carboxylic Acids

$$R-\underset{\underset{\text{Aldehyde}}{}}{\overset{\overset{O}{\|}}{C}}-H \quad + \quad [\,O\,] \quad \xrightarrow[H_2SO_4]{K_2Cr_2O_7} \quad R-\underset{\underset{\text{Carboxylic acid}}{}}{\overset{\overset{O}{\|}}{C}}-OH$$

Ketones do not oxidize easily. When ketones are heated with strong oxidizing agents, a mixture of different organic acids is formed.

A Mixture of Organic Acids is Formed When Attempting to Oxidize Ketones

$$R^1-\underset{\underset{\text{Ketone}}{}}{\overset{\overset{O}{\|}}{C}}-R^2 \quad + \quad 3\,[\,O\,] \quad \xrightarrow[H_2SO_4]{K_2Cr_2O_7} \quad R^1-\underset{\underset{\text{Carboxylic acid -1}}{}}{\overset{\overset{O}{\|}}{C}}-OH \quad + \quad R^2-\underset{\underset{\text{Carboxylic acid - 2}}{}}{\overset{\overset{O}{\|}}{C}}-OH$$

3. *Reduction Reactions:* Reduction of aldehydes by a mild reducing agent, like lithium aluminum hydride (LiAlH$_4$), sodium borohydride (NaBH$_4$), or sodium amalgam (NaHg), produces primary alcohols, whereas ketones are reduced to secondary alcohols.

Reduction Reactions of Aldehydes and Ketones

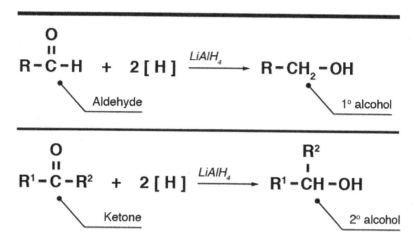

Reduction of carbonyl compounds with a strong reducing agent, like zinc amalgam and concentrated hydrochloric acid (ZnHg + HCl), causes conversion of the =C=O group into a –CH$_2$– group, resulting in formation of saturated hydrocarbons. This reaction is termed a *Clemmensen reduction.*

A Clemmensen Reduction Produces a Saturated Hydrocarbon

$$R-\overset{\overset{\displaystyle O}{\|}}{C}-H \ + \ 4\,[\,H\,] \ \xrightarrow[HCl]{Zn/Hg} \ R-CH_2-H \ + \ H_2O$$

Aldehyde → Alkane

$$R^1-\overset{\overset{\displaystyle O}{\|}}{C}-R^2 \ + \ 4\,[\,H\,] \ \xrightarrow[HCl]{Zn/Hg} \ R^1-CH_2-R^2 \ + \ H_2O$$

Ketone → Alkane

4. *Aldol Condensation Reaction:* In the presence of dilute acid or alkali, two molecules of an aldehyde or ketone containing α-hydrogen can undergo a condensation reaction to form a β-hydroxy aldehyde or β-hydroxy ketone. This reaction is termed "aldol condensation," as the product carries both the <u>ald</u>ehyde (–CHO) and alcoh<u>ol</u> (–OH) functional groups. Aldol condensation helps form a new C–C bond. For example, in the presence of dilute NaOH, two molecules of acetaldehyde condense to from an aldol, 3-hydroxybutanal (or β-hydroxybutanal).

An Aldol Condensation Reaction

When aldol is heated in the presence of an acid (e.g. HCl), a water molecule is removed causing the formation of α-β unsaturated aldehyde or ketone.

Heating an Aldol in the Presence of an Acid

Carbonyl compounds that do not have α-hydrogen cannot participate in aldol condensation. For example, formaldehyde (H-CHO) and trimethyl-acetaldehyde [$(CH_3)_3$C–CHO] do not have an α-hydrogen, so they cannot undergo aldol condensation.

5. *Cannizzaro Reaction:* In the presence of a concentrated base (e.g. NaOH), aldehydes *lacking* an α-hydrogen undergo oxidation-reduction reaction. In such reactions, one aldehyde molecule is oxidized to form a salt of carboxylic acid and the other is reduced to form an alcohol. For example, when treated

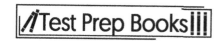

with NaOH, two molecules of formaldehyde undergo an oxidation-reduction reaction to form methanol and sodium formate.

An Example of an Aldehyde Undergoing the Cannizzaro Reaction

$$H-CHO \ + \ H-CHO \ \xrightarrow{NaOH} \ CH_3-OH \ + \ H-COONa$$

Formaldehyde   Methanol   Sodium formate

Biomolecules

Humans, animals, and plants are built with different biomolecules and vital life functions rely on them. *Biomolecules* are organic polymers that perform various functions in the human body. Among their many functions in the human body, biomolecules do the following: provide structure, provide nutrition and energy to cells, perform various enzymatic reactions, regulate the body's defense mechanism, and control genetic functions through heredity.

Classes of important biopolymers include:

1. Carbohydrates, such as starch (in animals) and cellulose (in plants).

2. Proteins, such as nucleoprotein, plasma protein, hormones, enzymes, and antibodies.

3. Nucleic acids, such as ribonucleic acid (RNA) and deoxyribonucleic acid (DNA).

4. Triglycerides, which are formed from fatty acids with a glycerol backbone.

All biomolecules are polymers, which are formed from repeated units of basic monomers. These polymers can be broken down (hydrolyzed) into their respective monomers.

Proteins can be hydrolyzed under acidic condition to form amino acids.

Hydrolyzing a Protein

Proteins can be hydrolyzed under acidic condition to form amino acids

$$Protein \ + \ nH_2O \ \xrightarrow{H^+} \ Amino \ acids$$

Nucleic acids are structurally associated with some proteins to form nucleoproteins. The hydrolysis of nucleic acids and nucleoproteins is illustrated below:

Nucleoprotein and Nucleic Acid Hydrolysis

Nucleoprotein $+$ nH_2O $\xrightarrow{H^+}$ Nucleic acid $+$ Protein

Nucleic acid $+$ nH_2O $\xrightarrow{H^+}$ Base $+$ Pentose sugar $+$ H_3PO_4
purine
pyrimidine

Carbohydrates are represented by a general formula,

$$(C_6H_{10}O_5)_n,$$

where $40 \leq n \leq 3000$. When hydrolyzed, starch produces the monosaccharide glucose ($C_6H_{12}O_6$).

Hydrolyzing a Starch

$(C_6H_{10}O_5)_n$ $+$ nH_2O $\xrightarrow{H^+}$ $C_6H_{12}O_6$

However, glucose remains stored as glycogen in the liver and muscle. Glycogen stores energy in the body.

Storing Glucose as Glycogen in the body

$n\ C_6H_{12}O_6$ $\xrightarrow{Glycogen\ Synthase}$ $(C_6H_{10}O_5)_n$

Where n = approx. 6000 - 30000

Carbohydrates are classified into three categories:

1. *Monosaccharides*: Monosaccharides are the monomers of carbohydrates. They are represented by a general formula $(CH_2O)_n$, where $n = 3 - 6$. Examples of monosaccharides are glucose (dextrose), fructose, and galactose.

They are named based on the number of C atoms in the molecule, such as triose (C3), tetrose (C4), pentose (C5) and hexose (C6). A monosaccharide with an aldehyde group is called an "aldose," and one with a ketone group is called a "ketose" (e.g. fructose).

Monosaccharides

D-ribose	D-Glucose	D-Fructose
	an aldohexose	a ketohexose

D-ribose:

$$H\diagdown_{C}\!\diagup^{O}$$
$$|$$
$$H - C - OH$$
$$|$$
$$H - C - OH$$
$$|$$
$$H - C - OH$$
$$|$$
$$CH_2OH$$

D-Glucose:

$$H\diagdown_{C}\!\diagup^{O}$$
$$|$$
$$H - C - OH$$
$$|$$
$$HO - C - H$$
$$|$$
$$H - C - OH$$
$$|$$
$$H - C - OH$$
$$|$$
$$CH_2OH$$

D-Fructose:

$$H$$
$$|$$
$$H - C - OH$$
$$|$$
$$C = O$$
$$|$$
$$HO - C - H$$
$$|$$
$$H - C - OH$$
$$|$$
$$H - C - OH$$
$$|$$
$$CH_2OH$$

2. *Disaccharides*: Disaccharides are two monosaccharides joined together. When hydrolyzed, disaccharides produce two monosaccharide molecules. See the examples below:

Hydrolysis of Disaccharides

$$C_{12}H_{22}O_{11} \ + \ H_2O \ \xrightarrow{H^+} \ C_6H_{12}O_6 \ + \ C_6H_{12}O_6$$

    Sucrose   Glucose   Fructose

$$C_{12}H_{22}O_{11} \ + \ H_2O \ \xrightarrow{H^+} \ C_6H_{12}O_6 \ + \ C_6H_{12}O_6$$

    Lactose   Glucose   Galactose

$$C_{12}H_{22}O_{11} \ + \ H_2O \ \xrightarrow{H^+} \ C_6H_{12}O_6 \ + \ C_6H_{12}O_6$$

    Maltose   Glucose   Glucose

3. *Polysaccharides*: Polysaccharides are high molecular weight carbohydrates. When hydrolyzed, they produce many molecules of monosaccharides. Examples of polysaccharides are starch, glycogen, and cellulose.

Carbonyl compounds, aldehydes and ketones, can react with –OH group of an alcohol to form hemiacetal and acetal.

The Reaction of Carbonyl Compounds with Alcohols

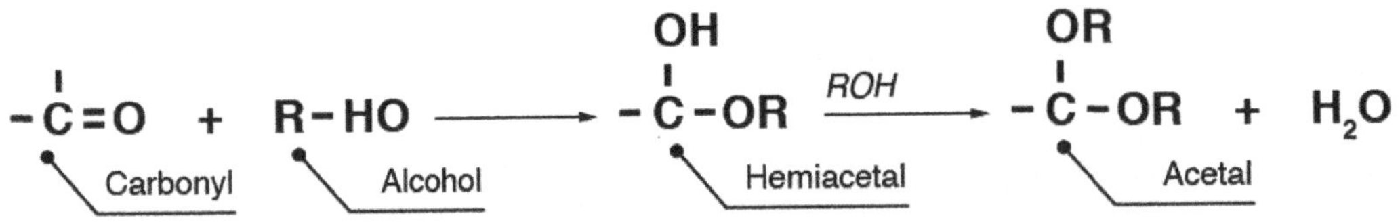

$$-\overset{\mid}{C}=O \ + \ R-HO \ \longrightarrow \ -\overset{\overset{\textstyle OH}{\mid}}{C}-OR \ \xrightarrow{ROH} \ -\overset{\overset{\textstyle OR}{\mid}}{C}-OR \ + \ H_2O$$

   Carbonyl   Alcohol   Hemiacetal   Acetal

Glucose and fructose contain both carbonyl and hydroxyl groups, and therefore can form intramolecular hemiacetal to produce a cyclic structure. This hemiacetal formation takes place between C1 and C5 carbons to form a stable heterocyclic structure. The cyclic structure forms a pyranose ring, which is a six-membered ring consisting of five C atoms and one O atom. In a cyclic structure, C1 might have an –OH group at the right or left side, and therefore may be termed α-D-glucose and β-D-glucose, respectively. View the structures below as an example:

The Formation of a Stable Cyclic Structure

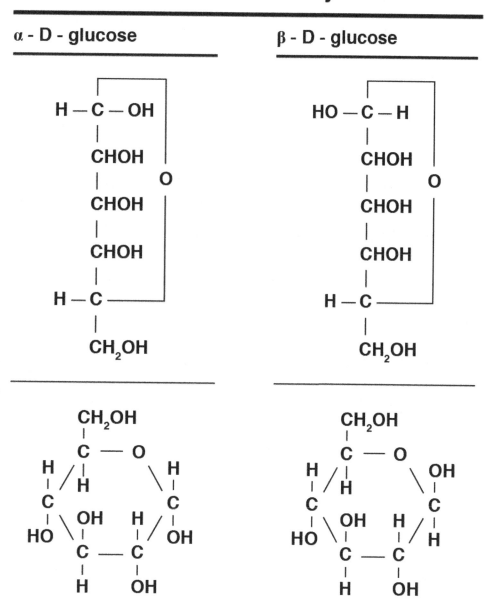

Reducing sugars are ones with a free aldehyde or ketone groups, and they can act as reducing agents. All monosaccharides, including glucose, fructose, and galactose, are reducing sugars. Many disaccharides, including lactose and maltose (except sucrose), are also reducing sugars. The reducing sugars are able to reduce Fehling's solution and Tollens' reagent.

When Fehling's solution is treated with a reducing sugar, the deep blue color of Fehling's solution fades and then forms a reddish precipitate. Fehling's solution is prepared by mixing a copper sulphate solution with potassium sodium tartrate in NaOH. See the reaction below as an example:

Fehling's Solution Treated with a Reducing Sugar

$$CH_2OH(CHOH)_4CHO + 2CuSO_4 \xrightarrow[\text{Heat}]{\text{NaOH}} Cu_2O + CH_2OH(CHOH)_4COONa + 3H_2O$$

Glucose   Deep blue   Reddish   Sodium gluconate

When a solution of reducing sugar is heated with Tollens' reagent, silver is precipitated and forms silver mirror on the inner surface of the reaction vessel. See the reaction that follows:

Tollens' Reagent Treated with a Reducing Sugar

$$CH_2OH(CHOH)_4CHO + 2Ag(NH_3)_2OH \longrightarrow 2Ag + CH_2OH(CHOH)_4COONH_4 + 3NH_3 + H_2O$$

Glucose   Tollens' reagent   Silver mirror

Reactions of Organic Compounds

In inorganic compounds, chemical reactions occur through the formation of electrically-charged ions and their polarization. However, organic compounds contain covalent bonds, and the chemical reactions of organic compounds involve simultaneous breakage and formation of these covalent bonds. The functional groups present in the molecule are important in determining the nature and rate of reactions. In addition to the functional groups, the reactions are also determined by the type of bonds present in the molecule (i.e. σ-bonds and π-bonds), and the nature of other reactants present.

Organic chemical reactions occur when an attacking reagent interacts with an organic molecule (substrate) to form a new organic compound (product). This can be explained with the reaction between chloromethane (CH_3—Cl) and sodium hydroxide (Na^+OH^-). In this reaction, CH_3—Cl is the organic

substrate, OH⁻ is the attacking reagent, and CH_3-OH (methanol) is the product. There are three steps in this chemical reaction:

The Steps in the Reaction Between Chloromethane and Sodium Hydroxide

1   NaOH ⟶ Na⁺ + OH⁻

2

$$H-\underset{\underset{H}{|}}{\overset{\overset{H}{|}}{\underset{\delta^+}{C}}}-\underset{\delta^-}{Cl} + OH^- \longrightarrow H-\underset{\underset{H}{|}}{\overset{\overset{H}{|}}{C}}-OH + Cl^-$$

Substrate   Attacking agent   Product

3   Na⁺ + Cl⁻ ⟶ NaCl

These chemical reactions take place through fission or breakdown of the C—C bonds and formation of carbon radical intermediates. There are two types of carbon radicals that are formed in chemical reactions: carbocation and carbanion.

A *carbocation* is a positively-changed carbon ion (i.e. cation). Examples are methyl carbocation ($^+CH_3$) and ethyl carbocation (CH_3—$^+CH_2$). Carbocation are formed due to an unequal distribution of the shared electrons from a bond cleavage. For example, in chloromethane (CH_3—Cl), the chlorine (Cl) atom is more electronegative than the C atom; therefore, electrons are drawn towards the Cl. Fission of the bond causes the Cl atom to take the shared electron of the C atom, which results in the formation of an electropositive carbocation.

The Formation of an Electropositive Carbocation

$$H-\underset{\underset{H}{|}}{\overset{\overset{H}{|}}{\underset{\delta^+}{C}}}\cdot\cdot\underset{\delta^-}{Cl} \longrightarrow H-\underset{\underset{H}{|}}{\overset{\overset{H}{|}}{C^+}} + \cdot\dot{C}l^-$$

Chloromethane   Carbocation

Carbocation groups are named after the parent alkyl group by "ium" (i.e. alkyl + ium = alkylium). For example, $^+CH_3$ is called methyl carbocation or "methylium." Similarly, $CH_3-^+CH_2$ is called ethyl carbocation ion or "ethylium."

The positively-charged carbocations are highly reactive and can readily bind with the electron-donating atoms or groups to form new organic compounds. Carbocations are stabilized by nearby electron donating groups, including alkyl groups. In terms of stability, the primary (1°) carbocation ion is less stable than secondary (2°) carbocation, which is less stable than tertiary (3°) carbocation. A 1° carbon is one that is bound to one other carbon, a 2° carbon is bound to two carbon atoms and a 3° carbon is bound to three carbon atoms. See the stability order below of carbocation where "R" refers to alkyl group(s).

The Stability Order of Carbocations

Tertiary Carbocation	Secondary Carbocation	Primary Carbocation

A *carbanion* is a negatively-charged carbon atom that carries an unshared pair of electrons. Examples of carbanions are methyl carbanion ($^-CH_3$) and ethyl carbanion ($CH_3-^-CH_2$). A carbanion is formed through fission of an σ-bond and concentration of an unshared electron pair in a C atom. This type of fission takes place when the C atom carries relatively higher electronegativity compared to the attached atom or group.

A Carbanion

The stability of a carbanion is increased in the presence of an electron-withdrawing (attracting) group that pulls the electron density toward itself and stabilizes the carbanion (e.g. $-NO_2$, $-CO$, $-CN$ etc.). On

the other hand, the stability is decreased in the presence of the electron-donating (repelling) alkyl (R) groups, which push the electron cloud toward the carbanion

(e.g. $-CH_3$, $-C_2H_5$ etc.)

Therefore, the stability order is as follows:

The Stability Order of Carbanions

Tertiary Carbanion	Secondary Carbanion	Primary Carbanion

$$\underset{\displaystyle R}{\overset{\displaystyle R}{R-\overset{\displaystyle |}{\underset{\displaystyle |}{C}}:^-}} \quad < \quad \underset{\displaystyle H}{\overset{\displaystyle R}{R-\overset{\displaystyle |}{\underset{\displaystyle |}{C}}:^-}} \quad < \quad \underset{\displaystyle H}{\overset{\displaystyle R}{H-\overset{\displaystyle |}{\underset{\displaystyle |}{C}}:^-}}$$

In chemical reactions, a carbocation or carbanion binds with the attacking reagent to form a new organic compound. The nature of the reaction depends significantly on the attacking reagent. The mechanism of interaction between the organic substrate and the attacking reagent can be illustrated in the steps below:

The Steps of the Interaction Between an Organic Substrate and an Attacking Reagent

$$Z \; + \; X\!:\!Y \longrightarrow \left[\begin{array}{l} :X^- \; + \; Y^+ \xrightarrow{Z^+} XZ \; + \; Y^+ \\[2ex] X^+ \; + \; :Y^- \xrightarrow{Z} XZ \; + \; Y^- \end{array} \right.$$

Z = Attacking reagent   X:Y = Substrate

Depending on the nature, attacking reagents (Z) can be categorized into an electrophile or electrophilic reagent and a nucleophile or nucleophilic reagent.

Electrophiles are attacking species that have a strong attraction to electrons. Electrophiles are naturally electron-deficient, and therefore, they have electron-accepting characteristics. There are two types of electrophiles: positive (or charged) electrophiles and neutral electrophiles.

Positive electrophiles (E^+) are positively-charged, so they have a high affinity for electrons. Examples are proton (H^+), alkylium (R^+), nitronium (NO_2^+), nitrosonium (^+NO) etc., where R = alkyl groups.

Neutral electrophiles (E) do not have an electric charge but have a strong affinity for electrons to complete the octet in their valence shell. Examples of neutral electrophiles are aluminum chloride $(AlCl_3)$, boron trifluoride (BF_3), and ferric chloride $(FeCl_3)$.

Neutral Electrophiles *(E)*

Aluminum Chloride	Boron Trifluoride

$$Cl \ddot{:}\underset{}{Al}\ddot{:}\,Cl \quad \text{with } Cl \text{ above}$$

$$F\ddot{:}B\ddot{:}F \quad \text{with } F \text{ above}$$

Nucleophiles are rich in electrons, so they act as electron donors. Nucleophiles are also categorized into negative (or charged) nucleophile and neutral nucleophile.

Negative nucleophiles (Nu^-) are negatively-charged attacking reagents. Examples include methyl carbanion $(^-CH_3)$, chloride (Cl^-), bromide (Br^-), hydroxide ion (^-OH), and cyanide ion (^-CN).

Neutral nucleophiles (Nu) carry an unshared pair of electrons, but do not carry any negative charge. Examples are ammonia (NH_3), water (H_2O), and alcohol (R-OH).

Neutral Nucleophile *(Nu)*

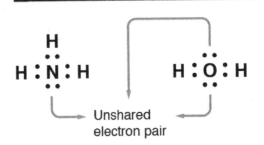

Unshared electron pair

Based on the nature of the attacking agents, organic reactions can also be classified into following categories:

1. An *electrophilic addition* is an addition reaction in which an electrophile (electron-attracting species) is added to a molecule that is mostly carrying unsaturated bonds. For example, the electrophilic proton (H^+) attacks the double bond of ethene ($CH_2=CH_2$) to form ethane ($CH_3—CH_3$).

2. An *electrophilic substitution* is an elimination reaction in which an electrophile is eliminated from an organic molecule, e.g. elimination of protons from ethane (CH_3-CH_3) causes the formation of ethene ($CH_2=CH_2$).

3. A *nucleophilic addition* is an addition reaction in which an electron-rich reactant (i.e. nucleophile) attaches to an electrophile to form a new bond.

4. A *nucleophilic substitution* is an elimination reaction that causes the removal of a nucleophile from an organic molecule.

Oxidation and Reduction

Oxidation/reduction, or redox reactions, are reactions in which the oxidation state of an atom changes. This oxidation state change relates to the number of electrons lost or gained during the reaction. In the oxidation half of the reaction, an electron is lost. In the reduction half of the reaction, an electron is gained. These two parts are referred to as "half-reactions" because oxidation is always accompanied with reduction. Usually, the change in the oxidation state of the C atom determines if an organic redox reaction is termed an oxidation or a reduction reaction.

1. *Alkanes*: Alkanes are generally stable and unreactive. Therefore, alkanes rarely participate in chemical redox reactions. The most common redox reaction of alkanes is combustion, in which the burning of hydrocarbon chains in the presence of oxygen produces carbon dioxide, water, and energy. In the absence of adequate oxygen, oxidative combustion of hydrocarbons produces carbon monoxide, water, and energy. This is considered an oxidation reaction because the C atom loses electrons. In this equation, the O atoms are reduced. Linear and small chain alkanes are more readily oxidized than larger and branched chain alkanes. The figure below shows the oxidation of methane and is labeled with the oxidation numbers of each atom to illustrate the change in oxidation state. Here, carbon is oxidized and oxygen is reduced.

The Oxidation of Methane Showing the Oxidation Numbers

$$\overset{+1}{H}-\overset{\overset{+1}{H}}{\underset{\underset{+1}{H}}{\overset{-4}{C}}}-\overset{+1}{H} \ + \ 2 \overset{0}{O}=\overset{0}{O} \ \longrightarrow \ \overset{-2}{O}=\overset{+4}{C}=\overset{-2}{O} \ + \ 2 \overset{+1}{H}-\overset{-2}{O}-\overset{+1}{H}$$

2. *Alkenes:* In the presence of a weak oxidizing agent such as dilute potassium permanganate ($KMnO_4$) in alkaline (KOH) solution, alkenes undergo oxidation to produce glycols. This reaction is utilized to detect unsaturation (double and triple bonds) in organic compounds. While observing this test, called *Baeyer's Test,* the pink color of potassium permanganate gradually fades.

Oxidation of an Alkene and Using the Baeyer's Test

$$R^1-CH=CH-R^2 \quad + \quad H_2O \quad + \quad [\,O\,] \quad \xrightarrow[KOH]{KMnO_4} \quad R^1-HOCH-CHOH-R^2$$

Alkene   Glycol

However, in the presence of a strong oxidizing agent such as concentrated $KMnO_4$ in acidic (H_2SO_4) media, alkenes are oxidized to produce organic acids.

Oxidizing an Alkene to an Organic Acid

$$R-CH=CH_2 \quad + \quad 5\,[\,O\,] \quad \xrightarrow[KOH]{KMnO_4} \quad R-COOH \quad + \quad CO_2 \quad + \quad H_2O$$

Alkene   Carboxylic acid

In the presence of catalysts like platinum (Pt), palladium (Pd), or nickel (Ni), alkenes are reduced to alkanes.

Reducing an Alkene to an Alkane

$$R^1-CH=CH-R^2 \quad + \quad H_2 \quad \xrightarrow{Pt \ or \ Pd} \quad R^1-CH_2-CH_2-R^2$$

Alkene   Alkane

3. *Alkynes:* When alkynes are treated with an oxidizing agent like alkaline potassium permanganate ($KMnO_4$) in the presence of a high temperature, the π-bonds undergo fission and oxidation to form carboxylic acids. The type of carboxylic acids formed depends on the position of the triple bond in the C chain.

Alkynes can be reduced by H_2 at room temperature in the presence of Pt or Pd, or at a high temperature ($150° – 180°C$) in the presence of Ni.

Reduction of an Alkyne

$$R-C \equiv CH \ + \ H_2 \ \xrightarrow[180°\ C]{Ni} \ R-CH=CH_2 \ \xrightarrow[180°\ C]{H_2,\ Ni} \ R-CH_3-CH_3$$

Alkyne   Alkene   Alkane

4. *Aromatic Compounds:* The alkyl side chain on the benzene is susceptible to oxidation. For example, the $-CH_3$ group of toluene could be oxidized by $K_2Cr_2O_7/H_2SO_4$ or $KMnO_4$ or dilute heated HNO_3 to form benzoic acid.

Oxidizing the Alkyl Side Chain of an Aromatic Compound

CH$_3$ COOH

$$+ \ 3\,[\,O\,] \ \xrightarrow[120°\ C]{HNO_3} \ \ + \ H_2O$$

Toluene   Benzoic acid

When treated with a weak oxidizing agent like chromyl chloride (CrO_2Cl_2), $-CH_3$ group of toluene is oxidized to form benzaldehyde. This is called the *Étard Reaction.*

The Étard Reaction

In the presence of nickel (Ni) catalyst and at temperature 200°C, benzene (C_6H_6) is reduced by H_2 to form cyclohexane (C_6H_{12}). In this reaction, three molecules of H_2 attach with a molecule of benzene.

Under similar conditions, toluene is reduced to hexahydrotoluene.

The Reduction of Toluene to Hexahydrotoluene

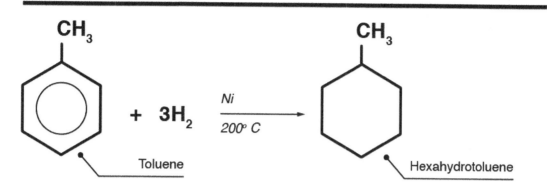

5. *Alcohols:* Alcohols are oxidized to produce carbonyl compounds like aldehydes and ketones. The type of carbonyl compounds formed depends on the nature of alcohol (1°, 2°, or 3°) and oxidizing agent.

6. *Carbonyl Compounds:* Carbonyl compounds can be oxidized to organic acids, whereas reduction of carbonyl compounds produces alcohols.

Hydration and Dehydration
Hydration refers to the addition of water to a compound. Conversely, dehydration is the removal of water from a molecule.

1. Alkenes: Alkenes can be hydrated in the presence of concentrated sulfuric acid to form alcohol. In such reactions, an unstable intermediate, alkyl hydrogen sulphate, is formed. This is further degraded to yield an alcohol.

The Hydration of an Alkene

$$R-CH = CH_2 \ + \ H_2SO_4 \longrightarrow \underset{\text{Alkyl hydrogen sulphate}}{R-CH-CH_3} \overset{H_2O}{\underset{Heat}{\longrightarrow}} \underset{\text{Alcohol}}{R-CH-CH_3}$$

with OSO_3H group on the middle carbon, and OH group on the middle carbon respectively.

The above reaction follows *Markovnikov's Rule*. According to this rule, when an asymmetric alkene interacts with an asymmetric reagent, the H atom of the reagent attaches with the C atom that carries

the greater number of H atoms. An example below shows how propene reacts with hydrogen bromide to yield 2-bromopropane as the principal yield in the reaction:

An Example of a Reaction that Follows Markovnikov's Rule

$$2CH_3 - CH = CH_2 \xrightarrow{2HBr} \begin{cases} CH_3 - CHBr - CH_3 \quad \text{2 - bromopropane } 90\% \\ CH_3 - CH_2 - CH_2Br \quad \text{1 - bromopropane } 10\% \end{cases}$$

2. *Alkynes*: When treated with sulfuric acid and in the presence of mercuric sulphate catalyst, alkynes are converted into carbonyl compounds. The reaction proceeds through the formation of an unstable intermediate that undergoes rearrangement to form the final carbonyl product.

The Conversion of Alkynes to Carbonyl Compounds

$$R - C \equiv CH + H_2O \xrightarrow[H_2SO_4]{Hg^{++}} R - \underset{\text{Intermediate compound}}{CH = CH_2} \xrightarrow{Rearrangement} R - \underset{\text{Ketone}}{\overset{O}{\overset{\|}{C}} - CH_3}$$

Alkyne

2. *Alcohols*: Alcohols undergo *dehydration* reactions in the presence of a *Lucas reagent*. This reaction causes the elimination of –OH group of alcohol, resulting in the formation of organic halide.

3. *Carbonyl Compounds*: Aldehydes and ketones can be hydrated to produce alcohols. Hydration of aldehydes produces 1° and 2° alcohols, whereas that of ketones produces 3° alcohols.

Hydrolysis

Hydrolysis refers to chemical reactions that involve cleavage of bonds in a molecule by the addition of water. In such reactions, a chemical bond in the organic molecule undergoes fission. The -OH group of water attaches to one part of the organic molecule, and proton (H) attaches to the other. The reactions are catalyzed by an acid or alkali. Hydrolytic bond cleavage is limited to certain classes of organic compounds, including amides, esters, ethers, and alkyl halides.

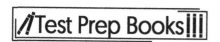

1. *Amides*: A primary amide is hydrolyzed to form a carboxylic acid and ammonia, whereas a secondary amide produces a carboxylic acid and primary amine.

Hydrolysis of Amides

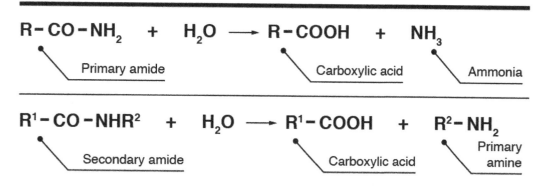

2. *Esters*: Hydrolysis of an ester produces a carboxylic acid and an alcohol.

Hydrolysis of Esters

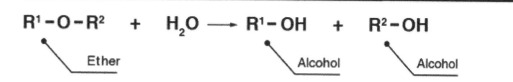

3. *Ethers*: Ethers are hydrolyzed to form alcohols.

Hydrolysis of Ethers

R¹ – O – R² + H₂O ⟶ R¹ – OH + R² – OH

Ether   Alcohol   Alcohol

4. *Alkyl Halides*: Hydrolysis of alkyl halides produces alcohol.

The Hydrolysis of Alkyl Halides

$$R - X \quad + \quad H_2O \longrightarrow R - OH \quad + \quad HX$$

Alkyl halide   Alcohol   Hydrogen halide

Organic compounds also undergo a variety of reactions that involve addition, substitution, and elimination of atom(s) or group(s).

Addition, Substitution, and Elimination Reactions

Addition Reactions

In *addition reactions*, two different molecules are combined to form a new organic compound. Generally, unsaturated compounds with π-bonds undergo addition reactions. These reactions involve breaking a π-bond to form more stable σ-bond. An example is the reaction between ethene and bromine, which causes the formation of 1,2-dibromo ethane.

The Addition Reaction Between Ethene and Bromine Breaks a π-Bond to Form a More Stable σ-Bond

$$\underset{\text{Ethene}}{H-C=C-H} \quad + \quad \underset{\text{Bromine}}{Br_2} \quad \longrightarrow \quad \underset{\substack{\text{Br} \quad \text{Br} \\ \text{1, 2 - dibromo ethane}}}{H-C-C-H}$$

Addition reactions are chemical reactions that involve the addition of a molecule to an organic compound that contains a double bond. Alkenes can undergo addition reactions with hydrogen (H_2), halogens (X_2), hydrogen halide (HX), hypohalous acid (HOX), and sulfuric acid (H_2SO_4). In such reactions,

the above electrophilic reagents interact with π-electrons to form new compounds. Below are examples of addition reactions of alkenes with various reagents:

Addition Reactions of Alkenes

1. Reaction with H_2

$$CH_2 = CH_2 \ + \ H_2 \longrightarrow CH_3 - CH_3$$

Ethene   Ethane

2. Reaction with X_2

$$CH_2 = CH_2 \ + \ Cl_2 \longrightarrow ClCH_2 - CH_2Cl$$

Ethene   1,2-dichloro ethane

3. Reaction with HX

$$CH_2 = CH_2 \ + \ HCl \longrightarrow CH_3 - CH_2Cl$$

Ethene   Ethyl chloride

4. Reaction with HOX

$$CH_2 = CH_2 \ + \ HOCl \longrightarrow ClCH_2 - CH_2OH$$

Ethene   2 - chloroethanol

5. Reaction with H_2SO_4

$$CH_2 = CH_2 \ + \ H_2SO_4 \longrightarrow CH_3 - CH_2 - OSO_3H \ + \ H_2O \longrightarrow CH_3 - CH_2 - OH \ + \ H_2SO_4$$

Ethene   Ethyl hydrogen sulphate   Ethanol

The π-bond in alkenes is readily oxidized by different oxidizing agents. In such reactions, the different compounds formed include glycols, ketones, carboxylic acid, and carbon dioxide. The reaction pattern, however, depends on the nature of the oxidizing agent. For example, in the presence of a weak oxidizing

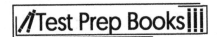

agent, such as dilute potassium permanganate (KMnO4), in alkaline (KOH) solution, ethene undergoes oxidation to produce an alcohol, ethylene glycol.

An Example of Oxidizing the π-Bond in in Alkene

$$CH_2=CH_2 \ + \ 2MnO_4^- \ + \ 2OH^- \longrightarrow \underset{\underset{OH \quad OH}{|\qquad|}}{CH_2-CH_2} \ + \ 2MnO_4^-$$

dark green solution

However, in the presence of a strong oxidizing agent, such as concentrated KMnO4, in acidic (H2SO4) conditions, propene undergoes oxidation to produce carboxylic acid, acetic acid.

$$CH_3-CH=CH_2 + 5[O] \rightarrow CH_3-COOH + CO_2 + H_2O$$

Like alkenes, alkynes can undergo addition reactions with hydrogen (H_2), halogens (X_2), hydrogen halide (HX), hypohalous acid (HOX), and sulfuric acid (H_2SO_4). In such reactions, these electrophilic reagents interact with π-electrons to form new compounds. Examples of addition reactions are illustrated below:

Alkyne Addition Reactions

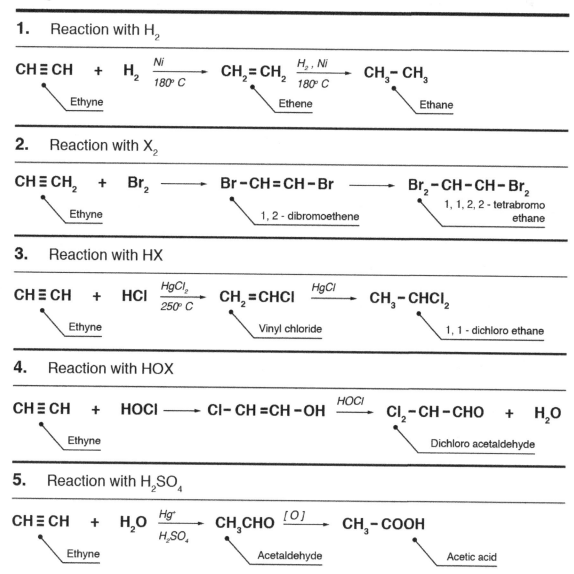

1. Reaction with H_2

2. Reaction with X_2

3. Reaction with HX

4. Reaction with HOX

5. Reaction with H_2SO_4

Benzene also undergoes addition reactions including hydrogen addition and halogen addition.

Hydrogen Addition: In the presence of a nickel (Ni) catalyst and at temperature 200°C, hydrogen (H_2) undergoes an addition reaction with benzene (C_6H_6) to form cyclohexane (C_6H_{12}). In this reaction, three molecules of H_2 attach with a molecule of benzene.

Hydrogen Addition

Benzene + 3H₂ → Cyclohexane

Halogen Addition: In the presence of ultraviolet radiation, three molecules of chlorine (Cl_2) add with benzene (C_6H_6) to form hexachlorocyclohexane ($C_6H_6Cl_6$).

Halogen Addition

Benzene + 3Cl₂ → Hexachlorocyclohexane

Substitution Reactions

A *substitution reaction* is when an atom or group is substituted by another atom or group to form a new compound. An example is a reaction between chloroethane and sodium hydroxide, where the chloride group of chloroethane is replaced by the hydroxyl group to form ethanol.

A Substitution Reaction

The chloride group of chloroethane is replaced by the hydroxyl group to form ethanol

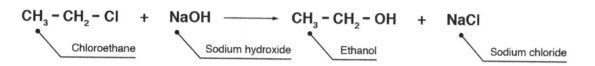

$$CH_3 - CH_2 - Cl \quad + \quad NaOH \quad \longrightarrow \quad CH_3 - CH_2 - OH \quad + \quad NaCl$$

Chloroethane   Sodium hydroxide   Ethanol   Sodium chloride

In substitution reactions, one functional group is replaced with another. For alkanes, H atoms (R-H) are replaced by attacking reagents such as halogens (X) or nitro ($-NO_2$) groups. Introduction of a halogen is called *halogenation*, and introduction of a nitro group is called *nitration*.

Halogenation of alkanes requires UV light and high temperatures ($300°–400°$ C) because alkanes are very stable. Halogenation can occur at every available H atom. For example, halogenation of alkanes by chlorine (Cl_2) will produce methyl chloride (CH_3Cl), dichloromethane (CH_2Cl_2), trichloromethane or chloroform ($CHCl_3$), and carbon tetrachloride (CCl_4), respectively. The steps of reaction are illustrated below.

     1. $CH_4 + Cl_2 \rightarrow CH_3Cl + HCl$

     2. $CH_3Cl + Cl_2 \rightarrow CH_2Cl_2 + HCl$

     3. $CH_2Cl_2 + Cl_2 \rightarrow CHCl_3 + HCl$

     4. $CHCl_3 + Cl_2 \rightarrow CCl_4 + HCl$

The chlorination reaction takes place in three steps:

     1. Chlorine free-radical is formed in the presence of high temperature and UV radiation.

     2. A chlorine free-radical attacks a C—H bond of alkane to form an alkyl (R) free-radical and HCl.

3. The other chlorine free-radical binds with the alkyl free-radical to form chlorinated (halogenated) alkane (R—Cl).

The Steps of the Chlorination (Halogenation) Reaction of an Alkane

1.
$$Cl:Cl \longrightarrow Cl\cdot + Cl\cdot$$

2.
$$Cl\cdot + H:R \longrightarrow R\cdot + HCl$$

3.
$$Cl\cdot + R\cdot \longrightarrow R\text{-}Cl$$

Because a carbon free-radical is formed, the rate and extent of halogenation depends on the stability of the carbon free-radical. For example, a H atom attached to a 3° carbon is more reactive than that attached with 2° and 1° carbon, respectively. Therefore, in the following reaction, the yield of 2-chloro propane (55%) is more than that of propyl chloride (45%).

The Rate and Extent of Halogenation Depends on the Stability of the Carbon Free-Radical

$$\underset{\text{Propane}}{\overset{2^0 \quad 1^0}{CH_3 - CH_2 - CH_3}} + \underset{\text{Chlorine}}{Cl_2} \longrightarrow \underset{\underset{45\%}{\text{Propyl chloride}}}{CH_3 - CH_2 - CH_2} + \underset{\underset{55\%}{\text{2 - chloro propane}}}{\overset{\overset{Cl}{|}}{CH_3 - CH - CH_3}}$$

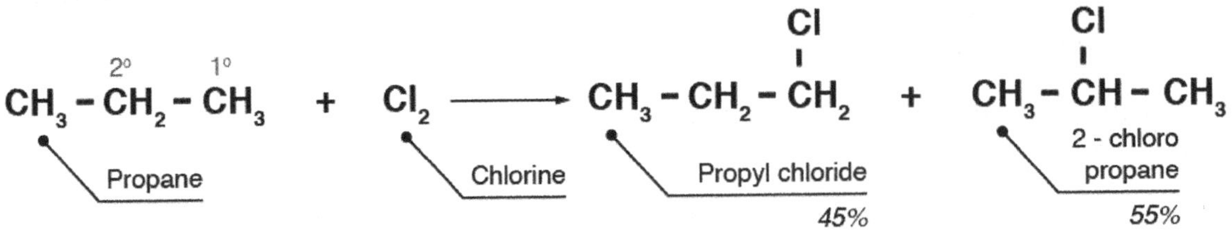

Halogenation with any halogen occurs through the same general steps. However, the different halogens have different reactivities. The order of reactivity is as follows: $F_2 > Cl_2 > Br_2 > I_2$.

Benzene and aromatic compounds undergo substitution reactions. Since benzene is quite stable, each H atom is substituted sequentially. The substituent groups/atoms take specific positions on the benzene ring during those substitution reactions.

1. *Halogenation:* Halogen substitution is a catalyst-dependent reaction, commonly using ferric chloride ($FeCl_3$), ferric bromide ($FeBr_3$) and aluminum chloride ($AlCl_3$) as catalysts. Halogenation is an electrophilic

addition reaction. For example, in the presence of anhydrous $AlCl_3$, chlorine (Cl_2) substitutes an H atom to form chlorobenzene (see below).

Halogenation of Benzene is an Electrophilic Addition Reaction

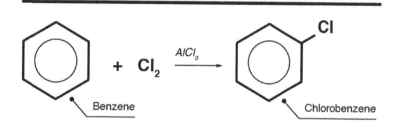

These substitutions can eventually replace all of the H atoms. In the example above, continued reaction would form dichlorobenzene, trichlorobenzene, tetrachlorobenzene, pentachlorobenzene and hexachlorobenzene, respectively.

Subsequent Halogen Substitutions

1. C_6H_6 + Cl_2 $\xrightarrow{AlCl_3}$ C_6H_5Cl + HCl
   Benzene   Chlorobenzene

2. C_6H_5Cl + Cl_2 $\xrightarrow{AlCl_3}$ $C_6H_4Cl_2$ + HCl
   Chlorobenzene   Dichlorobenzene

3. $C_6H_4Cl_2$ + Cl_2 $\xrightarrow{AlCl_3}$ $C_6H_3Cl_3$ + HCl
   Dichlorobenzene   Trichlorobenzene

4. $C_6H_3Cl_3$ + Cl_2 $\xrightarrow{AlCl_3}$ $C_6H_2Cl_4$ + HCl
   Trichlorobenzene   Tetrachlorobenzene

5. $C_6H_2Cl_4$ + Cl_2 $\xrightarrow{AlCl_3}$ C_6HCl_5 + HCl
   Tetrachlorobenzene   Pentachlorobenzene

6. C_6HCl_5 + Cl_2 $\xrightarrow{AlCl_3}$ C_6Cl_6 + HCl
   Pentachlorobenzene   Hexachlorobenzene

2. *Nitration:* This electrophilic substitution reaction replaces H atoms with a nitro (–NO$_2$) group. For example, when benzene is treated with concentrated sulfuric acid (H$_2$SO$_4$) and concentrated nitric acid (HNO$_3$) at around 50°C, one H atom is substituted by a –NO$_2$ group to produce nitrobenzene and water.

Nitration of Benzene is an Electrophilic Substitution Reaction

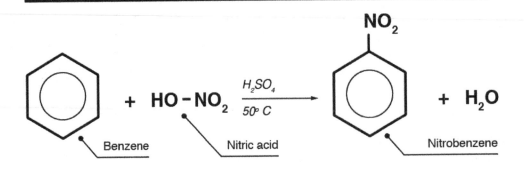

When the reaction is carried out at 100°C in the presence of concentrated H$_2$SO$_4$ and HNO$_3$, 1,3-dinitrobenzene (meta-dinitrobenzene) is formed.

Nitration of Benzene at 100°C in the Presence of Concentrated H$_2$SO$_4$ and HNO$_3$

NO$_2$

+ HO – NO$_2$ $\xrightarrow[100° C]{H_2SO_4}$

NO$_2$

NO$_2$

+ H$_2$O

Nitrobenzene   Nitric acid

1, 3 - dinitrobenzene

Meta-dinitrobenzene

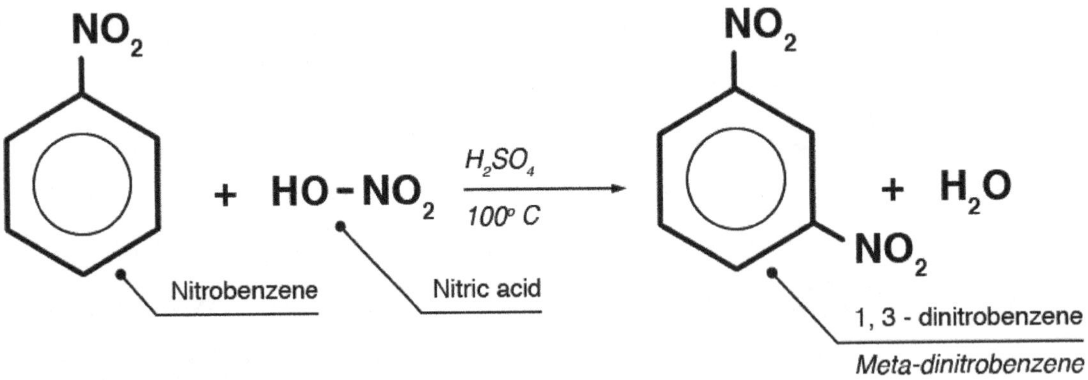

3. *Friedel-Crafts Alkylation:* When benzene is treated with an alkyl halide (R—X) in the presence of a Lewis acid catalyst (e.g. anhydrous $AlCl_3$ or $FeCl_3$), one H atom of the ring is substituted by the alkyl group to form alkyl benzene. This is called Friedel-Crafts alkylation. For example, in a reaction between benzene and methyl chloride, methyl benzene (toluene) is formed.

Friedel-Crafts Alkylation

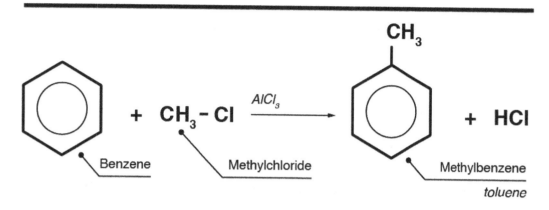

4. *Friedel-Crafts Acylation:* When benzene reacts with an acyl halide (RCO—X) in the presence of a Lewis acid, like $AlCl_3$ or $FeCl_3$, an acyl group (RCO-) is introduced in the benzene ring to form an aromatic ketone. For example, the reaction between benzene and acetyl chloride produces acetophenone (methyl phenyl ketone).

Friedel-Crafts Acylation

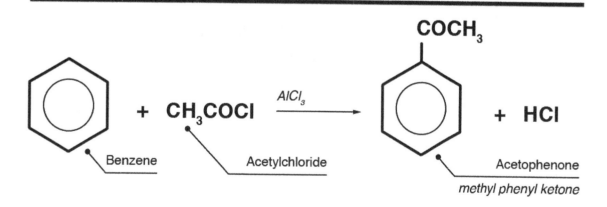

When a single H atom is substituted from the benzene ring, it makes mono-substituted benzene derivatives. Examples are methylbenzene (toluene), chlorobenzene, phenol, phenyl amine (aniline), benzaldehyde, and benzoic acid.

Mono-Substituted Benzene Derivatives

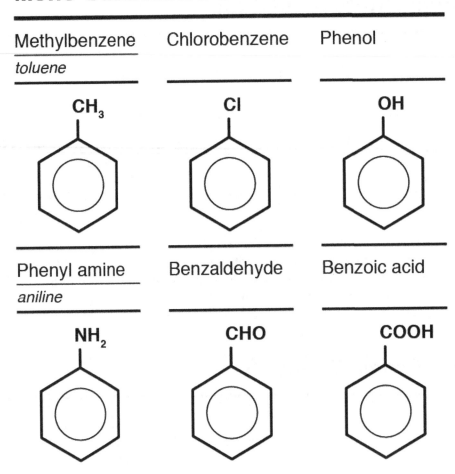

When a single H atom is substituted, there are still five H atoms that could be substituted by different atoms or groups. The C atom with first substitution is numbered 1, and other C atoms in the ring are numbered and named accordingly. Since benzene has a cyclic, symmetrical structure, 1:2 and 1:6 positions are equivalent. Similarly, 1:3 and 1:5 positions are equivalent. The 1:2 and 1:6 positions are

nearest, and called "ortho" or "o." The 1:3 and 1:5 positions are termed "meta" or "m", and the 1:4 position is the farthest, and termed "para" or "p."

Naming Mono-Substituted Benzene Derivatives

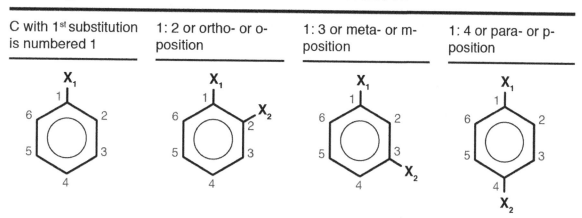

| C with 1st substitution is numbered 1 | 1: 2 or ortho- or o-position | 1: 3 or meta- or m-position | 1: 4 or para- or p-position |

The following example shows three possible isomers of dichlorobenzene based on the relative positions of the substituent atoms.

Possible Isomers of Dichlorobenzene

1, 2 - dichlorobenzene   **1, 3 - dichlorobenzene**   **1, 4 - dichlorobenzene**

ortho - dichlorobenzene   meta - dichlorobenzene   para - dichlorobenzene

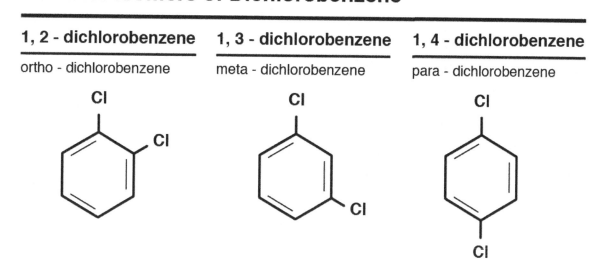

The substituents on benzene ring could have two types of effects: an inductive effect or a mesomeric effect.

1. *Inductive Effect (I):* The electronegativity of substituent atoms or groups can cause polarization of the bond between the C atom in benzene ring and the substituent group. When the electronegativity of the substituent (e.g. halogens, X: F, Cl, B, I) is higher than carbon, it pulls the σ-electrons and causes polarization of the bond. This is called a negative (−) inductive effect. In contrast, if the electronegativity

of substituent atoms or groups (e.g. alkyl groups, R: –CH₃, –C₂H₅) is lower than the ring C atom, it pushes the σ-electrons toward C atom, and is called a positive (+) inductive effect.

The Inductive Effect (I) Involves σ-Electrons

Negative inductive effect	Positive inductive effect
$C \longrightarrow X$	$C \longleftarrow R$

2. *Mesomeric Effect (M):* The electron cloud in a π-bond can be influenced by the nature of substituent atoms or groups. The delocalization of π-electrons toward the relatively higher electronegative atom or group is called a negative mesomeric effect. The opposite effect by an electron repelling/pushing atom or group is called a positive mesomeric effect.

The Mesomeric Effect (M) Involves a π-Bond

$$>C = O \longrightarrow >C^+ - O^-$$

Mesomeric effect

The nature of substituent group or atom determines the position of further substitutions in the ring. Therefore, the substituent groups/atoms can be categorized into two types: ortho-para directing and meta directing.

1. *Ortho-Para Directing Atoms/Groups:* Electron-donating substituents like -CH₃, -OH, and -NH₂ donate electrons to the benzene ring and make it more reactive. Due to the positive mesomeric effect, they push into the electron cloud and the electron density gets relatively higher at ortho-para positions. Therefore, the ortho-para position becomes more reactive, allowing further substitutions on those positions. In the case of halogens, the effect is relatively complicated. X atoms have a negative inductive effect, but they provide a positive mesomeric effect by pushing unpaired electrons to the benzene ring. Therefore, halogens are ortho-para directing substituents.

2. *Meta Directing Atoms/Groups:* Certain atoms/groups, due to their negative mesomeric effects, pull the electron cloud from the benzene ring. These atoms and groups make benzene ring less reactive. Examples are –NO₂, -CHO, -COOH, and –SO₃H. By withdrawing the electron cloud, these groups decrease the electron density significantly at the ortho-para positions. However, meta position is less affected, and thus, the second electrophile attacks the meta position. These groups are called meta-directing substituents.

Elimination Reactions

In *elimination reactions*, atoms or groups are eliminated from two adjacent C atoms to form π-bonds. This is opposite to addition reactions. As an example, in the presence of concentrated sulfuric acid, a water molecule is removed from ethanol to form a C-C π-bond, i.e. ethene.

An Elimination Reaction:
Ethene is formed from Ethanol and a Water Molecule is Removed

$$H-\underset{\underset{H}{|}}{\overset{\overset{H}{|}}{C}}-\underset{\underset{OH}{|}}{\overset{\overset{H}{|}}{C}}-H \xrightarrow[160°C]{Conc.\ H_2SO_4} H-\overset{\overset{H}{|}}{C}=\overset{\overset{H}{|}}{C}-H\ +\ H_2O$$

Ethanol   Ethene

Isomerization

Isomerization refers to a rearrangement of atoms or groups in an organic molecule to form an isomer. The newly formed compound has a similar molecular formula, but has a different structural arrangement than the parent compound. As an example, in the presence of aluminum chloride and hydrochloric acid, butane undergoes atomic rearrangement to form 2-methyl propane.

An Isomerization

$$CH_3-CH_2-CH_2-CH_3 \xrightarrow[300°C]{AlCl_3,\ HCl} CH_3-\overset{\overset{CH_3}{|}}{CH}-CH_3$$

Butane   2 - methyl propane

Alkanes can undergo *isomerization* to produce isomers. In such reactions, straight chain alkanes can be converted into corresponding branched-chain isomers. For example, in the presence of aluminum chloride ($AlCl_3$) and hydrochloric acid (HCl) and at 250°–300°C, butane is converted into iso-butane (2-methyl propane).

$$CH_3-CH_2-CH_2-CH_3 \rightarrow CH_3-CH(CH_3)-CH_3$$

Butane   Isobutane

Polymerization

In the presence of a high temperature, pressure, and a catalyst, molecules of a same alkene can join together to form a large molecule. This process is called *polymerization* and it results in the formation a polymer from a monomer (single unit). Polymerization is used widely in industries to synthesize compounds with modified physical properties and shear withstanding capacities. For example, polymerization of ethene (ethylene) causes the formation of polyethylene, which is widely used as a plastic material in manufacturing industries. Note that *n* equals the number of monomers.

$$nCH_2=CH_2 \rightarrow (-CH_2-CH_2-)n$$

Like alkenes, alkynes also undergo polymerization to produce larger molecules. Alkynes can produce both aliphatic and aromatic polymers during this process. The type of polymer formed depends on the reaction conditions.

1. *Formation of Aliphatic Polymers:* In the presence of cuprous chloride (CuCl) and ammonium chloride (NH$_4$Cl), ethyne (or acetylene) undergoes polymerization to first form vinylacetylene, and the further addition of a molecule forms divinylacetylene.

Complete Polymerization of Ethyne to Form Aliphatic Polymers

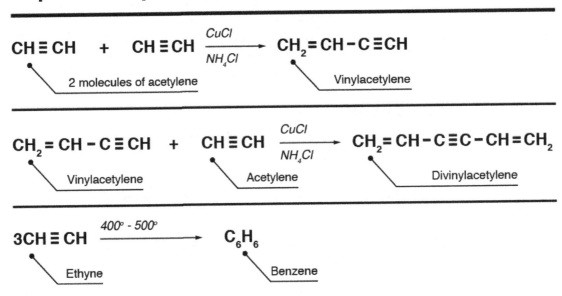

2. *Formation of Aromatic Polymers:* When ethyne (or acetylene) is passed through a heated iron pipe at a temperature of 400°–500°C, then three molecules join together to form aromatic compound benzene.

Biochemistry

Basic units of organic compounds are often called *monomers*. Repeating units of linked monomers are called *polymers*. The most important polymers found in all living things can be divided into just four categories: carbohydrates, lipids, proteins, and nucleic acids. This may be surprising since there is so much diversity in the outward appearances and functions of living things present on Earth. Carbon (C), hydrogen (H), oxygen (O), nitrogen (N), sulfur (S), and phosphorus (P) are the major elements of most biological molecules. Carbon is a common backbone of large molecules because of its ability to form four covalent bonds.

DNA and RNA

Nucleic acids have two important duties in the body. As monomers, they are crucial for energy transfer. As polymers, they are a fundamental component of genetic material. Nucleotides are the monomer units that assemble to form nucleic acids. Nucleotides have three components: a nitrogenous base and a phosphate functional group, both of which are attached to a five-carbon (pentose) sugar. There are two classes of nitrogenous bases, purines and pyrimidines. The two types of purines are guanine (G) and adenine (A), while the three types of pyrimidines are thymine (T), cytosine (C), and uracil (U). The two types of pentose sugars are deoxyribose and ribose. Nucleotides containing deoxyribose are termed deoxyribonucleic acid (DNA). DNA utilizes guanine, adenine, cytosine, and thymine as its nitrogenous bases. Nucleotides containing ribose are termed ribonucleic acid (RNA). RNA utilizes guanine, adenine, cytosine, and uracil as its nitrogenous bases.

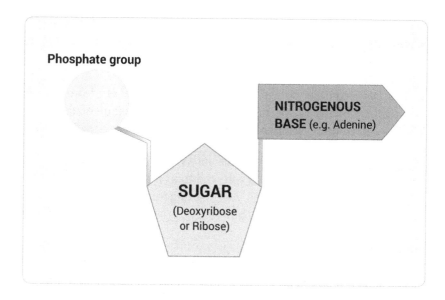

Chromosomes, Genes, Proteins, RNA, and DNA

Chromosomes are composed of hundreds to thousands of genes. Human cells contain 23 pairs of chromosomes for a total of 46 chromosomes. Genes are inherited in pairs, one from each parent.

Proteins are made of long chains of amino acids. In total, there are 20 amino acids, 11 of which humans can synthesize on their own and the remaining 9 of which are procured through diet. DNA contains the information for the synthesis of proteins, but that information on DNA has to undergo transcription and translation by RNA in order to produce proteins.

Codons

A *codon* represents a sequence of three nucleotides, which codes for either one specific amino acid or a stop signal during protein synthesis. Codons are found on messenger RNA (mRNA).

Twenty essential amino acids are utilized in the process of protein synthesis. The full set of codons encompasses 64 possible combinations and is termed the *genetic code*. In the genetic code, 61 codons represent amino acids and three codons are stop signals. The genetic code is redundant due to the fact that a single amino acid may be produced by multiple codons. For example, the codons AAA and AAG produce the amino acid lysine. The codons UAA, UAG, and UGA are stop signals. The codon AUG codes for both the amino acid methionine and the start signal. As a result, AUG when found in mRNA, marks the initiation point of protein translation.

RNA

Ribonucleic acid (RNA) plays crucial roles in protein synthesis and gene regulation. RNA is made of nucleotides consisting of ribose (a sugar), a phosphate group, and one of four possible nitrogen bases—adenine (A), cytosine (C), guanine (G), and uracil (U). RNA utilizes the nitrogenous base uracil in place of the base thymine found in DNA. Another difference between RNA and DNA is that RNA is typically found as a single-stranded structure, while DNA typically exists in a double-stranded structure.

RNA can be categorized into three major groups—messenger RNA (mRNA), ribosomal RNA (rRNA), and transfer RNA (tRNA). Messenger RNA (mRNA) transports instructions from DNA in the nucleus of a cell to the areas responsible for protein synthesis in the cytoplasm of a cell. This process is known as *transcription*. Transfer RNA (tRNA) deciphers the amino acid sequence for the construction of proteins found in mRNA. Both tRNA and ribosomal RNA (rRNA) are found in the ribosomes of cells. Ribosomes are responsible for protein synthesis. The process is known as *translation*, and both tRNA and rRNA play crucial roles. Both translation and transcription are further described below.

DNA

Deoxyribonucleic acid, or DNA, contains the genetic material that is passed from parent to offspring. It contains specific instructions for the development and function of a unique eukaryotic organism. The great majority of cells in a eukaryotic organism contains the same DNA.

The majority of DNA can be found in the cell's nucleus and is referred to as *nuclear DNA*. A small amount of DNA can be located in the mitochondria and is referred to as *mitochondrial DNA*. Mitochondria provide the energy for a properly functioning cell. All offspring inherit mitochondrial DNA from their mother. James Watson, an American geneticist, and Frances Crick, a British molecular biologist, first outlined the structure of DNA in 1953.

The structure of DNA visually approximates a twisting ladder and is described as a double helix. DNA is made of nucleotides consisting of deoxyribose (a sugar), a phosphate group, and one of four possible nitrogen bases—thymine (T), adenine (A), cytosine (C), and guanine (G). It is estimated that human DNA contains three billion bases. The sequence of these bases dictates the instructions contained in the DNA making each species singular. The bases in DNA pair in a particular manner—thymine (T) with adenine (A) and guanine (G) with cytosine (C). Weak hydrogen bonds between the nitrogenous bases ensure easy uncoiling of DNA's double helical structure in preparation for replication.

Transcription

Transcription refers to a portion of DNA being copied into RNA, specifically mRNA. It represents the first crucial step in gene expression. The process begins with the enzyme RNA polymerase binding to the

promoter region of DNA, which initiates transcription of a specific gene. RNA polymerase then untwists the double helix of DNA by breaking weak hydrogen bonds between its nucleotides. Once DNA is untwisted, RNA polymerase travels down the strand reading the DNA sequence and adding complementary nitrogenous bases. With the assistance of RNA polymerase, the pentose sugar and phosphate functional group are added to the nitrogenous base to form a nucleotide. Lastly, the weak hydrogen bonds uniting the DNA-RNA complex are broken to free the newly formed mRNA. The mRNA travels from the nucleus of the cell out to the cytoplasm of the cell where translation occurs.

Translation

Translation refers to the process of ribosomes synthesizing proteins. It represents the second crucial step in gene expression. The instructions encoding specific proteins to be made are contained in codons on mRNA, which have previously been transcribed from DNA. Each codon represents a specific amino acid or stop signal in the genetic code.

Amino acids are the building blocks of proteins. Ribosomes contain transfer RNA (tRNA) and ribosomal RNA (rRNA). Translation occurs in ribosomes located in the cytoplasm of cells and consists of the following three phases:

1. Initiation: The ribosome gathers at a target point on the mRNA, and tRNA attaches at the start codon (AUG), which is also the codon for the amino acid methionine.

2. Elongation: A new tRNA reads the next codon on the mRNA and links the two amino acids together with a peptide bond. The process is repeated until a polypeptide, or long chain of amino acids, is formed.

3. Termination: The ribosome disengages from the mRNA when it encounters a stop codon (UAA, UAG, or UGA). The event releases the polypeptide molecule. Proteins are made of one or more polypeptide molecules.

Lipids

Lipids are a class of biological molecules that are hydrophobic, meaning they don't mix well with water. They are mostly made up of large chains of carbon and hydrogen atoms, termed hydrocarbon chains. When lipids mix with water, the water molecules bond to each other and exclude the lipids because they are unable to form bonds with the long hydrocarbon chains. The three most important types of lipids are fats, phospholipids, and steroids.

Fats are made up of two types of smaller molecules: glycerol and fatty acids. Glycerol is a chain of three carbon atoms, with a hydroxyl group attached to each carbon atom. A hydroxyl group is made up of an oxygen and hydrogen atom bonded together. Fatty acids are long hydrocarbon chains that have a backbone of sixteen or eighteen carbon atoms. The carbon atom on one end of the fatty acid is part of a carboxyl group. A *carboxyl group* is a carbon atom that uses two of its four bonds to bond to one oxygen atom (double bond) and uses another one of its bonds to link to a hydroxyl group. Fats are made by joining three fatty acid molecules and one glycerol molecule.

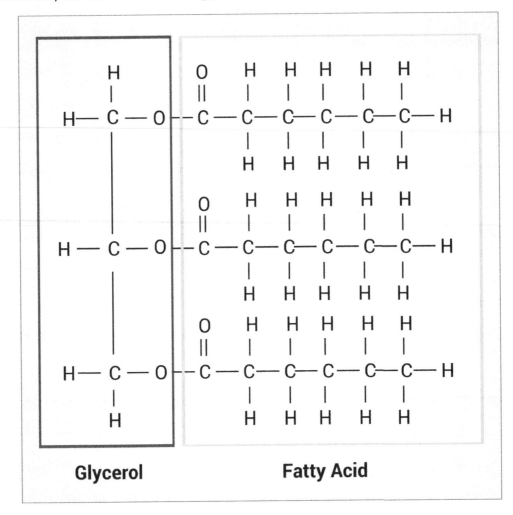

Glycerol **Fatty Acid**

Phospholipids are made of two fatty acid molecules linked to one glycerol molecule. A phosphate group is attached to a third hydroxyl group of the glycerol molecule. A phosphate group has an overall negative charge and consists of a phosphate atom connected to four oxygen atoms.

Phospholipids have an interesting structure because their fatty acid tails are hydrophobic, but their phosphate group heads are hydrophilic. When phospholipids mix with water, they create double-layered structures, called bilayers, that shield their hydrophobic regions from water molecules. Cell membranes are made of phospholipid bilayers, which allow the cells to mix with aqueous solutions outside and inside, while forming a protective barrier and a semi-permeable membrane around the cell.

Steroids are lipids that consist of four fused carbon rings. The different chemical groups that attach to these rings are what make up the many types of steroids. Cholesterol is a common type of steroid found in animal cell membranes. Steroids are mixed in between the phospholipid bilayer and help maintain the structure of the membrane and aids in cell signaling.

Proteins

Proteins are essential for most all functions in living beings. The name protein is derived from the Greek word *proteios*, meaning *first* or *primary*. All proteins are made from a set of twenty amino acids that are linked in unbranched polymers. The combinations are numerous, which accounts for the diversity of proteins. Amino acids are linked by peptide bonds, while polymers of amino acids are called *polypeptides*. These polypeptides, either individually or in linked combination with each other, fold up to form coils of biologically functional molecules, called proteins.

There are four levels of protein structure: primary, secondary, tertiary, and quaternary. The primary structure is the sequence of amino acids, similar to the letters in a long word. The secondary structure is beta sheets, or alpha helices, formed by hydrogen bonding between the polar regions of the polypeptide backbone. Tertiary structure is the overall shape of the molecule that results from the interactions between the side chains linked to the polypeptide backbone. Quaternary structure is the overall protein structure that occurs when a protein is made up of two or more polypeptide chains.

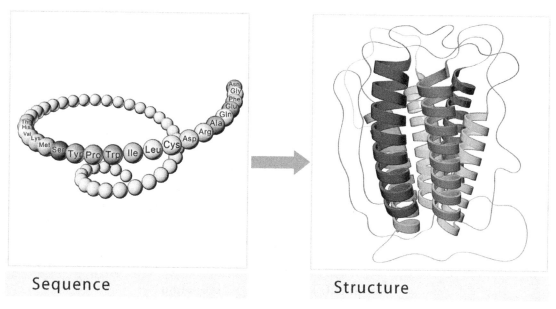

Sequence   Structure

Electrochemistry

Balancing Oxidation-Reduction Reactions

When two chemical species react with one another in solution, one species may lose electrons while the other gains electrons. **Oxidation-reduction** or **redox chemical reactions** involve the loss and gaining of electrons between two different reactants or reacting species. Therefore, a change in oxidation state for a particular ion within the chemical species will change when transitioning from a reactant to a product. For a given species or ion, **oxidation** refers to the loss of electrons or an increase in oxidation number (e.g., Fe^{2+} to Fe^{3+}), and **reduction** refers to gaining of electrons or decrease in oxidation state or number (e.g., Fe^{3+} to Fe^{2+}). Chemical reactions can occur in either an acidic or basic solution. The **half-reaction method** of balancing requires that a redox reaction be separated into two half-reactions, an oxidation half and a reduction half, to help balance the charge using electrons. Hydronium ions and water are used to balance an acidic reaction, and hydroxide ions are used to balance a reaction in basic solution. The half-reactions are then combined to give the overall reaction equation. For example, consider the following reaction that occurs in an acidic solution. First, identify the oxidation states for each species.

$$\text{Step 1: } Cu(s) + NO_3^-(aq) \rightarrow Cu^{2+}(aq) + NO_2(g)$$

Solid copper is oxidized because it loses electrons and increases in oxidation state (Cu to Cu^{2+}). The nitrate ion is the reactant species that is reduced because the gaseous nitrogen dioxide product has a charge of zero. Specifically, it is the nitrogen atom that changes or reduces in oxidation state because:

$$\overbrace{NO_3^-}^{x+3(-2)=-1} \rightarrow \overbrace{NO_2}^{x+2(-2)=0}$$

The x term refers to the oxidation number of nitrogen, and -2 is the known/standard oxidation number of oxygen. For the nitrate ion, the term x must be five ($x = +5$) because $5 + 3(-2) = -1$ and for nitrogen dioxide, x must be four ($x = +4$) because $4 + 2(-2) = 0$. The change in oxidation number for nitrogen is N^{5+} to N^{4+}. The reaction can now be separated into an oxidizing and reducing half as follows:

$$\text{Step 2: } Cu(s) \rightarrow Cu^{2+}(aq) \quad \text{Oxidation}$$

$$NO_3^-(aq) \rightarrow NO_2(g) \quad \text{Reduction}$$

Now balance each half-reaction based on the mass, if needed, using water molecules and hydronium ions, because the reaction occurs in aqueous solution. The copper half-reaction does not need mass balancing, but the second half-reaction does.

$$\text{Step 3: } Cu(s) \rightarrow Cu^{2+}(aq)$$

$$2\,H^+(aq) + NO_3^-(aq) \rightarrow NO_2(g) + H_2O(l)$$

Note that one water molecule is added first on the product side to balance the oxygen atoms, followed by the addition of hydronium ions on the reactant side to balance hydrogen. Next balance each half-

reaction based on charge by adding negatively charged electrons either to the reactant or product's side.

$$\text{Step 4: } Cu(s) \rightarrow Cu^{2+}(aq) + 2\ e^-$$

$$e^- + 2\ H^+(aq) + NO_3^-(aq) \rightarrow NO_2(g) + H_2O(l)$$

For the copper half-reaction, two electrons are added to the product side because oxidation results in the loss of electrons. The reactants side has an overall charge of zero, and the products side is also neutral because two electrons are needed to make the copper(II) ion neutral. For the second half-reaction containing nitrogen, the product side is neutral in charge. So, one electron must be added to make the reactant side neutral in charge: –1 from the electron, +2 from two protons, and –1 from the nitrate ion give $-1 + 2 - 1 = 0$. Before the half-reactions are combined, each half-reaction must be multiplied by a small whole number such that each half-reaction has an equal number of electrons. The second half-reaction must be multiplied by two.

$$\text{Step 5: } Cu(s) \rightarrow Cu^{2+}(aq) + 2\ e^-$$

$$2\left(e^- + 2\ H^+(aq) + NO_3^-(aq) \rightarrow NO_2(g) + H_2O(l)\right)$$

In the last step, the half-reactions can now be combined.

$$\text{Step 6: } Cu(s) \rightarrow Cu^{2+}(aq) + 2\ e^-$$

$$2\ e^- + 4\ H^+(aq) + 2\ NO_3^-(aq) \rightarrow 2\ NO_2(g) + 2\ H_2O(l)$$

$$Cu(s) + \cancel{2\ e^-} + 4\ H^+(aq) + 2\ NO_3^-(aq) \rightarrow Cu^{2+}(aq) + \cancel{2\ e^-} + 2\ NO_2(g) + 2\ H_2O(l)$$

The final balanced equation is:

$$Cu(s) + 4\ H^+(aq) + 2\ NO_3^-(aq) \rightarrow Cu^{2+}(aq) + 2\ NO_2(g) + 2\ H_2O(l)$$

Each side of the equation has a total charge of +2, and all elements are balanced according to mass. Copper is oxidized and acts as a **reducing agent**, which is a species that reduces another substance, thereby losing its electrons to that substance. The nitrate ion acts as the **oxidizing agent** because it is the species that oxidizes another substance and gains electrons from that substance.

Voltaic Cells

Voltaic cells are electrochemical cells that can be used to generate electricity through a spontaneous reaction. An electrical current or flow of electrical charge, created by the movement of electrons, is created. Figure 27 shows a schematic of a typical voltaic cell, which consists of two **half-cells**, with each half-cell containing an electrode (e.g., a metal plate) and an ionic solution where oxidation and reduction occur.

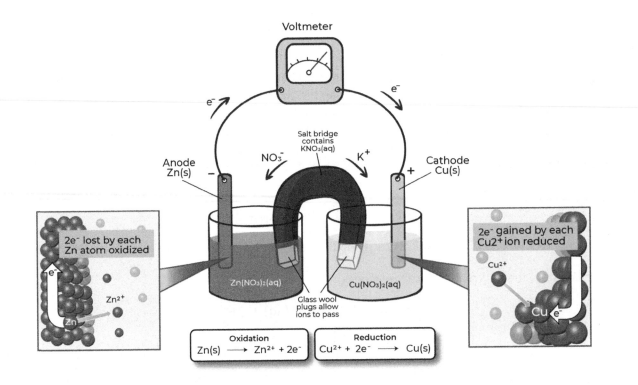

Figure 27. A galvanic or voltaic cell

Electrodes or conductive surfaces, e.g., metal strips, allow the flow of electrons and either undergo oxidation or reduction. The **anode** is the electrode or metal plate that undergoes oxidation, and the **cathode** is the electrode where reduction occurs. The anode carries a negative charge and the cathode a positive charge. An electrical wire connects the anode and cathode. Based on Le Chatelier's principle, solid zinc converts to zinc(II) ions, which increase in concentration over time. Likewise, in the reduction cell, copper(II) ions are converted to solid copper, which will collect around the plate over time. Electrons flow from the anode to the cathode (along a wire), which spontaneously creates a current that is opposite to electron flow. A **salt bridge** containing a strong electrolyte (e.g., potassium nitrate) is found in a U-shaped tube, with a permeable stopper at the end of each tube and allows for the passage of positive ions to the cathode and negative ions to the anode. The salt bridge prevents a buildup of charge through charge neutralization and therefore balances the charge of the cell, allowing the redox reaction to continue. The reaction is described as follows:

$$Zn(s) \rightarrow 2\,e^- + Zn^{2+}(aq) \quad \text{Oxidation}$$

$$Cu^{2+}(aq) + 2\,e^- \rightarrow Cu(s) \quad \text{Reduction}$$

$$Zn(s) + Cu^{2+}(aq) \rightarrow Cu(s) + Zn^{2+}(aq)$$

The reaction lies mostly to the right because zinc has a greater tendency to ionize or lose electrons than copper, which causes the zinc electrode to become negatively charged. The reaction may take on the following electrochemical cell notation:

$$Zn(s)|Zn^{2+}(aq)||Cu(s)|Cu^{2+}(aq)$$

The salt bridge is indicated by $||$, which separates the two half-reactions. The anode is to the left of the salt bridge where the oxidation half takes place, and the cathode is on the right side where the reduction process takes place. The vertical line $|$ separates the different phases for a particular species.

Stoichiometry of Redox Reactions

The electrical current is measured in **amperes** or **amps (A)**, which has units of coulombs per second ($1\ A = 1\ C/s$). An electron has a charge of:

$$1.602 \times 10^{-19}\ C$$

For the reduction of Cu^{2+}(aq) to Cu(s), which occurs at the cathode in Figure 27, if there is a constant current of 1.50 A (flowing through the circuit), how long will it take for 1.00 g of Cu(s) to accumulate on the copper plate (cathode)? To solve this problem, consider the following roadmap or relationship regarding unit conversions:

$$\text{grams of Cu} \rightarrow \text{moles of Cu} \rightarrow \text{moles of } e^- \rightarrow \text{coulombs} \rightarrow \text{time in minutes}$$

The moles of copper are related to the moles of electrons needed for reduction, which is two moles.

$$Cu^{2+}(aq) + 2\ e^- \rightarrow Cu(s)$$

One mole of electrons has a specific charge of 96,485 C or 1 faraday (C/mol) because:

$$\frac{1\ \text{mole of electrons}}{6.022 \times 10^{23}\ \text{electrons}} \times \frac{1\ \text{electron}}{1.602 \times 10^{-19}\ C} \approx \frac{1\ \text{mole of electrons}}{96{,}485\ C}$$

From the definition of an amp, 1.50 A is equivalent to 1.50 C/s. Therefore, the amount of time it takes for 1.00 g of Cu(s) to collect on the cathode is given by the following calculation:

$$1.00\ \text{g of Cu} \times \frac{1\ \text{mol Cu}}{63.55\ \text{g}} \times \frac{2\ \text{mol } e^-}{1\ \text{mol Cu}} \times \frac{96{,}485\ C}{1\ \text{mol } e^-} \times \frac{1\ s}{1.50\ C} \times \frac{1\ \text{min}}{60\ s} = 33.7\ \text{min}$$

Predicting the Spontaneous Direction of a Redox Reaction Using Standard Electrode Potentials

In a redox reaction, the **potential difference** (in units of volts, V), refers to an electrical current that is created by a difference in potential energy, in joules per coulomb: $1\ V = 1\ J/C$. The force created by this potential difference is called the **electromotive force (emf)** or **cell potential (E_{cell})**. The **standard emf** or **standard cell potential (E°_{cell})** refers to the standard cell potential at 25 °C and 1 M concentration for any reaction in solution or at 1 atm for a gaseous substance. The greater the value of E°_{cell} the more likely the reaction will occur. A positive E°_{cell} indicates a spontaneous forward reaction and a negative E°_{cell} refers to a nonspontaneous forward reaction. The E°_{cell} is also the difference between two standard electrode potentials or the standard half-cell potentials (E°_{anode} and $E^{\circ}_{cathode}$). The E°_{cell} for the copper and zinc reaction (Figure 27) is 1.10 V and is the difference of each half-cell potential between the cathode (final state) and anode (initial state). To find E_{cell}°, one must consult a table of standard electrode potentials (given at 25.0 °C), which lists several half-cell reduction reactions with corresponding half-cell potential (E°) values. Values for the half-cell potentials will range from 3.00 V to −3.00 V.

The more positive the half-cell potential is for the reduction of a species, the greater the oxidation strength (stronger oxidizing agent), and the more negative the half-cell potential, the greater the reducing strength (stronger reducing agent). The greater the negativity of an electrode potential, the more likely that the electrode will repel electrons and undergo oxidation. In other words, electrons will spontaneously flow from the negative electrode (anode) to the positive electrode (cathode).

Conversely, the more positive the electrode potential (e.g., $E_{cathode}^{\circ}$) is, the more likely that the electrode will attract electrons to undergo reduction. A useful mnemonic is: more Negative Is Oxidation (NIO), and more Positive Is Reduction (PIR). The standard half-cell potentials for zinc and copper, from positive to negative E° are listed below.

$$Cu^{2+}(aq) + 2\,e^- \rightarrow Cu(s) \quad E^{\circ} = +0.34\text{ V}$$

$$Zn^{2+}(aq) + 2\,e^- \rightarrow Zn(s) \quad E^{\circ} = -0.76\text{ V}$$

Because copper has a greater half-cell potential, it is more likely to be reduced at the electrode (as the stronger oxidizing agent), meaning that the copper electrode will be the cathode or the positively charged electrode. Zinc has a more negative half-cell potential and is more easily oxidized (as the stronger reducing agent). When combining the half-cell reactions, the zinc reaction must be written to indicate that oxidation is occurring. The value of the half-cell potential will also change sign.

$$\text{Oxidation: } Zn(s) \rightarrow 2\,e^- + Zn^{2+}(aq)$$

Electrons flow from the anode (initial) to the cathode (final), so E_{cell}° is:

$$E_{cell}^{\circ} = E_{final}^{\circ} - E_{initial}^{\circ} = E_{cathode}^{\circ} - E_{anode}^{\circ} = 0.34\text{ V} - (-0.76\text{ V}) = 1.10\text{ V}$$

Note that the half-cell potential value of copper changes sign, which is consistent with the new half-cell potential for zinc. For a standard reduction table, half-reactions at the top (positive electrode potentials) attract electrons and tend to proceed in the forward direction. Half-reactions at the bottom (negative electrode potentials) repel electrons and will move in the reverse direction. A reduction reaction is spontaneous if it's paired with the reverse of any reaction listed below it in a standard reduction table. For example, consider the following reaction:

$$Zn(s) + Ni^{2+}(aq) \rightarrow Zn^{2+}(aq) + Ni(s)$$

The standard half-cell potentials, from positive to negative, are:

$$Ni^{2+}(aq) + 2\,e^- \rightarrow Ni(s) \quad E^{\circ} = -0.23$$

$$Zn^{2+}(aq) + 2\,e^- \rightarrow Zn(s) \quad E^{\circ} = -0.76$$

The half-cell reaction for zinc is listed below nickel, so reversing the zinc reaction ($E° = +0.76$ V) and pairing it with the forward reaction for nickel, will make the reaction spontaneous.

$$\text{Reduction: } Ni^{2+}(aq) + 2\,e^- \rightarrow Ni(s)$$

$$\text{Oxidation: } Zn(s) \rightarrow Zn^{2+}(aq) + 2\,e^-$$

$$Ni^{2+}(aq) + \cancel{2\,e^-} + Zn(s) \rightarrow Ni(s) + Zn^{2+}(aq) + \cancel{2\,e^-}$$

The resulting cell potential ($E°_{cell}$) is calculated the same way as before:

$$E°_{cell} = E°_{final} - E°_{initial} = E°_{cathode} - E°_{anode} = -0.23 \text{ V} - (-0.76 \text{ V}) = 0.53 \text{ V}$$

The Gibbs Standard Free Energy Change $\Delta G°$ and $E°_{cell}$

The relation between the standard cell potential, $E°_{cell}$, and the **Gibbs standard change in free energy** ($\Delta G°$) is given by the following equation:

$$\Delta G° = nFE°_{cell}$$

The term n refers to the number of mole electrons that are transferred in the redox reaction. The term **F is Faraday's constant**, which is equal to 96,485 C/(mol e^-). $E°_{cell}$ is also related to the equilibrium constant, K, through the following equation:

$$E°_{cell} = \frac{0.0592 \text{ V}}{n} \log K$$

For a spontaneous redox reaction where the products are favored in the standard state, $\Delta G°$ is negative (< 0), $E°_{cell}$ is positive (> 0), and K is greater than one (> 1). In contrast, for a nonspontaneous reaction, where the reactants are favored in their standard states, $\Delta G°$ is positive (> 0), $E°_{cell}$ is negative (< 0), and K is less than one (< 1). For nonstandard or nonequilibrium conditions, the relationship between E_{cell} and the reaction quotient, Q, is given by the **Nernst equation**:

$$E_{cell} = E°_{cell} - \frac{0.0592 \text{ V}}{n} \log Q$$

Under standard conditions, if Q is equal to one ($Q = 1$) then $E_{cell} = E°_{cell}$ because $\log 1 = 0$. If $Q < 1$, then the concentrations of the reactants are greater than the products, so the reaction proceeds to the right and $E_{cell} > E°_{cell}$. If $Q > 1$, then the concentrations of the products are greater than the reactants, so the reaction proceeds to the left and $E_{cell} < E°_{cell}$. At equilibrium, $Q = K$, the reaction will not move in any particular direction, so $E_{cell} = 0$. When a battery discharges and becomes depleted, it's because the reaction proceeds toward equilibrium such that the reactants are depleted and $Q = K$. The cell potential becomes zero.

Nuclear Chemistry

Nuclear Chemistry

Radioactivity describes the emission of subatomic particles (e.g., nuclei, protons, electrons) that can occur for specific atoms. The French scientist **A. H. Becquerel** discovered radioactivity in 1896 when he exposed a photographic plate to sunlight containing uranium. He hypothesized that the plate emitted X-rays and exhibited **phosphorescence**, which is the emission of light due to light absorption. The plate was shown to glow in the dark. However, when Becquerel repeated the results without exposing the plate to light, he discovered that the uranium crystals were emitting uranic rays even without phosphorescence. Shortly after, a doctoral student by the name of **Marie Curie** discovered that the elements polonium (named after Curie's home country Poland) and radium also emitted uranic rays.

The term "uranic rays" was eventually referred to as radiation or radioactivity because other elements besides uranium also emitted these rays. In 1903, Marie Curie, Pierre Curie, and Becquerel were awarded the Nobel Prize in physics for the discovery of radioactivity. In 1911, Marie Curie received the Nobel Prize in chemistry for her discovery of polonium and radium. Since the discovery of radioactive atoms, many scientists such as **Ernest Rutherford** focused on nuclear chemistry and radioactivity. **Nuclear chemistry** is the study of reactions in which the nuclei of atoms are transformed whereby the identities of the elements are changed. The unstable nuclei spontaneously decompose, emitting certain sub-particles, in order to become more stable. These reactions can involve large changes in energy—much larger than the energy changes that occur when chemical bonds between atoms are made or broken. Nuclear chemistry is also used to create electricity.

Radioactivity Types: Alpha, Beta, and Gamma Radiation

In some cases, and for specific atoms, the nucleus of an atom constantly emits particles due to its instability. These atoms are described as radioactive, and the isotopes are referred to as **radioisotopes.** Specific isotopes of an element are called **nuclides.** Recall that the atomic number (Z) refers to the number of protons and that the mass number (A) indicates the number of protons and neutrons for an element, X. The atomic number is shown as a subscript and the mass number as a superscript, which is both placed to the left of the chemical symbol, X. Argon has three stable isotopes (Ar-36, Ar-38, and Ar-40), which have a different number of neutrons with respect to one another:

$$\ce{^{A}_{Z}X} \quad \ce{^{36}_{18}Ar} \quad \ce{^{38}_{18}Ar} \quad \ce{^{40}_{18}Ar}$$

The equation $N = A - Z$ gives the number of neutrons (N). The number of neutrons for each nuclide is $N = 36 - 18 = 18$ (Ar-36), $N = 38 - 18 = 20$ (Ar-38), and $N = 40 - 18 = 22$ (Ar-40). The notation for sub-particles such as the proton, neutron, and electron is:

$$\ce{^{1}_{1}p} \text{ (proton)} \quad \ce{^{1}_{0}n} \text{ (neutron)} \quad \ce{^{0}_{-1}e} \text{ (electron)}$$

The proton and neutron have a mass number (A) of one, whereas the electron is given a value of zero. The subscripts for the proton and neutron indicate one and zero protons. A value of −1 is given to the electron and reflects the charge of the electron. The notation doesn't follow the naming convention for an atomic number, although it's useful to note the atomic number is typically equal to the number of electrons.

Several types of natural radioactive decay occur and include alpha (α), beta (β), and gamma (γ) ray emission, and positron emission. Certain unstable nuclei will even absorb an electron from one of its orbitals to become more stable, a process called **electron capture**. **Alpha radiation** is emitted or released when a nucleus decomposes to produce two protons and two neutrons, which are collectively called an **alpha (α) particle**. Helium-4 nuclei contain two protons and two neutrons, so an alpha particle also takes the same symbol: $_2^4$He. **Nuclear equations**, a type of equation showing a nuclear reaction or process such as radioactivity, can be used to show how specific elements emit an alpha particle, which shows how the number of protons within a nucleus changes. Nuclear equations have different notations than regular chemical equations. For example, consider the alpha decay of uranium (U-238). The nuclear equations are written as follows, where the top number or superscript is the atomic mass number (A), and the bottom number or subscript is the atomic number (Z) for each element:

$$\overset{\text{parent}}{\overset{\text{nuclide}}{\overbrace{_{92}^{238}\text{U}}}} \rightarrow \overset{\text{daughter}}{\overset{\text{nuclide}}{\overbrace{_{90}^{234}\text{Th}}}} + \overset{\text{alpha}}{\overset{\text{particle}}{\overbrace{_2^4\text{He}}}}$$

The equation describes the spontaneous decomposition of uranium-238 (U-238) into thorium-234 (Th-234) and helium-4 (He-4) via alpha decay. When this happens, the process is referred to as **nuclear decay**, and results in a change in chemical identity, unlike chemical reactions which retain their identity. However, similar to chemical equations, nuclear equations must be balanced on each side; the sum of the mass numbers and the sum of the atomic numbers should be equal on both sides of the equation. Table 12 shows how the mass number A decreases by four, and how the atomic number Z lowers to two with respect to the conversion of Uranium to Thorium.

Table 19. Decomposition of U-238 to Th-234 and He-4	
Reactants: Uranium	Products: Thorium and Helium
$A = 238$	$A = 234 + 4 = 238$
$Z = 92$	$Z = 90 + 2 = 92$

Based on the law of conservation of mass, the nuclear equation is balanced. Alpha particles have relatively low penetrating power, or the ability to move through or penetrate matter. For example, a substance that emits alpha particles will not go through a piece of paper and will even have trouble moving through air.

Beta radiation or **Beta (β) decay** occurs when an unstable nucleus emits a stream of high-speed electrons. The beta particles are often noted as β^-. Specifically, a neutron will transform into a proton as an electron is emitted. The general reaction is:

$$\text{Beta } (\beta) \text{ decay: neutron } (_0^1\text{n}) \rightarrow \text{proton } (_1^1\text{p}) + \overset{\text{beta particle}}{\overbrace{\text{electron } (_{-1}^0\text{e})}}$$

In a beta reaction, the parent nuclide will change its atomic number by one due to the addition of a proton. For example, consider the following beta reaction:

$$\overset{\text{parent}}{\overset{\text{nuclide}}{\overbrace{_{97}^{249}\text{Bk}}}} \rightarrow \overset{\text{daughter}}{\overset{\text{nuclide}}{\overbrace{_{98}^{249}\text{Cf}}}} + \overset{\text{beta}}{\overset{\text{particle}}{\overbrace{_{-1}^0\text{e}}}}$$

A neutron from berkelium-249 (Bk-249) transforms into a proton to form californium-249 (Cf-249), which has one extra proton. One beta particle is emitted in the process. A beta particle will have less ionizing power because it is much smaller than an alpha particle but will have more penetrating power. A sheet of metal can be used to stop a beta particle.

Gamma (γ) radiation is a form of electromagnetic radiation and occurs when the nucleus emits high-energy or short-wavelength photons. The following notation represents gamma rays:

$$\text{Gamma (γ) ray: } {}^{0}_{0}\gamma$$

Because gamma rays have no mass or charge, both the superscript and subscript have a value of zero. Consequently, the parent nuclide will not change in mass or atomic number during gamma ray emission. This type of radiation represents a rearrangement of an unstable nucleus into a more stable one and often accompanies other types of radioactive emissions. For example, the alpha decay of U-238 to Th-234 and He-4 additionally produces a gamma ray. Gamma rays have the highest penetrating power but the lowest ionizing power. Biologically, gamma rays are hazardous because they can pass through the body. Thick slabs of concrete or lead can serve as a shield against gamma radiation, which can be produced from nuclear explosions or nuclear reactors (e.g., decomposition of U-235 to U-236).

In **positron (β⁺) emission**, a proton is converted to a neutron within an unstable nucleus, and a positron is emitted. The **positron** is referred to as the antiparticle with respect to an electron (i.e., an antielectron) because it has the same mass as an electron but with a +1 charge. If an electron and positron collide with one another, each particle is annihilated, and energy is released in the form of a gamma ray. The general nuclear mechanism is:

$$\text{Positron emission: proton } ({}^{1}_{1}\text{p}) \rightarrow \text{neutron } ({}^{1}_{0}\text{n}) + \text{positron } ({}^{0}_{+1}\text{e})$$

During positron emission, the atomic number (Z) for the parent nuclide will decrease by one. The daughter nuclide will have one less proton compared to the parent nuclide. For instance, consider the decomposition of carbon-10 to boron-10 in the following nuclear reaction:

$$\overset{\text{parent}}{\underset{\text{nuclide}}{\overbrace{{}^{10}_{6}\text{C}}}} \rightarrow \overset{\text{daughter}}{\underset{\text{nuclide}}{\overbrace{{}^{10}_{5}\text{B}}}} + \overset{\text{positron}}{\overbrace{{}^{0}_{+1}\text{e}}}$$

The mass number remains the same because a proton is converted to a neutron. Positrons (β⁺) have similar penetrating and ionizing power like beta (β⁻) particles.

Electron capture differs from other types of radioactive decay, and instead of emitting a particle or radiation, the process involves the absorption of a particle. Specifically, when a nucleus acquires an electron from one of its inner orbitals within its electron cloud, the process is called **electron capture**. The process is similar to positron emission and involves the conversion of a proton to a neutron.

$$\text{Electron capture: } {}^{1}_{1}\text{p} + {}^{0}_{-1}\text{e} \rightarrow {}^{1}_{0}\text{n}$$

Notice how the subscripts will denote the charge, which is conserved on both sides of the equation. For an atom that undergoes electron capture, its atomic number (Z) will decrease by one with its mass number (A) remaining the same. Consider the following example where iodine-111 decomposes to tellurium-111:

$$^{111}_{53}\text{I} + {}^{0}_{-1}\text{e} \rightarrow {}^{111}_{52}\text{Te}$$

Determining Radioactivity Type

Protons and neutrons, called **nucleons**, are attracted to each other by the **strong force**, a force that acts at short distances and keeps the positively charged protons in the nucleus together. The nuclear stability of an isotope is determined by the ratio of neutrons to protons (N/Z). If the number of neutrons was plotted as a function of the atomic number, it would allow one to determine which isotopes are stable and which ones would undergo decay. Figure 28 shows a number of stable isotopes.

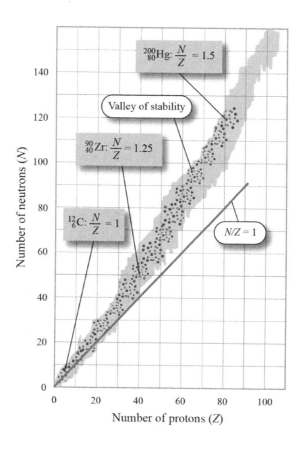

Figure 28. Valley of stability

The isotopes carbon-12, zirconium-90, and mercury-200 are among many stable isotopes that are found in the valley or island of stability (scattered dots). The neutron to proton ratio for these elements is between 1 and 1.5. If the N/Z ratio is too high (i.e., lies above the shaded region) or over 1.5, then that nuclide/isotope has too many neutrons and will tend to beta decay. In other words, that nuclide will undergo beta decay where a neutron is transformed into a proton, thereby reducing the number of neutrons in the daughter nuclide. The daughter nuclide will be closer to the valley of stability. For example, consider the following reaction:

$$\overbrace{^{249}_{97}\text{Bk}}^{\substack{\text{parent}\\\text{nuclide}}} \rightarrow \overbrace{^{249}_{98}\text{Cf}}^{\substack{\text{daughter}\\\text{nuclide}}} + \overbrace{^{0}_{-1}\text{e}}^{\substack{\text{beta}\\\text{particle}}}$$

In this example, Bk-249 has an N/Z ratio of 1.6, and Cf-249 has an N/Z ratio of 1.5. If the N/Z ratio is too low, then there are too many protons, and some of these protons will convert to neutrons either through electron capture or positron emission. For example, consider the positron emission of C-10 to B-10, shown previously. The N/Z ratio for C-10 is 0.67, but for B-10, it's 1.0. The new nuclide moves up or closer to the valley of stability. To determine whether a given nuclide is stable, it's best to find its N/Z value, and see if it lies along the valley of stability. For example, magnesium-28 or $^{28}_{12}Mg$ has 12 protons and 16 neutrons. Its N/Z ratio is $16/12 = 1.3$. Based on Figure 28, this value is not between 1 and 1.25. The N/Z ratio is too high, meaning that there are too many neutrons, and it will undergo beta decay:

$$^{28}_{12}Mg \rightarrow {}^{28}_{13}Al + \overset{\text{beta particle}}{\overbrace{{}^{0}_{-1}e}}$$

A neutron is converted to a proton, and an electron or beta particle is emitted. The N/Z ratio for aluminum-28 is 1.2.

Atoms that have an atomic number greater than 83 ($Z > 83$) are **radioactive** and can decay in one or multiple steps either through alpha, beta, or gamma decay, such as the alpha decay of uranium-238 to thorium-234. The daughter nuclide, thorium-234, can also decay to protactinium-234 via beta decay.

Kinetics of Radioactive Decay

Radioactive decay is often described in terms of its **half-life ($t_{1/2}$)**, which is the time that it takes for half of the radioactive substance to react. Radioactive decay follows first-order kinetics and is similar to chemical reactions. The relationship between the half-life of a nuclide and its rate constant (k) is given by:

$$t_{1/2} = \frac{0.693}{k}$$

The larger the rate constant is for the nuclide, the shorter the half-life is.

For example, the radioisotope strontium-90 has a half-life of 28.8 years. If there are 10 grams of strontium-90 to start with, after 28.8 years, there would be 5 grams left. If you were only given the initial amount (10 g) and a final amount at time t (5 grams) and were asked to find the half-life at time t, then you would need to find the rate constant.

$$k = \frac{0.693}{t_{1/2}} = \frac{0.693}{28.8 \text{ years}} = 0.02406 \text{ years}^{-1}$$

Now solve for t using the following equation below:

$$\ln[B]_t = -kt + \ln[B]_0$$

$$\ln[B]_t - \ln[B]_0 = -kt$$

$$t = -\ln\left(\frac{[B]_t}{[B]_0}\right)\frac{1}{k}$$

The term $[B]_t$ is the amount at a time which happens to be the half-life because 5 g is 50 percent of the initial amount (10 g). $[B]_0$ represents the initial amount at time zero. Therefore, $[B]_0 = 10$ g, so $[B]_t = 0.50(10$ g$) = 5$ g. The half-life is:

$$t = t_{1/2} = -\ln\left(\frac{5 \text{ g}}{10 \text{ g}}\right)\frac{1}{0.02406 \text{ years}^{-1}} = 28.8 \text{ years}$$

If asked to calculate the time given the half-life, where the final amount is half the initial amount, the answer will be the half-life (28.8 years). However, suppose you were asked to find the time at which strontium-90 decayed to 40 percent of its initial amount (4.0 g), then the time would be:

$$t = -\ln\left(\frac{0.40([B]_0)}{[B]_0}\right)\frac{1}{0.02406 \text{ days}^{-1}} = 38.1 \text{ days}$$

Note that 40 percent in decimal form is just $40/100 = 0.40$ and that the final time is now longer than the half-life. It's also important to recognize how to rearrange the previous equation if asked to calculate the amount of substance at a time t, given the initial amount and the half-life:

$$[B]_t = [B]_0 e^{-kt}$$

For instance, how much of strontium-90 is left after $t = 38.1$ days if it has $t_{1/2} = 28.8$ days and an initial starting amount of 10 g? Solving for $[B]_t$ gives:

$$[B]_t = [B]_0 e^{-kt} = (10 \text{ g})e^{-0.02406 \text{ days}^{-1} \times 38.1 \text{ days}} = 4.0 \text{ g}$$

Note that half-life was given to solve for k, which was shown above.

Fission and Fusion

There are two distinct types of nuclear reactions: fission and fusion reactions. Both involve a large energy release. In **fission** reactions, a large atom is split into two or more smaller atoms. The nucleus absorbs slow-moving neutrons, resulting in a larger nucleus that is unstable. The unstable nucleus then undergoes fission. Nuclear power plants depend on nuclear fission reactions for energy. The bombardment of neutrons toward uranium-235 atoms results in the formation of barium-140 and krypton-93 in addition to several neutrons.

$$^{235}_{92}\text{U} + ^{1}_{0}\text{n} \rightarrow ^{140}_{56}\text{Ba} + ^{93}_{36}\text{Kr} + 3\,^{1}_{0}\text{n} + \text{energy}$$

Scientists later realized that this reaction could undergo a **chain reaction**, a process where neutrons produced from one fission reaction induce a fission reaction in other uranium nuclei. The energy produced is the basis for how atomic bombs work.

Fusion reactions involve the combination of two or more lighter atoms into a larger atom. Fusion reactions do not occur on Earth naturally due to the extreme temperature and pressure conditions required to make them happen. Fusion products are generally not radioactive. Fusion reactions are responsible for the energy that is created by the sun. Nuclear fusion, the basis of modern nuclear bombs, employs the following type of reaction:

$$^{2}_{1}\text{H} + ^{3}_{1}\text{H} \rightarrow ^{4}_{2}\text{He} + ^{1}_{0}\text{n}$$

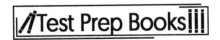

Hydrogen bombs are created by first detonating a fission bomb, which creates temperatures high enough to allow fusion to occur whereby deuterium (hydrogen-2) and tritium (hydrogen-3) combine.

Practice Test

Some of the following questions may require use of the Periodic Table, so one is provided below for your convenience.

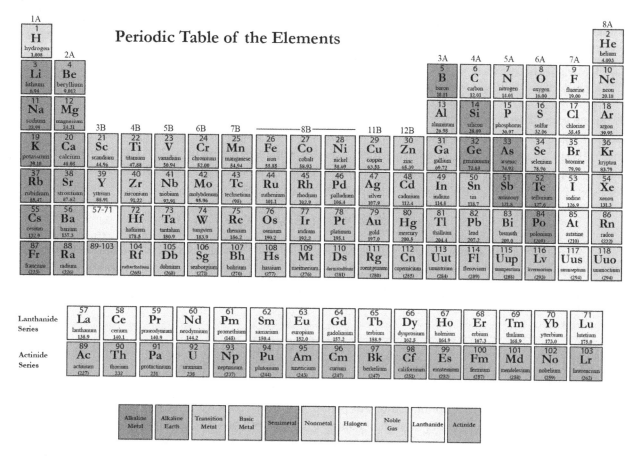

Periodic Table of the Elements

1. When visiting Mono Lake in California, a lake that is rich in toxic arsenic, you analyzed two samples of mud that both contained sulfur and oxygen, two of the basic building blocks of life on Earth.

	Sulfur	Oxygen
Sample A	2.420 g	3.662 g
Sample B	3.566 g	3.558 g

Based on the law of multiple proportions, what is the mass ratio of oxygen to sulfur for the two compounds A and B?

    a. 2:1
    b. 1:4
    c. 3:2
    d. 5:2
    e. 3:7

2. Yellow crazy ants are some of the most destructive and invasive ants that spray formic acid into the air and ground to subdue their victims. The acid even makes it painful for people to breathe. These ants are classified as predatory scavengers and will even eat a large variety of animal tissue, which has resulted in the population reduction of Hawaiian seabirds. Given the molecular formula, CH_2O_2, what is the approximate mass percentage of each element in the compound?

    a. Mass % C = 26, Mass % H = 4, Mass % O = 70
    b. Mass % C = 31, Mass % H = 4, Mass % O = 65
    c. Mass % C = 28, Mass % H = 5, Mass % O = 67
    d. Mass % C = 29, Mass % H = 6, Mass % O = 65
    e. Mass % C = 31, Mass % H = 6, Mass % O = 63

3. During your expedition to the Sahara Desert in Africa in search of the desert hedgehog, your car tires become flat after going over a hidden quartz rock bed. You are low on water, and help from the nearest town, 100 miles away, won't come for another two days. You have plenty of fuel and a carbon water filter to purify the water. You realized that turning on your car will produce water based on the following reaction:

$$2C_8H_{18}(g) + 25O_2(g) \rightarrow 16CO_2(g) + 18H_2O(g)$$

Out of desperation, you started your car every two hours at night and attempted to collect the water, through condensation, from the car's tailpipe. If you had five gallons of gasoline, approximately 20.0 kg, how much water was possibly produced?

    a. 16.2 kg
    b. 30.3 kg
    c. 14.1 kg
    d. 28.4 kg
    e. 19.9 kg

4. The ground state electronic configuration for helium, He, is $1s^2$. When a relatively low-energy photon bombards He, the electronic configuration becomes $1s^12s^1$, where one electron from the 1s shell moves to the 2s subshell in accordance with the Aufbau principle. Based on Aufbau's principle, what is one possible excited state electronic configuration of calcium?

    a. $1s^22s^22p^63s^23p^64s^2$
    b. $1s^22s^22p^63s^23p^63d^14s^1$
    c. $1s^22s^22p^63s^23p^63d^14s^2$
    d. $1s^22s^22p^63s^23p^64s^13d^1$
    e. $1s^22s^22p^63s^23p^63s^24s^1$

5. Which of the following is true regarding the ionization energy between selenium and bromine?

    a. The ionization energy of bromine is greater than selenium because the effective nuclear charge or Coulomb force is less.
    b. The ionization energy of selenium is greater than bromine because the effective nuclear charge or Coulomb force is greater.
    c. The ionization energy of bromine is greater than selenium because the outer valence shell is farther away from the nucleus.
    d. The ionization energy of selenium is greater than bromine because the effective nuclear charge or Coulomb force is less.
    e. The ionization energy of bromine is greater than selenium because the effective nuclear charge or Coulomb force is greater.

6. Orbital diagrams are often used to depict the arrangement and number of electrons in each subshell. For example, the ground state electronic configuration of carbon is represented by the orbital diagram below. Each electron is represented by an up or down arrow, with a maximum of two electrons or arrows per orbital. Which of the following orbital diagrams is NOT a possible representation for the electron configuration of carbon?

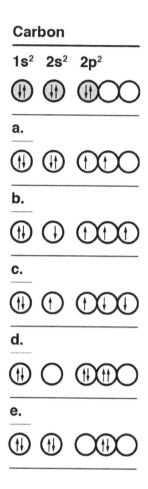

7. The following table shows the isotopic masses and fractional abundances of silicon, Si. Based on the information provided, what is the average atomic mass of silicon correct to four significant figures?

Silicon Isotope	Atomic Mass (amu)	Relative Intensity %	Fractional Abundance
Si-28, $^{28}_{14}Si$	27.9769271	100.0	0.9223
Si-29, $^{29}_{14}Si$	28.9764949	5.08	0.0468
Si-30, $^{30}_{14}Si$	29.9737707	3.35	0.0309

a. 28.0914
b. 28.0804
c. 28.0854
d. 28.0824
e. 28.0845

8. A simulated infrared spectrum of sulfur dioxide (SO_2) is shown below. The absorbance line representing A at 518 cm^{-1} represents the wavenumber at which SO_2 undergoes a scissoring-type motion. Absorbance lines B and C correspond to a symmetric and asymmetric stretch at 1151 and 1362 cm^{-1}, respectively. What is the amount of energy corresponding to the symmetric stretch? Note that $h = 6.626 \times 10^{-34} J \cdot s$.

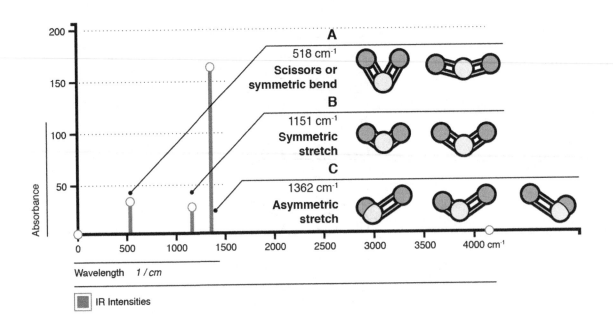

a. $2.288 \times 10^{-22} J$
b. $2.288 \times 10^{-20} J$
c. $1.727 \times 10^{-28} J$
d. $2.288 \times 10^{-24} J$
e. $1.988 \times 10^{-25} J$

9. Lead (Pb^{2+}) is a heavy water-soluble metal ion that can contaminate water and result in adverse health effects when consumed. Lead can attack the brain or central nervous system, resulting in problems with thinking and dizziness and fatigue. Lead, in particular, can enter faucet water when certain plumbing materials, such as pipes containing lead, begin to erode. Older homes, built before 1986, are more likely to have lead pipes from the water main to their homes. Due to the Safe Drinking Water Act, pushed by Congress, and with new standards set by the Environmental Protection Agency, lead-free pipes were implemented such that pipes or pipe fittings did not contain more than 8.0% lead. Suppose you moved to a home built in the 1970s and were told by the realtor that most pipes in the home were replaced. In an attempt to verify this claim, you filled a 1.000 L bottle with faucet water and dumped sodium chromate (VI) salt in the bottle. Not to your surprise, a yellow precipitate formed, indicating lead (II) ion was present:

$$Pb^{2+}(aq) + Na_2CrO_4(aq) \rightarrow PbCrO_4(s) + 2Na^+(aq)$$

If you recover 0.550 gram of the yellow solid, lead (II) chromate (VI), how many lead (II) ions are present? Is the concentration of lead (II) in the bottle more than 8.0%? Note that the concentration of water in an aqueous solution is approximately 55.5 M.
  a. 1.02×10^{21} molecules of Pb^{2+}, no
  b. 2.22×10^{22} molecules of Pb^{2+}, yes
  c. 1.12×10^{23} molecules of Pb^{2+}, no
  d. 4.02×10^{24} molecules of Pb^{2+}, no
  e. 5.12×10^{25} molecules of Pb^{2+}, yes

10. Ice is a crystalline solid made up of repeating units of water molecules. Which choice does NOT describe a property of an ice cube?
  a. Has an exact melting point
  b. Breaks across a plane
  c. Becomes soft before it melts
  d. Cannot be compressed
  e. Bonds break at the same time when it melts

11. At a temperature of 300 K and pressure of 1.3 atm, what volume will 2.5 moles of CO_2 gas occupy, where the ideal gas constant is equal to 0.082 L atm mol^{-1} K^{-1}?
  a. 47.3 L
  b. 4.7 L
  c. 26.9 L
  d. 2.5 L
  e. 300 L

12. Scientist A needs to make 100 mL of a 6 M solution of aluminum chloride, $AlCl_3$. She has a 60 M $AlCl_3$ solution on hand. Which combination will give her what she needs?
  a. 6 mL 60 M $AlCl_3$ + 94 mL water
  b. 10 mL 60 M $AlCl_3$ + 90 mL water
  c. 60 mL 60 M $AlCl_3$ + 40 mL water
  d. 10 mL 60 M $AlCl_3$ + 100 mL water
  e. 60 mL 60 M $AlCl_3$ + 60 mL water

13. Which of the following shows the correct order of strength of intermolecular bonds?
    a. London dispersion > hydrogen > dipole-dipole
    b. Hydrogen > London dispersion > dipole-dipole
    c. London dispersion > dipole-dipole > hydrogen
    d. Hydrogen > dipole-dipole > London dispersion
    e. Dipole-dipole > hydrogen > London dispersion

14. Which property of liquids describes how water defies gravity to move upward from the root of a plant through the stem and into the leaves?
    a. Capillary action
    b. Surface tension
    c. High vapor pressure
    d. High boiling point
    e. Solubility

15. Which element's atoms have the greatest number of electrons?
    a. Hydrogen
    b. Iron
    c. Copper
    d. Magnesium
    e. Iodine

16. Given that the energy of an ionic bond is calculated by $\Delta H = (C \times Z^+ \times Z^-)/R_0$, what would increase the bond energy between two ions?
    a. Increasing the positive charge while decreasing the negative charge by the same magnitude
    b. Changing one ion into a neutral atom
    c. Decreasing the distance between the ionic nuclei
    d. Decreasing the positive charge while increasing the negative charge by the same magnitude
    e. Increasing the distance between the ionic nuclei

17. What is the molecular formula for the molecule represented in the following Lewis diagram?

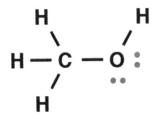

    a. CH_3-H_2O
    b. CH_3
    c. CH_3OH^{-4}
    d. CH_3O
    e. CH_3OH

18. Given the atomic radii of the following metals, which two could be mixed together to form an interstitial alloy?
    a. Copper- 128 pm and zinc- 139 pm
    b. Carbon- 70 pm and iron- 126 pm
    c. Aluminum- 143 pm and copper- 128 pm
    d. Iron-126 pm and aluminum- 143 pm
    e. Aluminum- 143 pm and zinc- 139 pm

19. What type of solid is a semiconductor such as silicon?
    a. Metallic solid
    b. Neutral solid
    c. Ionic solid
    d. Covalent network solid
    e. Molecular solid

20. What type of reaction occurs when fossil fuels react with oxygen gas?
    a. Combustion
    b. Combination
    c. Decomposition
    d. Vaporization
    e. Solidification

21. Consider the following reaction: $2AgI + Na_2S \rightarrow Ag_2S + 2NaI$. If you start with six moles of AgI and three moles of Na_2S, how many moles of NaI are produced by the reaction if you have a 90 percent yield?
    a. 3
    b. 6
    c. 5.4
    d. 90
    e. 10

22. Which coefficient would fill in the blank in the following equation to correctly balance the chemical reaction between iron oxide and carbon monoxide: $FeO_3 + 3CO \rightarrow Fe + \underline{}CO_2$?
    a. 1
    b. 3
    c. 2
    d. 6
    e. 4

23. Which characteristic is NOT a property of water?
    a. If it acts as an acid, it donates a proton.
    b. It is amphoteric and can act as an acid or a base.
    c. If it donates a proton, it becomes a conjugate base in the form of hydroxide ions.
    d. If it accepts a proton, it becomes a conjugate acid in the form of hydronium ions.
    e. If it acts as a base, it donates a proton.

24. In the following reaction, which elements experience changes in oxidation numbers, indicating an oxidation-reduction reaction?

Reaction	$I_2O_5 + 5CO_2 \rightarrow I_2 + 5CO$						
Elements	I_2	O_5	$+ 5C$	O_2	$\rightarrow$	I_2	$+ 5C$	O
Oxidation #s	+5	-2	-2	+2		0	+4	-2

a. Iodine and carbon
b. Carbon and oxygen
c. Oxygen and Iodine
d. Carbon monoxide
e. Carbon dioxide

25. Scientist A is observing an unknown substance in the lab. Which observation describes a chemical property of the substance?
a. She sees that it is green in color.
b. She weighs it and measures the volume and finds the density to be 10 g/L.
c. She applies pressure to it and finds that it breaks apart easily.
d. She passes it through a flame and finds that it burns.
e. She smells it and finds it odorless.

26. Which statement is accurate about the following endothermic reaction: $CaCO_3 \rightarrow CaO + CO_2$?
a. The enthalpy of the reaction, ΔH, is -178 kJ/mol.
b. Heat energy was released from the system to form more, smaller molecules.
c. The products have less energy than the reactants.
d. There was a decrease in enthalpy in the system.
e. The enthalpy of the reaction, ΔH, is +178 kJ/mol.

27. At standard state conditions, Scientist A constructs a galvanic cell with a silver cathode and a zinc anode. Given the following two half-reactions and their energy potentials, what is the energy potential for the cell?

$$Ag^+ (aq) + e^- \rightarrow Ag (s) \quad E^o = 0.80 \text{ V}$$

$$Zn^{2+} (aq) + 2e^- \rightarrow Zn (s) \quad E^o = -0.76 \text{ V}$$

a. 0.04 V
b. 0.08 V
c. 1.56 V
d. -0.76 V
e. -1.56 V

28. Given the free energy and cell potential of the system, how many moles of electrons are transferred for every mole of reactant in a galvanic cell with zinc and copper electrodes?

$$E°_{cell} = -1.2 \text{ Joules/Coulomb}$$

$$F = 95835 \text{ Coulombs/mole of electrons}$$

$$\Delta G = 230 \text{ kJ}$$

a. 1 mole
b. 2 moles
c. 230 moles
d. 20 moles
e. 1.2 moles

29. For the kinetic study of the chemical reaction $2X + Y \rightarrow 2Z$, the following data was recorded for the initial rates of disappearance of X and Y.

	Initial Concentration mol/L		Initial rate of reaction of X $mol/(L \cdot s)$
	X	Y	
Experiment 1	0.0250	0.0506	0.0560
Experiment 2	0.0500	0.0506	0.224
Experiment 3	0.0250	0.1012	0.112

Which choice below represents the correct rate law?
a. $Rate = k[X]$
b. $Rate = k[X]^2[Y]$
c. $Rate = k[X][Y]^2$
d. $Rate = k[X][Y]$
e. $Rate = k[Y]$

30. A piece of bone from an animal found at a site in New Mexico was dated using a radioactive carbon-14 method. It was determined that the bone contained 5.0 dpm (disintegrations per minute) per gram of carbon. If the half-life of carbon is 5730 years and the amount of carbon found in living animals is 15.0 dpm per gram of carbon, how old is the animal bone?
a. 6000 years
b. 5000 years
c. 20,000 years
d. 15,000 years
e. 9,000 years

31. The table below shows a set of rate constants at different temperatures for the decomposition of dinitrogen pentoxide (N_2O_5). Which of the following best represents the correct activation energy barrier for this decomposition reaction?

Rate constant for decomposition of N_2O_5 at different temperatures		
	$k\ (s^{-1})$	Temperature °C
Experiment 1	4.8×10^{-4}	45.0
Experiment 2	8.8×10^{-4}	50.0
Experiment 3	1.6×10^{-3}	55.0
Experiment 4	2.8×10^{-3}	60.0

    a. $2.00 \times 10^2 J/mol$
    b. $4.53 \times 10^4\ J/mol$
    c. $1.03 \times 10^5 J/mol$
    d. $3.92 \times 10^1\ J/mol$
    e. $2.81 \times 10^6\ J/mol$

32. For the decomposition or disappearance of nitrous oxide (N_2O) to nitrogen and an oxygen atom, the activation energy for the forward reaction is 60.0 kcal with a standard bond enthalpy ($\Delta H°$) of +167 kJ. What is the activation energy barrier (E_a) for the reverse reaction in kilojoules (1 kcal is equal to 4.184 kJ)?

    a. 227 kJ
    b. 107 kJ
    c. 84 kJ
    d. 65 kJ
    e. 418 kJ

33. The decomposition of nitryl chloride, NO_2Cl, is believed to take place in the following steps:

$$2NO_2Cl\ (g) + Cl(g) \rightarrow 2NO_2(g) + Cl_2(g) \quad Fast$$

$$NO_2Cl \overset{k_1}{\rightleftharpoons} NO_2 + Cl \quad Slow$$

$$NO_2Cl + Cl \overset{k_2}{\rightleftharpoons} NO_2 + Cl_2 \quad Fast$$

Which choice below represents the correct rate law for the disappearance of NO_2Cl?
    a. $Rate = k[\ NO_2][Cl_2]$
    b. $Rate = k[\ NO_2Cl]^2\ [Cl]$
    c. $Rate = k[\ NO_2Cl][Cl]$
    d. $Rate = k[\ NO_2Cl]$
    e. $Rate = k[\ NO_2Cl]^2$

34. The following chemical equations show a possible reaction mechanism for a reaction involving nitramide (NH_2NO_2) and a small amount of hydroxide ion:

$$NH_2NO_2(aq) + OH^-(aq) \leftrightarrow NHNO_2^-(aq) + H_2O \ (l)$$

$$NHNO_2^-(aq) \rightarrow N_2O(g) + OH^-(aq)$$

Which species are the reaction intermediate and the catalyst, respectively?
   a. NH_2NO_2 and $NHNO_2^-$
   b. $NHNO_2^-(aq)$ and $N_2O(g)$
   c. OH^- and N_2O
   d. H_2O and $NHNO_2^-$
   e. $NHNO_2^-$ and OH^-

35. Consider the following chemical reaction with the given rate law below:

$$CH_3COOCH_3(aq) + H_2O(l) \rightarrow CH_3COOH(aq) + CH_3OH(aq)$$

$$Rate = k[H_3O^+][CH_3COOCH_3]$$

For a one-liter solution, if the concentration of CH_3COOCH_3 is 0.40 M and 0.30 M HCl, which of the following would increase the rate of reaction?
   a. Adding 0.5 L of H_2O
   b. Increasing the amount of CH_3COOH
   c. Increasing the amount of ^-OH
   d. Adding the catalyst CH_3COOH to the solution
   e. Adding the catalyst H_3O^+ to the solution

36. The potential energy surface below corresponds to the dissociation of two atoms. Which of the following choices represents the correct bond dissociation energy and stable (equilibrium) bond length, respectively?

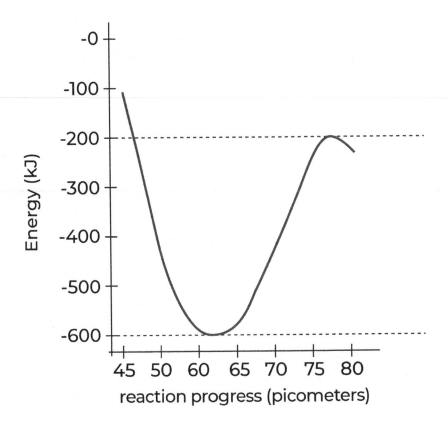

a. 600 kJ and 60 pm
b. 500 kJ and 62 pm
c. 400 kJ and 76 pm
d. 400 kJ and 62 pm
e. 200 kJ and 70 pm

37. The graph below corresponds to the potential energy surface for a synthesis reaction where reactants A and B combine to form product C. Which of the following choices is true regarding the graph below?

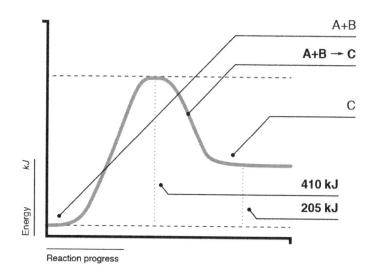

A+B

A+B → C

C

410 kJ

205 kJ

a. The reverse reaction is endothermic with a $\Delta H = +205\ kJ$.
b. The reverse reaction is exothermic and has a $\Delta H = -205\ kJ$.
c. The forward reaction is endothermic and has a $\Delta H = +410\ kJ$.
d. The reverse reaction is endothermic with a $\Delta H = -205\ kJ$.
e. The forward reaction is exothermic and has a $\Delta H = +205\ kJ$.

38. The specific heat capacity ($J \cdot g^{-1} \cdot {}^\circ C^{-1}$) for a substance is given by the following equation:

$$c = \frac{q}{m\Delta T}$$

The mass in grams is given by m, and the change in temperature (°C) is given by $\Delta T = T_f - T_i$. The amount of heat that is absorbed or transferred to a substance is given by q. Suppose you found a quarter in the desert that weighed about 5.67 g. What is the value of q for the quarter if it had a temperature of 50.0 °C, but was subsequently placed on your hand having a temperature of 37.0°C? Assume that $c = 0.400\ J \cdot g^{-1} \cdot {}^\circ C^{-1}$ for a quarter.
a. +34.5 J
b. -27.2 J
c. -34.5 J
d. +27.2 J
e. +68.1 J

39. Japanese ice coffee involves a relatively new brewing method that some craft coffee shops have adopted in the United States. It works by grinding fresh whole bean coffee and placing the grounds in a special coffee dripper and filter. As a fixed amount of hot water is poured over the coffee grounds, the sweet coffee flavors are extracted, which then drip into a cup containing a fixed amount of ice. What is the difference or change in internal energy of the extracted coffee if it loses 600 J of heat to the ice? Assume there is no volume change in the ice-to-coffee mixture.

    a. $\Delta U_{coffee} = -600J$
    b. $\Delta U_{ice} = -600J$
    c. $\Delta U_{coffee} = +600J$
    d. $\Delta U_{ice} = -600J$
    e. $\Delta U_{cup} = -600J$

40. When a gas expands inside a piston, the amount of negative work done by the gas is the product of the pressure (atm) and change in volume (liters) ($1\ atm \cdot L = 101.3$ J). If a gas loses 900 J of heat when it's compressed from 1.50 L to 0.50 L with an applied pressure of 1.20 atm from the piston, what is the change in internal energy for the surroundings?

    a. $\Delta U_{surr} = +899\ J$
    b. $\Delta U_{gas} = -778\ J$
    c. $\Delta U_{surr} = +778\ J$
    d. $\Delta U_{surr} = -778\ J$
    e. $\Delta U_{gas} = +778\ J$

41. Latent heat is often described as the product of the mass and change in enthalpy ($m\Delta H = q_p$), which is equal to the heat at constant pressure when the temperature does not change. In contrast, sensible heat is the type of heat that depends on the change in temperature ($q = mc\Delta T$). How much heat (in kJ) is needed to convert 870.0 g of ice at -15.0 °C to liquid water ($H_2O\ (l)$) at 15 °C? The specific heats of ice ($H_2O\ (s)$) and water ($H_2O\ (l)$) are 2.03 and 4.184 $J \cdot g^{-1} \cdot °C^{-1}$, respectively. The heat of fusion $\Delta H_{fus} = 0.334\ kJ \cdot g^{-1}$.

    a. 135 kJ
    b. 372 kJ
    c. 217 kJ
    d. 347 kJ
    e. 671 kJ

42. Suppose hypothetically that a mysterious moon rock with a mass of 1.47 g was taken from Apollo 18, and brought to a secure location in the Cheyenne Mountain Complex. The metal was heated to 60°C and dropped into a constant pressure calorimeter containing 100.0 g of water (heat bath) at 22.0°C. The final temperature taken from the thermometer that was inserted into the heat bath read 25.0°C. If the specific heat capacity of water, c_{water}, is 4.184 J/(g · °C), what is the specific heat capacity of the moon rock?

    a. 4.18 J/(g · °C)
    b. 0.901 J/(g · °C)
    c. 0.449 J/(g · °C)
    d. 2.41 J/(g · °C)
    e. 24.4 J/(g · °C)

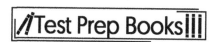

43. The table bellows lists enthalpies of formations, ΔH_f°, for various metals and metal oxides. The general reaction formation of a metal oxide (MO) from a metal (M) is:

$$2M(s) + O_2(g) \rightarrow 2MO(s)$$

Metal	ΔH_f° (kJ/mol)	Nonmetal	ΔH_f° (kJ/mol)	Metal Oxide	ΔH_f° (kJ/mol)
Mg (s)	0	O_2 (g)	0	MgO (s)	-601.2
Ca (s)	0	C (graphite)	0	CaO(s)	-635.1
Sr (s)	0	C (diamond)	1.896	SrO(s)	-592.0
Ba (s)	0	CO (g)	-110.5	BaO (s)	-548.1

Regarding the formation of a metal oxide, which reaction will be the MOST exothermic $(-\Delta H_{rxn}^\circ)$?
   a. MgO
   b. CaO
   c. SrO
   d. BaO
   e. CO

44. With exclusion to other types of stronger intermolecular forces (e.g., dipole-dipole), which of the following molecules will have the greatest dispersion forces?

a.

b.

c.

d.

e.

45. Which of the following processes can be SOLELY categorized as a chemical reaction process?
   a. The condensation of water vapor around the container of an ice cold beverage
   b. The dissolution of table sugar in water
   c. The shattering of a glass mason jar after falling on the floor
   d. The combination of sand (SiO_2) in water
   e. Baking brownies with chocolate chips

46. For the following reactions shown below, which of the following statements is true regarding the sign of $\Delta S°$?

$$i. \ (NH_4)_2Cr_2O_7(s) \rightarrow N_2(g) + 4H_2O(g) + Cr_2O_3(s)$$

$$ii. \ CO(g) + 2H_2(g) \rightarrow CH_3OH(l)$$

$$iii. \ CH_4(g) + 2O_2(g) \rightarrow CO_2(g) + 2H_2O(g)$$

a. $\Delta S°$ is negative for (i) and positive for (ii).
b. $\Delta S°$ is negative for (i) and (ii).
c. $\Delta S°$ is positive for (i) and negative for (ii).
d. $\Delta S°$ is positive for (i) and (iii).
e. $\Delta S°$ is negative for (ii) and positive for (iii).

47. Based on the reaction and information below, at what temperature will the following reaction be nonspontaneous? What is the standard Gibbs free energy $\Delta G°$ at 25.00°C?

$CH_4(g) + 2O_2(g) \rightarrow CO_2(g) + 2H_2O(g)$				
$\Delta H_f°$ (kJ/mol)	-74.87	0	-393.5	-241.826	$\Delta H_{rxn}° = -802.282$ kJ/mol
$S°$ (J/mol · K)	186.1	205.0	213.7	188.72	$\Delta S° = -4.96$ J/(mol · K)

a. Above $1.62 \times 10^5 K$ and $\Delta G° = -800.8$ kJ/mol
b. Above $1.32 \times 10^4 K$ and $\Delta G° = -2281$ kJ/mol
c. Above $1.19 \times 10^3 K$ and $\Delta G° = +800.8$ kJ/mol
d. Above $1.62 \times 10^5 K$ and $\Delta G° = -800.8$ kJ/mol
e. Above $1.62 \times 10^2 K$ and $\Delta G° = +675.8$ kJ/mol

48. During a mine craft excavation, you obtained iron ore and were attempting to extract pure iron from your load of samples to build a small rollercoaster track in the back of your home. However, the formation of iron from the ore sample is not thermodynamically favorable.

$$2Fe_2O_3(s) \rightarrow 4 \ Fe(s) + 3O_2(g); \quad \Delta G° = +1487 \ kJ$$

To obtain iron, you took a Yeti® ice cooler and drilled two holes in it to run two inlet gas pipes. By placing the iron ore sample inside the closed container and running carbon monoxide and oxygen gas through each pipe, it will react with the sample to produce iron. You are confident this will work because the gas mixture will be converted to carbon dioxide, which is a spontaneous process.

$$2CO(g) + O_2(g) \rightarrow 2CO_2(g); \quad \Delta G° = -514.4 \ kJ$$

Based on the reaction equations, what is the value of $\Delta G°$ for the production of iron?
a. -56 kJ
b. -48 kJ
c. +56 kJ
d. +48 kJ
e. -1543 kJ

49. Hydrogen peroxide (H_2O_2) is an antiseptic often used in toothpaste to whiten and clean teeth. The general reaction for the production of hydrogen peroxide is:

$$H_2(g) + O_2(g) \rightarrow H_2O_2(l)$$

Using Hess's Law, find the enthalpy change, ΔH, for the overall reaction given the following reaction stages for the production of hydrogen peroxide.

$$1.) \; 2H_2O_2(l) \rightarrow 2H_2O(l) + O_2(g); \qquad \Delta H_1 = -196.0 \text{ kJ}$$

$$2.) \; H_2(g) + \frac{1}{2}O_2(g) \rightarrow H_2O(l); \qquad \Delta H_2 = -285.8 \text{ kJ}$$

a. -187.5 kJ
b. 345.5 kJ
c. -375.0 kJ
d. +187.5 kJ
e. +375.0 kJ

50. For which concentration of S_2 would the reaction be favored to run in the forward direction and create more product, H_2S, as the reaction quotient, Q, would be less than the value of the equilibrium constant, K?

$$2H_2 + S_2 \rightarrow 2H_2S$$

$$[H_2] = 6 \text{ M}$$

$$[H_2S] = 3 \text{ M}$$

$$K = 15$$

a. 0.01 M
b. 0.3 M
c. 0.5 M
d. 1 M
e. 2.5 M

51. Using the information in the ICE table below, what is the value of K for the reaction $PCl_5 \rightarrow PCl_3 + Cl_2$?

Equation	PCl_5	PCl_3	Cl_2
Initial Moles	5	0	0
Relative Change in Moles	-X	+X	+X
Equilibrium Moles	3	2	2

a. 1.33
b. 0
c. 4
d. 5
e. 0.75

52. Which is NOT represented by particulate diagrams?
    a. The proportion of reactants needed to form products
    b. How bonds between reactants are broken to form the products
    c. The value of the equilibrium constant, K
    d. How bonds between the products are formed from the reactant elements
    e. The equilibrium mixture of reactants and products

53. Considering Le Chatelier's principle, if heat is taken away from the system running the reaction C (s) + H_2O (l) + heat $\leftrightarrow$ CO (g) + H_2 (g), which substance(s) would increase in concentration as equilibrium is restored?
    a. CO
    b. C
    c. H_2
    d. CO and H_2O
    e. C and H_2

54. The reaction N_2 (g) + O_2 (g) $\rightarrow$ 2NO (g) is at equilibrium when the system experiences an environmental change. The value of K was 11.5 but has changed to 15.6. Which type of stress disturbed the equilibrium of the system?
    a. A change in concentration of the reactants
    b. A change in concentration of the product
    c. A change in volume
    d. A change in temperature
    e. A change in pressure

55. What is the base ionization constant, K_b, for the reaction HCl + H_2O $\rightarrow$ H_3O^+ + Cl⁻ given that the pK_a of the reaction is -6.11?
    a. 1×10^{-14}
    b. 14
    c. 7.7×10^{-21}
    d. 1.3×10^6
    e. 20.11

56. What is the pH value at the halfway equivalence point of the reaction HBr + H_2O $\rightarrow$ H_3O^+ + Br⁻ given that the pK_a is 4 and the concentration of Br⁻ and HBr are both 3×10^7 moles?
    a. 4
    b. 7
    c. 1
    d. 14
    e. 10

57. What is the base-conjugate acid pair in the following chemical reaction: H_2O + NH_3 $\rightarrow$ NH_4^+ + OH⁻?
    a. H_2O and OH⁻
    b. H_2O and NH_4^+
    c. H_2O and NH_3
    d. NH_3 and NH_4^+
    e. NH_3 and OH⁻

58. Which is a TRUE statement about the current conditions of the following reaction given the conditions noted?

$$AgCl \ (s) \leftrightarrow Ag^+ \ (aq) + Cl^- \ (aq)$$

$$Ksp \ (\text{solubility product constant}) = [Ag^+][Cl^-] = 1.77 \times 10^{-10}$$

$$[Ag^+] = 1.33 \times 10^{-5} \ M$$

$$Q = 2.1 \times 10^{-10}$$

a. The reaction is at equilibrium.
b. Precipitation of AgCl occurs because Q > K, so there are more products than reactants present currently than would be at equilibrium.
c. The concentration of Cl⁻ at equilibrium is 3×10^{-5} M.
d. The reaction is being driven in the forward direction.
e. The solid AgCl is being dissociated into ions in aqueous solution because Q < K.

59. Under what condition would a chemical reaction run spontaneously?
a. Energy is absorbed by the system.
b. The value of K is less than one.
c. ΔG° has a negative value.
d. The reaction is endergonic.
e. ΔG° has a positive value.

60. The reaction of nitrogen and hydrogen to form ammonia is favored in the forward direction, but it is a reversible reaction. Which of the following conditions would cause the reaction to run in the reverse direction?

$$N_2 + 3H_2 \leftrightarrow 2NH_3$$

a. Adding heat to the system
b. Adding hydrogen to the system
c. Adding nitrogen to the system
d. Adding ammonia to the system
e. Adding equal proportions of reactants and products to the system

61. Water has many properties due to its unique structure. Which of the following does NOT play a role in water's properties?
a. Hydrogen bonding between molecules
b. Polarity within one molecule
c. Molecules held apart in solid state
d. Equal sharing of electrons

62. Which is the metric prefix for 10^{-3}?
a. Milli
b. Centi
c. Micro
d. Deci

63. Which of the following must have an equal mass number?
    I.   Isotopes
    II.   Isotones
    III.   Isobars
  a. I only
  b. I and II only
  c. II and III only
  d. III only

64. How are a sodium atom and a sodium isotope different?
  a. The isotope has a different number of protons.
  b. The isotope has a different number of neutrons.
  c. The isotope has a different number of electrons.
  d. The isotope has a different atomic number.

65. Which statement is true about nonmetals?
  a. They form cations.
  b. They form covalent bonds.
  c. They are mostly absent from organic compounds.
  d. They are all diatomic.

66. What is the basic unit of matter?
  a. Elementary particle
  b. Atom
  c. Molecule
  d. Element

67. Which particle is responsible for all chemical reactions?
  a. Electrons
  b. Neutrons
  c. Protons
  d. Orbitals

68. Which of these give atoms a negative charge?
  a. Electrons
  b. Neutrons
  c. Protons
  d. Orbital

69. How are similar chemical properties of elements grouped on the periodic table?
  a. In rows according to their total configuration of electrons
  b. In columns according to the electron configuration in their outer shells
  c. In rows according to the electron configuration in their outer shells
  d. In columns according to their total configurations of electrons

70. What volume of a 5.00 M HNO_3 solution should be added to water to make a 2.00 M HNO_3 solution?
  a. Take 350 mL of water and add 140 mL the concentrated acid.
  b. Take 210 mL of water and add 140 mL the concentrated acid.
  c. Take 350 L of water and add 140 L the concentrated acid.
  d. Take 210 L of water and add 140 L the concentrated acid.

71. What does the law of conservation of mass state?
    a. All matter is equally created.
    b. Matter changes, but is not created.
    c. Matter can be changed, and new matter can be created.
    d. Matter can be created, but not changed.

72. Which factor decreases the solubility of solids?
    a. Heating
    b. Agitation
    c. Large surface area
    d. Decreasing solvent

73. What is the term for a homogeneous mixture that does not have the Tyndall effect?
    a. Solvent
    b. Suspension
    c. Solution
    d. Colloid

74. How does adding salt to water affect its boiling point?
    a. It increases it.
    b. It has no effect.
    c. It decreases it.
    d. It prevents it from boiling.

75. What is the effect of pressure on a liquid solution?
    a. It decreases solubility.
    b. It increases solubility.
    c. It has little effect on solubility.
    d. It has the same effect as with a gaseous solution.

76. Nonpolar molecules must have what kind of regions?
    a. Hydrophilic
    b. Hydrophobic
    c. Hydrolytic
    d. Hydrochloric

77. Which of these is a substance that increases the rate of a chemical reaction?
    a. Catalyst
    b. Helium
    c. Solvent
    d. Inhibitor

78. Based on collision theory, what is the effect of temperature on the rate of chemical reaction?
    a. Increasing the temperature slows the reaction.
    b. Decreasing the temperature speeds up the reaction.
    c. Increasing the temperature speeds up the reaction.
    d. Collision theory and temperature are unrelated.

79. An atom of radium-226, $^{226}_{88}$Ra, contains:
    a. 88 neutrons, 138 protons, and 226 electrons
    b. 88 protons, 138 electrons, and 226 neutrons
    c. 88 protons, 88 electrons, and 138 neutrons
    d. 88 electrons, 88 protons, and 226 neutrons

80. Which type of bonding results from transferring electrons between atoms?
    a. Ionic bonding
    b. Covalent bonding
    c. Hydrogen bonding
    d. Dipole interactions

81. Which substance is oxidized in the following reaction?
    $$4Fe + 3O_2 \rightarrow 2Fe_2O_3$$
    a. Fe
    b. O
    c. O_2
    d. Fe_2O_3

82. Which statements are true regarding nuclear fission?
    I.   It involves the splitting of heavy nuclei
    II.   It is utilized in power plants
    III. It occurs on the sun
    a. I only
    b. II and III only
    c. I and II only
    d. III only

83. Which type of nuclear decay is occurring in the equation below?

    $$U^{236}_{92} \rightarrow He^4_2 + Th^{232}_{90}$$

    a. Alpha
    b. Beta
    c. Gamma
    d. Delta

84. Which statement is true about the pH of a solution?
    a. A solution cannot have a pH less than 1.
    b. The more hydroxide ions there are in the solution, the higher the pH will be.
    c. If an acid has a pH of greater than -1, it is considered a weak acid.
    d. A solution with a pH of 2 has ten times more hydrogen ions than a solution with a pH of 1

85. Which radioactive particle is the most penetrating and damaging and is used to treat cancer in radiation?
    a. Alpha
    b. Beta
    c. Gamma
    d. Delta

86. What coefficients are needed to balance the following combustion equation?

$$\_ C_2H_{10} + \_ O_2 \rightarrow \_ H_2O + \_ CO_2$$

    a. 1:5:5:2
    b. 1:9:5:2
    c. 2:9:10:4
    d. 2:5:10:4

87. If 100 mL of 0.050 M $AgNO_3$ is added to 50.0 mL of 0.050 M Na_2SO_4, which of the following choices is true? Note that K_{sp} of Ag_2SO_4 is 6.9×10^{-15}.
    a. A precipitate will form without any excess ions.
    b. A precipitate will form with excess Ag^+.
    c. There will be no precipitate.
    d. A precipitate will form with an excess of Ag^+ and SO_4^{2-}.

88. How many electrons are in the p subshell for the nitrogen atom?
    a. 2
    b. 3
    c. 4
    d. 5

89. Which of the following is NOT a unique property of water?
    a. High cohesion and adhesion
    b. High surface tension
    c. High density upon melting
    d. High freezing point

90. Salts like sodium iodide (NaI) and potassium chloride (KCl) have what type of bond?
    a. Ionic bonds
    b. Disulfide bridges
    c. Covalent bonds
    d. London dispersion forces

91. Which of the following is unique to covalent bonds?
    a. Most covalent bonds are formed between the elements H, F, N, and O.
    b. Covalent bonds are dependent on forming dipoles.
    c. Bonding electrons are shared between two or more atoms.
    d. Molecules with covalent bonds tend to have a crystalline solid structure.

92. Which of the following describes a typical gas?
    a. Indefinite shape and indefinite volume
    b. Indefinite shape and definite volume
    c. Definite shape and definite volume
    d. Definite shape and indefinite volume

93. Which of the following is the type of the reaction between ethene and hydrogen bromide?
    a. Addition reaction
    b. Substitution reaction
    c. Elimination reaction
    d. Polymerization

94. Which of the following alcohols has a chiral carbon?
    a. 2-methyl-2-butanol
    b. 2-butanol
    c. 2-methyl-1-butanol
    d. 3-methyl-2-buten-1-ol

95. Which of the following aldehyde(s) can undergo a Cannizzaro reaction?
    a. Formaldehyde and benzaldehyde
    b. Benzaldehyde and trimethyl acetaldehyde
    c. Formaldehyde and trimethyl acetaldehyde
    d. All of the above

96. Which of the following does NOT involve Clemmensen reduction?
    a. Conversion of ethanal (acetaldehyde) to ethane
    b. Conversion of benzaldehyde to toluene (methylbenzene)
    c. Conversion of ethanoic acid (acetic acid) to ethanol
    d. Conversion of propanone (acetone) to propane

97. Which of the following is a nucleophilic reagent?
    a. Br^+
    b. NO_2^+
    c. $AlCl_3$
    d. NH_3

98. How many enantiomers are possible for a glucose molecule?
    a. 4
    b. 8
    c. 16
    d. 32

99. What is the IUPAC name of $CH_3CH_2CH_2C(CH_3)_2CH_2CH(CH_3)_2$?
    a. 4,4,6,6-tetramethylhexane
    b. 1,1,3,3-tetramethylhexane
    c. 2,4,4-trimethylheptane
    d. 4,4-dimethyloctane

100. Which product forms from a reaction between propene and bromine?
    a. $CH_3–CHBr–CH_2Br$
    b. $BrCH_2–CH_2–CH_2Br$
    c. $CH_3–CH_2–CHBr_2$
    d. None of the above

101. Aspirin is used as a non-steroidal anti-inflammatory drug (NSAID). It is available as an over-the-counter (OTC) medication for the treatment of pain, inflammation, and fever. A low dose of aspirin is also used for prevention of cardiovascular events including heart attack and stroke, which occurs secondary to blood clot formation. However, aspirin is a prodrug that needs to be bio-converted into an active drug, salicylic acid. Given that aspirin is chemically termed as acetyl salicylic acid, and salicylic acid is ortho-hydroxybenzoic acid, how would you best describe the bio-conversion reaction?

    a. Oxidation
    b. Ester hydrolysis
    c. Hydrolysis
    d. Reduction

102. Which of the following statements is true about an aldol condensation reaction?

    I. Aldehydes with α-H can undergo the aldol condensation.
    II. CH_3CHO can undergo aldol condensation.
    III. Benzaldehyde can undergo aldol condensation.

    a. I and II
    b. I and III
    c. II and III
    d. I, II, and III

103. Which is NOT a disaccharide?

    a. Sucrose
    b. Lactose
    c. Maltose
    d. Mannose

104. Which is the most common reaction for alkenes and alkynes?

    a. Nucleophilic addition
    b. Electrophilic addition
    c. Nucleophilic substitution
    d. Electrophilic substitution

105. Which one of the following molecules has isomers that are optically active?

    a. $CH_3—CH_2—COOH$
    b. $C_5H_{10}O$
    c. $CH_3—CH(NH_2)—COOH$
    d. $CH_3—CH=CH—CH_3$

106. Which of the following compounds can be used to produce ethanol through an appropriate chemical reaction?

    I. $CH_2=CH_2$
    II. $C_2H_5—O—C_2H_5$
    III. $CH_3—CHO$

    a. I and II
    b. I and III
    c. II and III
    d. I, II, and III

107. C_nH_{2n-2} is the general formula for which of the following?
   a. Alkane
   b. Alkene
   c. Alkyne
   d. Aromatic compound

108. What is the electron group geometry of the NF_3 molecule?
   a. Trigonal planar
   b. Tetrahedral
   c. Trigonal bipyramidal
   d. Octahedral

109. Which of the following is an example of a Brønsted-Lowry base in water?
   a. H_3O^+
   b. CO_2
   c. NH_3
   d. HCl

110. What is the electrical charge of the nucleus?
   a. A nucleus always has a positive charge.
   b. A stable nucleus has a positive charge, but a radioactive nucleus may have no charge and instead be neutral.
   c. A nucleus always has no charge and is instead neutral.
   d. A stable nucleus has no charge and is instead neutral, but a radioactive nucleus may have a charge.

111. What is the temperature in Fahrenheit when it is 35 °C outside?
   a. 67 °F
   b. 95 °F
   c. 63 °F
   d. 75 °F

112. If sodium hydroxide (NaOH) is added to a dilute solution of potassium hydroxide (KOH), how will the NaOH equilibrium be affected, compared to NaOH dissolved in pure water?
   a. Sodium hydroxide always dissociates completely to form more ions.
   b. The equilibrium of sodium hydroxide is unaffected by KOH.
   c. The equilibrium of the reaction favors sodium hydroxide.
   d. Sodium hydroxide does not dissociate in the KOH solution.

113. Which of the following sets of quantum numbers is INCORRECT?
   a. $n = 2, l = 1, m_l = +1$
   b. $n = 2, l = 1, m_l = 0$
   c. $n = 1, l = 0, m_l = 0$
   d. $n = 3, l = 1, m_l = -2$

114. Suppose that you exhale half a liter (0.500 L) of carbon dioxide (CO_2) gas into a balloon. How many oxygen atoms are present inside the balloon if the oxygen atoms come from carbon dioxide? The molar volume of carbon dioxide is 22.4 L/mol.

    a. 1.34×10^{22} O atoms
    b. 2.69×10^{22} O_2 atoms
    c. 2.69×10^{22} O atoms
    d. 1.34×10^{22} O_2 atoms

115. The caffeine molecule, found in coffee and tea, has a molecular formula of $C_8H_{10}N_4O_2$. What is the mass in kilograms for one molecule of caffeine?

    a. 3×10^{-25} kg
    b. 4×10^{-25} kg
    c. 3×10^{-22} kg
    d. 1×10^{-22} kg

116. If a compound containing nitrogen and oxygen weighs a total of 2.04 g, what is the empirical formula if the substance contains 1.51 g of oxygen?

    a. N_2O_5
    b. N_2O
    c. NO
    d. NO_2

117. A compound containing sulfur and oxygen has the empirical formula of SO_2. If 0.162 grams of the actual compound, $(SO_2)_n$, contains 7.61×10^{20} molecules, what must the value of n be?

    a. 4
    b. 3
    c. 2
    d. 1

118. Find the mass of oxygen present in 1.50 mole of sucrose ($C_{12}H_{22}O_{11}$).

    a. 195 g
    b. 200 g
    c. 245 g
    d. 264 g

119. Carbon dioxide can be prepared by the reaction or combustion of carbon monoxide and oxygen:

$$2\ CO(g) + O_2(g) \rightarrow 2\ CO_2$$

If three moles of carbon monoxide react with one mole of oxygen gas, the amount of carbon dioxide that is produced is called the:

    a. Product yield
    b. Theoretical yield
    c. Limiting reactant
    d. Excess reactant

120. If carbon dioxide is passed over charcoal, it will produce carbon monoxide:

$$CO_2(g) + C(s) \rightarrow 2\ CO(g)$$

Suppose two moles of carbon dioxide react with three moles of carbon to produce a theoretical amount of carbon monoxide. If the excess reactant is consumed or depleted, how many more moles of carbon monoxide could be produced?

 a. 2 moles
 b. 3 moles
 c. 4 moles
 d. 5 moles

121. If 3.52 g of aluminum sulfate is mixed with 4.06 g of barium chloride in solution, what is the theoretical yield of barium sulfate (molar mass is 233.40 g/mol)?

$$Al_2(SO_4)_3(aq) + 3\ BaCl_2(aq) \rightarrow 3\ BaSO_4(s) + 2\ AlCl_3(aq)$$

 a. 3.15 g $BaSO_4$
 b. 1.95 g $BaSO_4$
 c. 3.09 g $BaSO_4$
 d. 4.55 g $BaSO_4$

122. If 3.96 grams of barium sulfate is collected through a separate filtration process, what is the percent yield for a mixture of aluminum sulfate and barium chloride that had a theoretical yield of 4.21 g?

 a. 106%
 b. 0.94
 c. 94%
 d. 1.06

123. How many grams of excess reactant is present if 3.52 g of aluminum sulfate is mixed with 4.06 grams of barium chloride?

 a. 2.00 g $Al_2(SO_4)_3$
 b. 0.50 g $Al_2(SO_4)_3$
 c. 1.30 g $Al_2(SO_4)_3$
 d. 2.50 g $Al_2(SO_4)_3$

124. A common way to express the concentration of a solution is in units of moles per liter, which is called the:

 a. Molality
 b. Molarity
 c. Molar volume
 d. Parts per mass

125. What mass of water is needed to dissolve 250.0 g of AgCl to make a 0.35 mol/kg AgCl aqueous solution?

    a. 5.0 kg
    b. 1.7 moles per liter
    c. 5.0 moles per liter
    d. 1.7 kg

126. What volume of a 5.00 M HNO_3 solution should be added to water to make a 2.00 M HNO_3 solution?

    a. Take 350 mL of water and add 140 mL the concentrated acid.
    b. Take 210 mL of water and add 140 mL the concentrated acid.
    c. Take 350 L of water and add 140 L the concentrated acid.
    d. Take 210 L of water and add 140 L the concentrated acid.

127. During an acid-base titration experiment, the endpoint was reached when 35.00 mL of a standardized 1.500 M potassium hydroxide was added to a 15.00 mL solution of sulfuric acid. What is the concentration of the acid assuming that the endpoint is near the equivalence point? The reaction of sulfuric acid and potassium hydroxide is shown below:

$$H_2SO_4(aq) + 2\,KOH(aq) \rightarrow K_2SO_4(aq) + 2\,H_2O(l)$$

    a. 0.5250 M H_2SO_4
    b. 0.750 M H_2SO_4
    c. 1.750 M H_2SO_4
    d. 3.500 M H_2SO_4

128. In anticipation of freezing weather, farmers may spray water over their plants and cover them with large plastic materials. As the temperature drops below 0 °C (32 °F), the following physical process will occur for the water found on the plants:

$$H_2O(l) \rightarrow H_2O(s)$$

If water is treated as the chemical system, which of the following is correct?

    a. The process is exothermic and $+q_{system} = -q_{surroundings}$.
    b. The process is endothermic and $+q_{system} = +q_{surroundings}$.
    c. The process is exothermic and $-q_{system} = +q_{surroundings}$.
    d. The process is endothermic and $-q_{system} = +q_{surroundings}$.

129. The standard enthalpies of formation for lithium bromide and lithium chloride are shown below.

Standard enthalpies of formation at 298K, ΔH_f° (kJ/mol)	
LiBr	−351.2
LiCl	−408.6

Which of the following statements is true?
a. ΔH_f° for LiCl is less exothermic than ΔH_f° for LiBr
b. ΔH_f° for LiCl is more exothermic than ΔH_f° for LiBr
c. ΔH_f° for LiCl is more endothermic than ΔH_f° for LiBr
d. ΔH_f° for LiCl is less endothermic than ΔH_f° for LiBr

130. The combustion of one mole of methane produces carbon dioxide and water with an amount of heat (in kJ) shown in the following thermochemical equation:

$$CH_4(g) + 2\,O_2(g) \rightarrow CO_2(g) + 2\,H_2O(g) \quad \Delta H_{rxn}^{\circ} = -802.3 \text{ kJ}$$

How much heat is produced if 1.5 kg of methane is combusted in oxygen gas?
a. -7.50×10^4 kJ/mol
b. $+7.50 \times 10^4$ kJ/mol
c. $+7.50 \times 10^4$ kJ
d. -7.50×10^4 kJ

131. For the reaction shown below, calculate the standard enthalpy of reaction ΔH_{rxn}° using the standard enthalpies of formation shown in the table below.

$$2\,Al(s) + Fe_2O_3(s) \rightarrow Al_2O_3(s) + 2\,Fe(s) \quad \Delta H_{rxn}^{\circ} = ?$$

Standard enthalpies of formation at 298K, ΔH_f° (kJ/mol)	
Al(s)	0
Fe_2O_3(s)	−824.2
Al_2O_3(s)	−1675.7
Fe(s)	0

a. −851.5 kJ
b. −2499.9 kJ
c. +851.5 kJ
d. +2499.9 kJ

132. A 43.0 g metal sample at a temperature of 100.0 °C is dropped into a 100.0 mL water bath with an initial temperature of 25.0 °C. The temperature of the water bath increased to a final temperature of 37.0 °C. What is the specific heat of the metal if the specific heat capacity of water is 4.184 J/(g °C)? The density of water is about 1 g/mL.
a. 2.95 J/(g °C)
b. 0.87 J/(g °C)
c. 1.85 J/(g °C)
d. 4.25 J/(g °C)

133. Which of the following processes involves an increase in entropy?
    a. The contraction of a gas.
    b. The freezing of water: $H_2O(l) \rightarrow H_2O(s)$
    c. $CaO(s) + CO_2(g) \rightarrow CaCO_3(s)$
    d. Adding sugar to water: $C_{12}H_{22}O_{11}(s) \xrightarrow{H_2O(l)} C_{12}H_{22}O_{11}(aq)$

134. The following equation gives the Gibbs free energy ΔG for a chemical reaction:

$$\Delta G = \Delta H - T\Delta S$$

If the Gibbs free energy of the system is negative, the reaction is spontaneous. If it is positive, the reaction is not spontaneous and will occur. If the free energy is zero, then the reaction is at equilibrium. Which of the following is true if both ΔH and ΔS are positive?
    a. The reaction is spontaneous when the temperature is increased.
    b. The reaction is spontaneous at any temperature.
    c. The reaction is spontaneous when the temperature is decreased.
    d. The reaction is nonspontaneous at any temperature.

135. Consider the following hypothetical reaction:

$$2\,A(g) \rightarrow A_2(g) \quad \Delta H° = -300 \text{ kJ}$$

At what conditions would this reaction be spontaneous? What is the sign of $\Delta S°_{rxn}$?
    a. Spontaneous at low temperatures and $+\Delta S°_{rxn}$
    b. Spontaneous at high temperatures and $+\Delta S°_{rxn}$
    c. Spontaneous at low temperatures and $-\Delta S_{rxn}$
    d. Spontaneous at high temperatures and $+\Delta S_{rxn}$

136. Based on the Aufbau principle, which of the following choices most likely represents the set of quantum numbers corresponding to the eighth electron in the oxygen atom?
    a. $n = 2, l = 0, m_l = 0, m_s = +1/2$
    b. $n = 2, l = 1, m_l = -1, m_s = +1/2$
    c. $n = 2, l = 0, m_l = 0, m_s = -1/2$
    d. $n = 2, l = 1, m_l = -1, m_s = -1/2$

137. Which of the following quantum numbers does not represent an electron in the nitrogen atom?
    a. $n = 2, l = 1, m_l = -2, m_s = -1/2$
    b. $n = 1, l = 0, m_l = 0, m_s = -1/2$
    c. $n = 2, l = 0, m_l = 0, m_s = -1/2$
    d. $n = 2, l = 1, m_l = -1, m_s = +1/2$

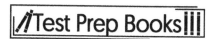

138. For the three structures below, which Lewis structure for nitrous oxide (N_2O) is the most likely the correct structure? Choose the right answer choice below which has the correct corresponding formal charges for the N, N, and O atoms.

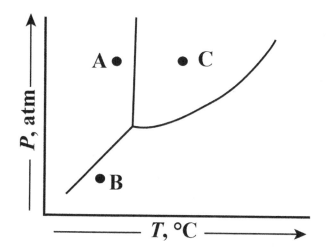

a. Structure 1, formal charges are −1, +1, 0
b. Structure 2, formal charges are 0, +1, −1
c. Structure 3, formal charges are −2, +1, +1
d. Structure 1, formal charges are +2. +1, 0

139. Which of the following statements is NOT true regarding the phase diagram below.

a. If a compound is in state A, with the temperature held constant, reducing the pressure will cause it to sublime.
b. Decreasing the pressure and temperature from point C to point B represents evaporation.
c. The slope of the solid and liquid equilibrium line is positive, similar to the phase diagram of water.
d. Increasing the temperature from point A to point C represents melting.

140. The vapor pressure of a solution is lowered when a nonvolatile solute is added to a pure solvent. As a result, the freezing point for the solution decreases, and the boiling point increases. The following equation gives the freezing point depression:

$$\Delta T_f = m \times K_f$$

The term $\Delta T_f = T_{solvent} - T_{solution}$ represents the temperature change or freezing point depression in degrees Celsius, and the terms m and K_f are the molality (units of molal or mol/kg) and freezing point depression constant for the solvent (units of $\frac{°C}{mol\ kg^{-1}}$). If 1.00×10^3 g of ethylene glycol ($C_2H_6O_2$, molar mass $= 62.07$ g/mol) is added to a radiator that contains 5.00×10^3 g of water, how much would the freezing point lower? Note that $K_f = 1.86$ °C/(mol kg^{-1}).

   a. −5.99 °C
   b. +5.99 °C
   c. −3.22 °C
   d. +3.22 °C

141. The activation energy for a chemical reaction involving the breakage of a carbon-oxygen single bond was estimated at 125 kJ/mol at 25.0 °C. If the temperature is increased to about 50.0 °C, by what factor does the rate constant increase?

   a. 10.5
   b. 32.1
   c. 49.5
   d. 29.6

142. The half-life for an unknown element called X is 7.95 days How many days would this sample need to reach 35.0 percent of its original amount?

   a. 19.5 days
   b. 12.0 days
   c. 11.5 days
   d. 15.0 days

143. Consider the following two-step reaction below:

$$2\ CO(g) + O_2(g) \rightleftharpoons 2\ CO_2$$

$$CO_2(g) + C(s) \rightleftharpoons 2\ CO(g)$$

What is the overall equilibrium expression?

   a. $K_{overall} = \frac{[CO_2]^2}{[CO]^2[O_2]}$

   b. $K_{overall} = \frac{[CO]^2}{[CO_2][C]}$

   c. $K_{overall} = \frac{[CO_2]^2}{[O_2][C]}$

   d. $K_{overall} = \frac{[CO_2]}{[O_2][C]}$

144. Consider the following exothermic reaction:

$$SO_2(g) + CaO(s) \rightleftharpoons CaSO_3(s) + \text{heat}$$

Which of the following changes will result in a decrease in pressure of $SO_2(g)$ at equilibrium?
a. Addition of $CaO(s)$
b. Removal of $CaSO_3(s)$
c. Increasing the volume of the reaction vessel
d. Decreasing the temperature of the chemical reaction system

145. What is the pH of a 0.35 M solution of ammonia in water? Note that $K_b = 1.8 \times 10^{-5}$
a. 10.20
b. 8.01
c. 11.25
d. 8.96

146. What is the equilibrium concentration of a reaction mixture that initially had 1.00 mol H_2 and 1.00 mol I_2 in a 1.0 L vessel?

$$H_2(g) + I_2(g) \rightleftharpoons 2HI(g) \quad K_c = 49.7 \text{ at } 458\,°C$$

a. $[H_2] = [I_2] = 0.22$ M, $[HI] = 1.6$ M
b. $[H_2] = [I_2] = 0.33$ M, $[HI] = 1.8$ M
c. $[H_2] = [I_2] = 0.44$ M, $[HI] = 2.0$ M
d. $[H_2] = [I_2] = 0.55$ M, $[HI] = 2.2$ M

147. For the following radioactive decay process, which of the following is the correct product shown in the nuclear equation below?

$$^{92}_{44}\text{Ru} + ^{0}_{-1}\text{e} \rightarrow ?$$

a. $^{92}_{45}\text{Rh}$

b. $^{92}_{43}\text{Rh}$

c. $^{92}_{43}\text{Tc}$

d. $^{4}_{2}\text{He} + ^{92}_{43}\text{Tc}$

148. Find the N/Z ratio of Pb-212 to predict whether beta decay or positron emission will occur, and then determine which choice below represents the N/Z ratio of the daughter nuclide.
a. 1.59
b. 1.55
c. 1.41
d. 1.37

149. For the following redox reaction, which of the following statements is true?

$$6 \, I^-(aq) + 4 \, H_2O(l) + 2 \, MnO_4^-(aq) \rightarrow 3 \, I_2(aq) + 2 \, MnO_2(s) + 8 \, OH^-(aq)$$

a. MnO_4^- is oxidized in the reaction.
b. MnO_4^- is the oxidizing agent.
c. I^- is reduced in the reaction.
d. I^- is the oxidizing agent.

150. Consider the standard reduction potentials for the reactions shown below. Calculate *the* standard cell potential, E_{cell}°, for the spontaneous reaction.

$$Ni^{2+}(aq) + 2 \, e^- \rightarrow Ni(s) \quad E^{\circ} = -0.23 \text{ V}$$

$$Al^{3+}(aq) + 3 \, e^- \rightarrow Al(s) \quad E^{\circ} = -1.66 \text{ V}$$

a. $E_{cell}^{\circ} = -1.43 \, V$
b. $E_{cell}^{\circ} = +1.43 \, V$
c. $E_{cell}^{\circ} = -2.63 \, V$
d. $E_{cell}^{\circ} = +2.63 \, V$

151. The solubility of silver chloride (AgCl) is approximately 1.9×10^{-3} g/L in water. Which of the following choices below most closely corresponds to the correct solubility product constant (K_{sp})?
a. 1.6×10^{-10}
b. 1.8×10^{-10}
c. 2.0×10^{-10}
d. 3.6×10^{-10}

152. Which of the following choices is NOT true regarding the equilibrium constant.
a. It can be used to predict the extent of the reaction.
b. It gives information on the direction of the reaction.
c. It can be used to determine the time needed to reach equilibrium.
d. It can be used to help calculate equilibrium concentrations.

Answer Explanations

1. C: Based on the law of definite proportions, the mass element ratio of each sample is:

$$Sample\ A\ ratio = \frac{3.662\ g\ O}{2.420\ g\ S} = 1.513\ O\ to\ S;\ \sim \frac{3}{2}$$

$$Sample\ B\ ratio = \frac{3.558\ g\ O}{3.566\ g\ S} = 0.9978\ O\ to\ S;\ \sim \frac{1}{1}$$

In sample A, the O:S mass element ratio is 3:2, and in sample B, it is 1:1. The law of multiple proportions says the ratio of two different compounds, or samples, will give a whole number mass element ratio. The O:S mass element ratio of sample A to B is:

$$Sample\ A\ to\ B\ ratio = \frac{\frac{3}{2}\ O\ to\ S}{\frac{1}{1}\ O\ to\ S} = \frac{3}{2}$$

Compound A contains "three times" as much oxygen compared to compound B. Compound A contains "twice" as much sulfur compared to compound B. In sample B, the O:S ratio is 1:1, and because the mass of oxygen is approximately 16.0 grams and sulfur is roughly 32.0 grams, this indicates that compound B may be SO_2, where the mass ratio of O:S is $[\frac{2 \times 16.0}{32.0}] \sim 1$ or 1:1.

The law of multiple proportions says that because compound A has three times more O and two times more sulfur than compound B, the chemical formula for A must be $S_{1 \times 2}O_{2 \times 3} = S_2O_6$ (dithionate). It is vital to note that the ratios listed above are mass element ratios and not necessarily related to the subscripts of the chemical formulas. In addition, the law of multiple proportions does not necessarily give us the exact molecular formula but rather just the mass element ratio, O:S, of sample A to B. For example, compound B may also be S_2O_4, where the mass element ratio of O:S is s1:1; therefore, compound A must be $S_{2 \times 2}O_{4 \times 3} = S_4O_{12}$, which has a mass ratio, O:S, of 3:2. If the identity of one molecular compound is known, the other can be deduced.

2. A: To find the mass percentage of each element, use the atomic mass from the periodic table. Carbon has an atomic mass of 12.01 amu; therefore, the mass in grams is 12.01. Similarly, hydrogen is 1.008 grams, and oxygen is 16.00 grams. The molecular formula, CH_2O_2, indicates there is more than one hydrogen and oxygen; therefore, the mass for each one must be multiplied by the number denoted in each subscript. There is only one carbon atom. The % mass of each element (C, 2H, 2O) in the compound is:

$$Mass\ \%\ element = \frac{total\ mass\ of\ element}{total\ mass\ of\ compound} \times 100\%$$

First, find the total mass or molar mass of the compound, which is:

$$Total\ mass\ of\ compound = (1 \times 12.01\ amu\ C) + (2 \times 1.008\ amu\ H) + (2 \times 16.00\ amu\ O)$$

$$= 46.025\ amu\ CH_2O_2$$

For each element and the compound, grams can be used in place of amu. In either case, units will still cancel out. The mass % of each element is:

$$Mass\ \%\ C = \frac{12.01\ g}{46.025\ g} \times 100\% = 26.09\%$$

$$Mass\ \%\ H = \frac{2 \times 1.008\ g}{46.025\ g} \times 100\% = 4.380\%$$

$$Mass\ \%\ O = \frac{2 \times 16.00}{46.025\ g} \times 100\% = 69.527\%$$

Choice A is the closest choice if the above percentages are rounded to the ones place: Mass % C = 26, Mass % H = 4%, and Mass % O = 70.

3. D: It's useful to consider the following conversion road map when converting the mass or moles of one substance to another:

$$\text{Mass A} \overset{\substack{\text{periodic} \\ \text{table}}}{\leftrightarrows} \text{moles A} \overset{\substack{\text{stoichiometric} \\ \text{ratio}}}{\leftrightarrows} \text{moles B} \overset{\substack{\text{periodic} \\ \text{table}}}{\leftrightarrows} \text{mass B}$$

When converting from the mass of A to moles of A, the periodic table must be used to obtain the molar mass of that substance. When converting the moles of substance A to B, the stoichiometric ratio, which gives the coefficients of each substance, must be used. Lastly, when converting from the moles of substance B to its mass, the periodic table is used again to acquire the mass. The order of the conversion road map will not change. The problem requires that the following conversions be carried out:

$$\text{Mass } C_8H_{18} \overset{\substack{\text{periodic} \\ \text{table}}}{\leftrightarrows} \text{moles } C_8H_{18} \overset{\substack{\text{stoichiometric} \\ \text{ratio}}}{\leftrightarrows} \text{moles } H_2O \overset{\substack{\text{periodic} \\ \text{table}}}{\leftrightarrows} \text{mass } H_2O$$

For every two moles of octane that react, eighteen moles of water will be produced:

$$2C_8H_{18}(g) + 25O_2(g) \rightarrow 16CO_2(g) + 18H_2O(g)$$

The following unit-type conversion is useful:

$$\left(\frac{18\ \text{moles } H_2O}{2\ \text{moles } C_8H_{18}}\right)$$

Moles of the reactant (e.g., octane) is placed at the bottom and moles of the product on top (e.g., water) if converting from a reactant to a product. Note that kilograms of octane must be converted to grams of octane because the periodic table gives the molar mass in terms of grams per mole:

$$20.0\ \text{kg } C_8H_{18} \times \left(\overset{}{\frac{1000g\ C_8H_{18}}{1\ \text{kg } C_8H_{18}}}\right) \times \left(\overset{\text{molar mass}}{\frac{1\ \text{mole } C_8H_{18}}{114.232\ g\ C_8H_{18}}}\right) \times \left(\overset{\text{stochiometric ratio}}{\frac{18\ \text{moles } H_2O}{2\ \text{moles } C_8H_{18}}}\right) \times \left(\overset{\text{molar mass}}{\frac{18.016\ g\ H_2O}{1\ \text{mole } H_2O}}\right)$$

$$= 28,388\ g\ H_2O \sim 28.4\ \text{kg } H_2O$$

This value is roughly close to eight gallons of water. However, experimentally, the combustion process will not be 100% efficient, and additional side products may be produced, e.g., carbon monoxide.

Furthermore, most of the water may be lost as water vapor if the outside air is at room temperature or higher. Starting the car at night, when the temperatures is lowest, would most likely result in the condensation of some of the water vapor that is produced.

4. B: The total number of electrons is determined by the atomic number of calcium (Ca), which is provided in any periodic table. Because calcium has an atomic number of Z = 20, there are twenty protons and twenty electrons. The order of the subshells, based on Aufbau's principle, is 1s, 2s, 2p, 3s, 3p, 3d, 4s, 4p. The s orbitals contain at most two electrons, the p orbitals six electrons, and the d orbitals up to ten electrons. The 4s subshell is always filled first before the 3d, and if there are more than twenty electrons, some of these will be filled in the 3d orbitals. Choice A represents the ground state electron configuration of calcium because the sum of the superscripts gives a total of twenty electrons ($2 + 2 + 6 + 2 + 6 + 2 = 20$). A shorthand notation of the ground state electron configuration for calcium is often written with a noble gas core, Z < 20, followed by a representation of the filled 4s subshells in calcium: $[Ar]4s^2$.

The 3d subshell is not shown because there are no extra electrons to fill one of the 3d orbitals. Choice C is incorrect because it has a total of twenty-one electrons ($2 + 2 + 6 + 2 + 6 + 1 + 2 = 21$), which is the electron configuration for scandium, not calcium. Choice D is incorrect because the "4s" subshell is placed before the "3d" subshell. Recall that the Aufbau principle predicts the electron configuration and shows which subshells are filled up first, with the "3d" subshell placed before "4s" in the final electron configuration for Z > 21. Choice E is incorrect because two electrons are found in the 3d subshell, and the total number of electrons is twenty-one.

One excited state electron configuration of calcium (Choice B) would require that one of the electrons from the outer valence shell, e.g., an electron in 4s, move to one of the 3d orbitals. The periodic table below is useful for predicting ground state electron configurations, which are separated into s, d, and p blocks. The row number in which the element resides indicates the highest possible subshell value. For example, calcium is in group 4 (n = 4) and in the s block. Because calcium is in group IIA, there are "two" valence electrons found in the 4s subshell; therefore, the electron configuration must be $[Ar]4s^2$.

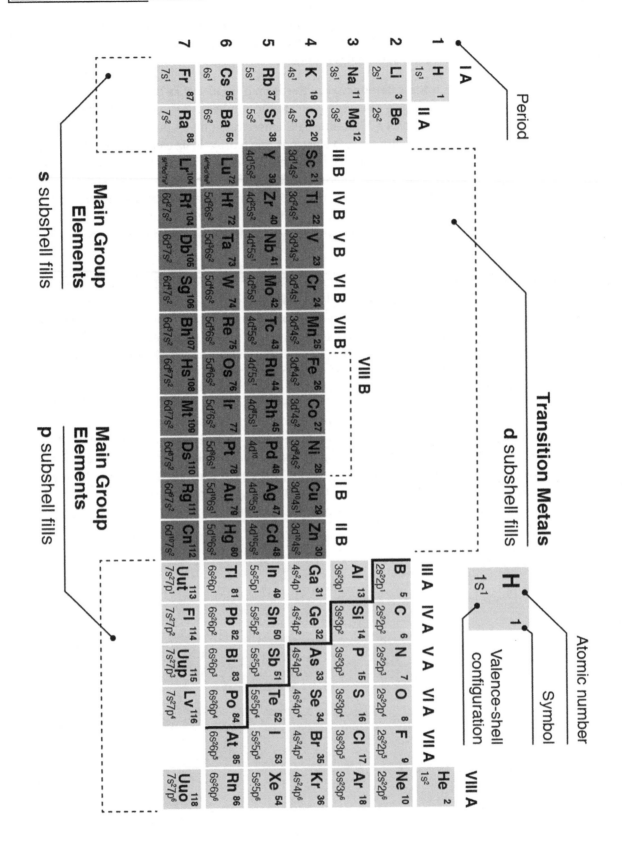

5. E: In general, the ionization energy will increase when moving left to right across a row in the periodic table. Because bromine is to the right of selenium, it is expected that the first ionization energy of bromine will be greater. The electron configuration of selenium and bromine are:

Se (Z = 34): $1s^2 2s^2 2p^6 3s^2 3p^6\ 3d^{10} 4s^2 4p^4$

Br (Z = 35): $1s^2 2s^2 2p^6 3s^2 3p^6\ 3d^{10} 4s^2 4p^5$

Even though bromine has an extra electron compared to selenium, the size of the 4p subshell does not increase drastically compared to the addition of a new subshell, e.g., 5s or 5p. Because the atomic number increases, the effective nuclear charge or the Coulomb force increases, which makes it more difficult to remove an electron in the 4p subshell.

6. D: Each answer choice, with the exception of Choice *D*, represents several possible higher energy or excited state electron configurations. For example, Choices *A, B,* and *C* represent other possible high-energy state configurations, which have orbitals containing either one electron or orbitals with two electrons that have opposite spin. Choice *E* represents the ground state electron configuration of carbon because one 2p orbital is completely filled with two electrons with opposite spin. Choice *D* is not a possible electron configuration because it shows one 2p orbital containing electrons with the same spin and violates the Pauli exclusion principle, which says that for any two electrons in one orbital, the electrons must have opposite spin.

7. C: The following formula can be used to find the average atomic mass for silicon:

$$Average\ atomic\ mass\ (amu) = \sum (atomic\ mass) \times (fraction\ abundance)$$

The sum symbol "Σ," refers to the summation for different terms. Because there are three isotopes, there are three terms. Each term is equal to the atomic mass for a given isotope multiplied by its fractional abundance. The average atomic mass is:

$$= \overbrace{(27.9769271) \times (0.9223)}^{\text{term 1:Si}-28} + \overbrace{(28.9764949) \times (0.0468)}^{\text{term 2:Si}-29} + \overbrace{(29.9737707) \times (0.0309)}^{\text{term 3:Si}-30}$$

$$= 28.0854\ amu$$

The average atomic mass of silicon to four significant figures is 28.0854 amu.

8. B: The equation that relates the energy to the wavelength of light is given by:

$$E = h\nu\ or\ E = h\frac{c}{\lambda};\quad c = 3.00 \times 10^8\ \text{m/s and}\ h = 6.626 \times 10^{-34}\ \text{J} \cdot \text{s}$$

The units for the wavelength must be in units of meters (m). Because the wavenumber is given in inverse centimeters (cm^{-1}), the value of 1151 cm^{-1} must be inverted and converted to meters:

$$Step\ 1: Take\ inverse\ to\ get\ cm\ units: (1151\ \text{cm}^{-1})^{-1} = 8.688 \times 10^{-4}\ \text{cm}$$

$$Step\ 2: Convert\ to\ meters: 8.688 \times 10^{-4}\ \text{cm} \times \frac{1\ \text{m}}{100\ \text{cm}} = 8.688 \times 10^{-6}\ \text{m}$$

In step 2, the unit conversion of $1\ cm = 1 \times 10^{-2}\ m$ could have been used alternatively. In the last step, the wavelength is inserted into the energy equation:

$$Step\ 3: E = \frac{hc}{\lambda} = \frac{(6.626 \times 10^{-34} J \cdot s)(3.00 \times 10^8\ m/s)}{8.688 \times 10^{-6}\ m} = 2.288 \times 10^{-20} J$$

9. A: To find the number of lead (II) ions, use the following conversion road map:

$$Mass\ of\ PbCrO_4 \overset{\substack{molar \\ mass}}{\Rightarrow}\ moles\ of\ PbCrO_4 \overset{stoichiometry}{\Rightarrow}\ moles\ of\ Pb^{2+} \overset{N_A/mole}{\Rightarrow}\ molecules\ of\ Pb^{2+}$$

The molar mass of lead (II) chromate is 323.2 g/mol. There is a one-to-one molar relationship between lead (II) chromate and lead (II) ion. The number of molecules can then be determined from the definition of the mole, which is equal to Avogadro's number, N_A:

$$Molecules\ of\ Pb^{2+} =$$

$$0.550\ g\ PbCrO_4 \times \left(\frac{1\ mole\ PbCrO_4}{323.2\ g\ PbCrO_4}\right) \times \left(\frac{1\ mole\ Pb^{2+}}{1\ mole\ PbCrO_4}\right) \times \left(\frac{6.02 \times 10^{+23}\ Pb^{2+} molecules}{1\ mole\ Pb^{2+}}\right)$$

$$= 1.02 \times 10^{21}\ molecules\ of\ Pb^{2+}$$

There are many ways to find out if the concentration of lead in the bottle is less than 8%. One qualitative method is to use the mass of the precipitate and mass of water. A 1.000 L bottle of water is approximately 1000 mL or 1000 g of water given that the water density is 1 g/mL. The percentage of Pb^{2+} ions is roughly less than the percentage of lead (II) chromate:

$$Percent\ Pb^{2+} in\ water < percent\ PbCrO_4 = \frac{0.550\ g}{0.550\ g + 1000\ g} \times 100\% \sim 0.0550\%$$

Therefore, the percentage of lead (II) ion is much less than 8%. Alternatively, the mass percent of lead in the 0.550 gram sample of lead (II) chromate could have been found, but the percent value will still be much less than 8%. The percentage of Pb^{2+} ions in the 1.000 L bottle can also be determined through the molarity:

$$0.550\ g\ PbCrO_4 \times \left(\frac{1\ mole\ PbCrO_4}{323.2\ g\ PbCrO_4}\right) \times \left(\frac{1\ mole\ Pb^{2+}}{1\ mole\ PbCrO_4}\right) = 1.70 \times 10^{-3}\ moles\ of\ Pb^{2+}$$

The concentration of Pb^{2+} is equal to $1.70 \times 10^{-3} \frac{moles}{1.000}$ L = 1.70×10^{-3} M. The concertation of water is typically 55.0 M in aqueous solutions:

$$Percent\ Pb^{2+} in\ water = \frac{1.70 \times 10^{-3}\ M}{1.70 \times 10^{-3}\ M + 55.5\ M} \times 100\% \sim 3.06 \times 10^{-3}\%$$

Another way to calculate the percentage is to use the number of water molecules instead of the molarity of water in the previous equation. Therefore, the number of water molecules is:

$$1000 \; g \; H_2O \times \left(\frac{1 \; \text{mole} \; H_2O}{18.016 \; \text{g} \; H_2O} \right) \times \left(\frac{6.02 \times 10^{+23} \; H_2O \; \text{molecules}}{1 \; \text{mole} \; H_2O} \right)$$

$$= 3.341 \times 10^{25} \; \text{molecules of} \; H_2O$$

$$percent \; Pb^{2+} in \; water = \frac{1.02 \times 10^{21}}{1.02 \times 10^{21} + 3.341 \times 10^{25}} \times 100\% \sim 3.05 \times 10^{-3}\%$$

The percent value determined through molarity and the number of molecules is approximately the same, with the percentage of lead (II) ion still much less than 8%.

10. C: The correct answer is Choice *C*. Crystalline solids do not become soft before they melt. Amorphous solids become soft and pliable before melting because some of the bonds between the nonidentical units break before others. In crystalline solids, the bonds all break at the same time, Choice *E*, because the bonds between the identical repeating units are all equal in strength. This also gives the crystalline solid an exact melting point, Choice *A*. Crystalline solids also display cleavage and break across a plane, Choice *B*. They also cannot be compressed, Choice *D*.

11. A: The correct answer is Choice *A*. The volume of the gas can be calculated using the ideal gas equation, PV = nRT. All of the required information for the equation is given in the question, except for the volume. Plugging in the known values, gives you (1.3 atm)V = (2.5 mol)(0.082 L atm mol^{-1} K^{-1})(300 K). Solving for V gives you 47.3 L of volume.

12. B: Scientist A needs to dilute her 60 M AlCl$_3$ solution 10-fold to get a 6 M AlCl$_3$ solution. The correct answer is Choice *B*, where 10 mL of 60 M AlCl$_3$ is diluted to a total volume of 100 mL by adding 90 mL water. In 10 mL of 60 M AlCl$_3$, there are 0.6 moles of AlCl$_3$. Putting 0.6 moles of AlCl$_3$ in 100 mL total volume gives the equivalent of 6 moles in 1 L of volume, which has a solution strength of 6 M. Choices *A* and *C* are not 10-fold dilutions, although the total volume is 100 mL. The total volume is not 100 mL and the dilution is not 10-fold for Choices *D* and *E*.

13. D: The correct answer is Choice *D*. London dispersion forces are the weakest type of intermolecular attraction. They form between induced dipoles that are temporary and break apart easily. Hydrogen bonds are the strongest type of weak intermolecular attraction. When hydrogen bonds to a highly electronegative atom, the interaction becomes exaggerated and creates a stronger bond than expected. Dipole-dipole interactions are stronger than London dispersion forces because the dipoles are permanently present in the molecular configuration, but are weaker than hydrogen bonds.

14. A: Capillary action is a combination of adhesion and cohesion and allows water to move up the stem of a plant, Choice *A*. Adhesion is the attraction of liquid to the container that it is in and cohesion is the attraction of liquid molecules to each other. These intermolecular forces combine to move water upward throughout the plant from the roots. Surface tension, Choice *B*, occurs when the molecules at the surface of a liquid are pulled down into the body of the liquid and a strong interface is created at the surface. Vapor pressure and boiling points, Choices *C* and *D*, involve the ability of a liquid to become a gas. Water does not dissolve the plant stem, so solubility is not involved in the travel of water up the stem, making Choice *E* incorrect.

15. E: Iodine has the greatest number of electrons at 53 electrons, Choice *E*. The number of electrons increases in elements going from left to right across the periodic table. Hydrogen, Choice *A*, is at the top left corner of the periodic table, so it has the fewest electrons (one electron). Iron has 26 electrons, copper has 29 electrons, and magnesium has 12 electrons, Choices *B*, *C*, and *D*, respectively.

16. C: Decreasing the value of the denominator of the equation, which is the distance between the nuclei, would increase the value of ΔH, the bond energy of the ionic bond, which is expressed in Choice *C*. Changing the charges of ions in opposite directions by the same magnitude would cancel out any changes in bond energy, so Choices *A* and *D* are incorrect. Changing an ion into a neutral atom would change the value of one of the Z variables to zero and would make the bond energy zero, meaning an ionic bond could not form, Choice *B*. Increasing the distance between the nuclei would increase the value of the denominator of the equation and would decrease the magnitude of the bond energy, so Choice *E* is incorrect.

17. E: The correct answer is Choice *E*; the molecule is methanol. There are three hydrogen atoms attached to the carbon atom, giving CH_3, and one hydrogen atom attached to an oxygen atom, giving OH, which makes the formula CH_3OH. Although there are free valence electrons from the oxygen, there is no charge on the molecule, so Choice *C* is incorrect. Choices *A*, *B*, and *D* do not have the correct number of carbon, hydrogen, or oxygen atoms as depicted in the Lewis diagram.

18. B: Interstitial alloys are formed by two metals that are different enough in size that one metal can fill in the spaces between the lattice formed by the larger metal. Carbon is much smaller than iron, so the correct answer is Choice *B*. Substitutional alloys are formed by two metals that are similar in size so that one metallic atom can replace the other within the lattice. The radii of copper, zinc, iron, and aluminum, Choices *A*, *C*, *D*, and *E*, are similar enough that they would form substitutional alloys rather than interstitial alloys.

19. D: Semiconductors are covalent network solids that are in between being good insulators and good conductors but are neither in entirety. Choice *D* is the correct answer. Covalent network solids are formed between nonmetals with strong covalent bonds. Metallic solids, Choice *A*, are good conductors because of the delocalized electrons that transfer electricity. The electrons in ionic solids, Choice *C*, are held tightly in place, so they cannot conduct electricity at all. Molecular solids, Choice *E*, are held together loosely by London dispersion forces with electrons shared in covalent bonds.

20. A: When fossil fuels react with oxygen gas, they are burning, which causes a combustion reaction. Combustion reactions create carbon dioxide and water vapor. Combination reactions, Choice *B*, take two smaller molecules and make one larger molecule. Decomposition reactions, Choice *C*, take a larger molecule and turn it into two or more smaller molecules. Choice *D*, vaporization, would indicate liquids becoming gases, and Choice *E*, solidification, would indicate liquids becoming solids, neither of which is occurring in this reaction.

21. C: The stoichiometry of the reaction tells you that for every two moles of AgI added to one mole of Na_2S, one mole of Ag_2S and two moles of NaI are produced. Therefore, if you start with six moles of AgI and three moles of Na_2S, six moles of NaI would be produced with a 100 percent yield. Because the yield is only 90 percent, the six moles should be multiplied by 0.9 and the result is 5.4 moles. With a 100 percent yield, three moles of Ag_2S would be produced, Choice *A*, and six moles of NaI, Choice *B*, would be produced. Choice *D*, 90, indicates the percent yield, not the number of moles. Choice *E*, 10, indicates the percent yield that is lost from the theoretical yield.

22. B: For a chemical equation to be balanced, the same quantity of each element must be present on each side of the equation. On the left side of the equation, there is one Fe atom, and there are six oxygen atoms and three carbon atoms. On the right side of the equation, there is one iron atom and the rest of the oxygen and carbon atoms must be accounted for by the carbon dioxide molecules. Having three carbon dioxide molecules would give three carbon atoms and six oxygen atoms, which is equivalent to the reactants' side of the equation.

23. E: Choice *E* is the correct answer because it is the only one that contains an incorrect statement about the properties of water. Water is amphoteric and can act as an acid or a base, as stated in Choice *B*. When it acts as an acid, it donates a proton to the molecule it is reacting with, Choice *A*, and then becomes hydroxide ions and acts as a conjugate base, Choice *C*. When it acts like a base, it accepts a proton, Choice *D*, and becomes hydronium ions, which act as a conjugate acid. When water acts as a base, it does not donate a proton, making Choice *E* the correct answer.

24. A: In redox reactions, exactly two elements must experience a change in oxidation number. Iodine starts with an oxidation number of +5 and ends with an oxidation number of 0. Carbon starts with an oxidation number of -2 and ends with an oxidation number of +4. Oxygen keeps an oxidation number of -2 throughout the reaction. Therefore, Choice *A* is the correct answer and Choices *B* and *C* are incorrect. Oxidation numbers are assigned to individual elements and not whole molecules, so Choices *D* and *E* are incorrect.

25. D: Chemical properties of a substance describe how they react with another substance, whereas physical properties describe the appearance of the substance by itself. Choice *D* is the correct answer because it describes how the substance reacts while burning and interacting with oxygen molecules. It is flammable because it does burn. The physical properties of color, density, fragility, and odor are described by Choices *A, B, C,* and *E*, respectively.

26. E: The reaction is noted to be endothermic, so heat must be absorbed into the system, and the enthalpy, ΔH, must have a positive value. Choice *A* and *D* are incorrect because they both describe decreased enthalpy for the system. Heat energy is added and absorbed in the system when larger molecules are broken and smaller molecules are formed, so Choice *B* is incorrect. In energy diagrams of endothermic reactions, the energy of the products is always higher than that of the reactants, so Choice *C* is incorrect. Choice *E* correctly describes an increase in heat energy being put into the system and a positive enthalpy for the system for the endothermic reaction.

27. C: The energy potential of the cell is calculated by the equation $E°_{cell} = E°_{red} + E°_{ox}$. The silver side of the galvanic cell is the cathode, so it is being reduced. The zinc side is the anode, so it is being oxidized. Because the zinc is being oxidized, its standard electrode reduction potential must be reversed. Filling in the equation, $E°_{cell} = 0.08 + (+0.76) = 1.56$ V, so Choice *C* is the correct answer. You would get Choice *A* if the oxidation potential was not reversed. Choice *B* is the potential of just the reduction reaction. Choice *D* is the potential of just the oxidation reaction. Choice *E* would be the answer if the reduction reaction was reversed instead of the oxidation reaction. The balanced redox reaction would read as follows:

$$2Ag^+(aq) + Zn(s) \rightarrow Zn^{2+}(aq) + 2Ag(s)$$

28. B: Gibb's free energy is calculated by the equation $\Delta G = -nFE°_{cell}$. The number of moles of electrons that are transferred per mole of reactant is represented by the "n" in the equation. Plugging in the given information yields the following:

$$230 \text{ kJ} = 230{,}000 \text{ J} = -n \times 95835 \frac{\text{Coulombs}}{\text{mole of electrons}} \times -1.2 \frac{\text{Joules}}{\text{Coulomb}}$$

$$230{,}000 \text{ J} = n \times 115002 \frac{\text{Joules}}{\text{mole of electrons}}$$

$$n = 2 \text{ moles of electrons}$$

So, n = 2 moles of electrons, Choice *B*. The number of moles of electrons does not always equal the number of moles of reactant in the equation, so Choice *A* is incorrect. It also does not equal the amount of free energy or the electric potential of the cell, so Choices *C* and *E* are incorrect. If the free energy has a positive value, the reaction is not favored and will not occur spontaneously. Because the free energy of this reaction is positive, it will not occur spontaneously and is not favored.

29. B: The initial rate method must be used to find the exponent, or superscript, for each reactant. Experiment 1 and 2 can be used to find the exponent for reactant *X* because the concentrations for *Y* do not change. First, write the general rate law in terms of the reactants:

$$Rate = k[X]^m[Y]^n$$

Now, write a rate expression for Experiments 1 and 2 using the values from the table.

$$Rate_1 = k[0.0250]^m[0.0506]^n$$

$$Rate_2 = k[0.0500]^m[0.0506]^n$$

Divide Rate 2 by Rate 1:

$$\frac{Rate_2}{Rate_1} = \frac{k[0.0500]^m[0.0506]^n}{k[0.0250]^m[0.0506]^n}$$

$$\frac{Rate_2}{Rate_1} = \frac{[0.0500]^m}{[0.0250]^m} = 2^m$$

Insert the rate values from the table provided in the question:

$$\frac{Rate_2}{Rate_1} = \left(\frac{0.0500}{0.0250}\right)^m = 2^m$$

$$\frac{0.224}{0.0560} = 2^m$$

$$4 = 2^m \; ; \; m = 2$$

Therefore, the order for the reactant X is 2. To find the exponent, or order, for Y, use Experiments 1 and 3 because the concentrations for X are constant, but vary for Y:

$$\frac{Rate_3}{Rate_1} = \frac{k[0.0250]^m[0.1012]^n}{k[0.0250]^m[0.0506]^n} = \left(\frac{0.1012}{0.0506}\right)^n = 2^n$$

$$\frac{Rate_3}{Rate_1} = \frac{0.112}{0.0560} = 2^n$$

$$2 = 2^n; \quad n = 1$$

The order for reactant Y is one. The final rate equation is:

$$Rate = k[X]^2[Y]^1 \quad \text{or} \quad Rate = k[X]^2[Y]$$

30. E: To find the age of the animal bone, the following equations must be used:

$$t_{\frac{1}{2}} = \frac{0.693}{k} \quad \text{and} \quad ln\frac{[A]_t}{[A]_0} = -kt$$

First, find the rate constant for carbon-14 which has a half-life of 5730 years:

$$k = \frac{0.693}{t_{\frac{1}{2}}} = \frac{0.693}{5730 \text{ years}} = 1.209 \times 10^{-4}/\text{years}$$

The initial concentration $[A]_0$ of carbon-14 corresponds the amount of this isotope found in present-day animals. Recall that carbon-14 is found in animals because they can ingest plants that consume carbon dioxide from the upper atmosphere. However, once the animal dies, the concentration of carbon-14 will gradually decrease over time and is represented by $[A]_t$. The latter is the concentration of carbon-14 at time t, or the age of the animal bone. Concentration in this problem is represented by the units dpm/(g carbon). $[A]_0$ is equal to 15.0 dpm/(g Carbon) and $[A]_t$ is 5.0 dpm/(g Carbon). Notice that $[A]_t$ should be less than the initial concentration. Solving for the age of the animal bone (t) using the calculated value of k gives:

$$t = -\frac{1}{k}ln\frac{[A]_t}{[A]_0} = -\frac{1}{1.209 \times 10^{-4}/\text{years}} ln\frac{5.0 \text{ dpm/(g carbon)}}{15.0 \text{ dpm/(g carbon)}}$$

$$t = 9,087 \text{ years or } 9.0 \times 10^3 \text{ years}$$

The age of the animal bone is approximately 9,000 years.

31. C: The following equation can be used to calculate the activation energy (E_a) because the rate constant and temperature are given:

$$ln\frac{k_2}{k_1} = \frac{E_a}{R}\left(\frac{1}{T_1} - \frac{1}{T_2}\right)$$

Any two temperatures and rate constants can be chosen because the slope ($m = -E_a/R$) will remain nearly the same.

$$\underbrace{\ln(k)}_{y} = \overbrace{\left(-\frac{E_a}{R}\right)}^{m}\overbrace{\left(\frac{1}{T}\right)}^{x} + \underbrace{\ln(A)}_{b}$$

The units for temperature are Kelvin because the constant $R = 8.314$ J/(mol K). The table below shows the temperature in Kelvin (K = °C + 273.15).

Rate Constant for the Decomposition of N_2O_5 at Different Temperatures		
	$k\ (s^{-1})$	Temperature (K)
Experiment 1	4.8×10^{-4}	318.15 K
Experiment 2	8.8×10^{-4}	323.15 K
Experiment 3	1.6×10^{-3}	328.15 K
Experiment 4	2.8×10^{-3}	333.15 K

The first equation, which contains two rate constants, shown above can be rearranged to:

$$E_a = \ln\frac{k_2}{k_1}\left(\frac{1}{T_1} - \frac{1}{T_2}\right)^{-1} R$$

Experiments 1 and 4 are chosen because they will give a broader estimate of E_a, although any two experiments chosen will provide roughly the same answer. If $k_1 = 4.8 \times 10^{-4}$, $T_1 = 318.15\ K$ and $k_2 = 2.8 \times 10^{-3}$, $T_2 = 333.15\ K$, then:

$$E_a = \ln\left(\frac{2.8 \times 10^{-3}}{4.8 \times 10^{-4}}\right) \times \left(\frac{1}{318.15\ \text{K}} - \frac{1}{333.15\ \text{K}}\right)^{-1} \times 8.314\ \text{J} \cdot \text{mol}^{-1}\text{K}^{-1}$$

$$E_a = 1.03 \times 10^5 \text{J/mol}$$

The table below lists estimates of E_a for values of the rate constant and temperature taken from different experiments.

Experiments 1 and 2	$E_a = 1.04 \times 10^5 \text{J/mol}$
Experiments 2 and 3	$E_a = 1.05 \times 10^5 \text{J/mol}$
Experiments 3 and 4	$E_a = 1.02 \times 10^5 \text{J/mol}$
Experiments 1 and 4	$E_a = 1.03 \times 10^5 \text{J/mol}$

Note that the average of E_a taken from the first three rows is $E_a = 1.03 \times 10^5 \text{J/mol}$. Choice C is the best answer choice.

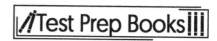

32. C: The potential energy for the decomposition of nitrous oxide is shown below:

$$N_2O\ (g) \rightarrow N_2(g) + O(g)$$

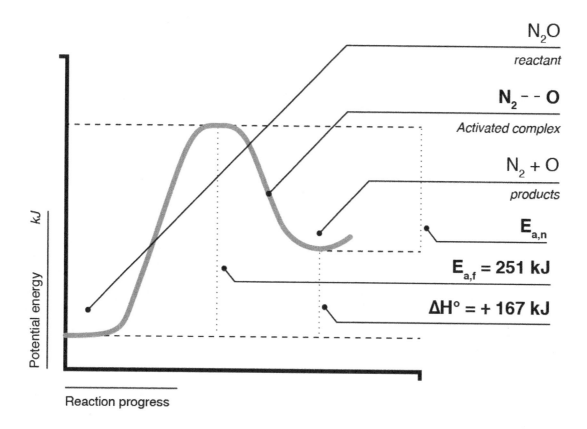

Reaction progress

The term $E_{a,f}$ is the forward activation energy and $E_{a,r}$ is the reverse activation energy. Typically, the units of energy are kilocalories (kcal) for small molecules. First convert $E_{a,f}$ from units of kcal to kJ:

$$60.0\ \text{kcal} \times \frac{4.184\ \text{kJ}}{1\ \text{kcal}} = 251\ \text{kJ}$$

To find $E_{a,r}$, subtract $\Delta H°$ from $E_{a,f}$:

$$E_{a,r} = E_{a,f} - \Delta H° = 251\ \text{kJ} - 167\ \text{kJ} = 84\ \text{kJ}$$

The reverse activation barrier should be less than the forward reaction because there is a positive bond enthalpy. The forward reaction is endothermic, while the reverse reaction is exothermic.

33. D: The rate law for the decomposition of nitryl chloride, NO_2Cl, will depend on the rate-determining step, which is the slowest step. The first elementary step of the reaction mechanism is the slowest:

$$NO_2Cl \overset{k_1}{\rightleftarrows} NO_2 + Cl \quad Slow, rate - Determining\ step$$

The rate will depend only on the disappearance of the reactant NO_2Cl and is expressed as:

$$Rate = k_1[\,NO_2Cl\,]$$

In the second step, NO_2Cl will only react once Cl is produced in the slow first step; however, this depends on the rate of decomposition of NO_2Cl because $k_1 < k_2$. So, the rate is proportional to the concentration of NO_2Cl. Therefore, the overall rate law is the rate law for the rate-determining step.

34. E: The reaction intermediate is a species that is present only briefly in the reaction mechanism but gets consumed in another elementary step. Writing the overall or net reaction equation is one way to confirm which species is a reaction intermediate. From the two given elementary steps the overall equation is:

$$NH_2NO_2(aq) + OH^-(aq) \leftrightarrow NHNO_2^-(aq) + H_2O\ (l)$$

$$\underline{NHNO_2^-(aq) \rightarrow N_2O(g) + OH^-(aq)}$$

$$NH_2NO_2(aq) + \cancel{OH^-(aq)} + \cancel{NHNO_2^-(aq)} \rightarrow \cancel{NHNO_2^-(aq)} + H_2O\ (l) + N_2O(g) + \cancel{OH^-(aq)}$$

$$NH_2NO_2(aq) \rightarrow H_2O\ (l) + N_2O(g) \quad \text{Net equation}$$

The overall equation indicates that either $NHNO_2^-$ or $OH^-(aq)$ is the reaction intermediate. However, based on the definition of a reaction intermediate, $NHNO_2^-(aq)$ is produced and then consumed, which makes it the intermediate structure. The hydroxide ion, $OH^-(aq)$, acts as a catalyst because it appears in the reactants for the first elementary reaction and then the products of the final elementary step. Even though it is consumed, it is also produced.

35. E: Adding 0.5 L of water might seem to have the effect of pushing the reaction to the right; however, water is not included in the rate law, so it would not increase the rate. In contrast, it would decrease the rate because the concentration of H_3O^+ is decreased, thereby decreasing the rate of reaction. Methanol, CH_3OH, is not included in the rate reaction so the rate will remain the same. Addition of hydroxide ion (⁻OH) through the addition of a base (e.g., NaOH) will decrease the reaction rate because the ion reacts with hydronium ion, H_3O^+.

Adding an acid produces a proton or hydronium ion:

$$HCl(aq) \rightarrow H^+(aq) + Cl^-(aq) \qquad \text{or} \qquad HCl(aq) \overset{H_2O}{\rightarrow} H_3O^+(aq) + Cl^-(aq)$$

These species, H^+ and H_3O^+, are the same and will react with a base:

$$H^+(aq) + OH^-(aq) \rightarrow H_2O(l) \qquad \text{or} \qquad H_3O^+(aq) + OH^-(aq) \rightarrow 2H_2O(l)$$

In either case, the rate of reaction will decrease because the concentration of H_3O^+ decreases when OH⁻ is added. The neutralization reaction produces more water and decreases the concentration of H_3O^+, which causes the solution to become more diluted. Acetic acid, CH_3COOH, is not a catalyst but a product that is formed in the reaction equation. Adding acetic acid will likely shift the equilibrium to the left if the reaction was reversible (e.g., ↔ specifies an equilibrium). The hydronium ion, H_3O^+, is a catalyst in the reaction and shows up in the rate law but not the general equation.

36. D: The potential energy graph plots the energy with respect to the distance between two atomic nuclei. Any point on the line refers to a specific bond distance. For example, an energy of -200 kJ corresponds to a bond distance of about 47 picometers. A stable chemical bond between two atoms will result in a lowering of the potential energy, which is often referred to as an energy minimum, where the bond is at equilibrium. Based on the given graph, that energy minimum occurs between 60–65 picometers. Therefore, the equilibrium or stable bond distance between the two atoms is about 62 picometers, which is the length, or distance, between each nucleus.

The energy corresponding to the equilibrium bond distance is about -600 kJ, which is the lowest energy value on the potential energy surface. If the bond distance between the two atoms becomes shorter than 62 picometers (e.g., 50 picometers), then that means that the atoms will be closer to one another and the energy will increase (~600 to 400 kJ) due to the repulsion between the nuclei (positively charged protons). A bond distance of 78 picometers corresponds to the transition state with an energy of -200 kJ. Typically, energy is added to the chemical system to break or stretch the bond. The bond dissociation energy (BDE or BE) is:

$$BDE = transition\ state\ bond\ energy - equilibrium\ bond\ energy$$

$$= -200\ kJ - (-600\ kJ) = 400\ k$$

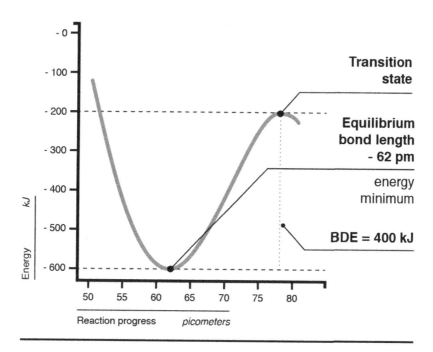

The bond dissociation energy (BDE or BE) is:

BDE = transition state bond energy - equilibrium bond energy

= - 200 kJ - (- 600 kJ)

= 400 kJ

Experimentally or realistically, a chemical bond will remain near its equilibrium bond distance because it's the most stable bond distance. For example, at room temperature, the chemical bond will vibrate between a distance of 59–65 picometers.

37. B: The forward reaction, $A + B \rightarrow C$, is an endothermic reaction because the reactants have lower energy compared to the products. Therefore, the reverse reaction, $C \rightarrow A + B$, must be an exothermic reaction because C has more energy than the products and will release energy to the surroundings when transitioning to $A + B$. This indicates that Choices A, D, and E are incorrect. This result gives two possible choices, either B or C. Although Choice C correctly indicates that the forward reaction is endothermic, the value of ΔH is not +410 kJ because the value corresponds to the activation energy barrier. The value of ΔH should reflect the energy difference between the reactants and the products. The value of ΔH will be +205 kJ in the forward direction, but -205 kJ in the reverse direction. Therefore, Choice B is correct. As the reaction progress from $C \rightarrow A + B$, the reaction will be exothermic, and C will lose -205 kJ to the surroundings. Although $A + B$ have less potential energy, these substances should have more kinetic energy compared to C.

38. B: Because the quarter is losing heat, the final answer should have a negative value. Rearranging the specific heat capacity equation and solving for q gives:

$$q = mc\Delta T$$

We know that the mass is m = 5.67 g and the specific heat capacity $c = 0.400$ J/(g° × C). The change in temperature is given by the difference in temperature between the quarter and the body temperature of the hand:

$$\Delta T = T_f - T_i = 37.0°C - 50.0°C = -12.0°C$$

Plugging these three values for the variables into the formula gives:

$$q = (5.67 \text{ g})(0.400 \text{ J} \cdot \text{g}^{-1} \cdot °\text{C}^{-1})(-12.0°C) = -27.2 \text{ J}$$

The quarter loses 27.2 J of heat, which is transferred to your hand. Note that if heat is lost from the quarter, is must be gained by the object that's holding it, which is consistent with the First Law of Thermodynamics:

$$-q_{quarter} = +q_{hand}$$

39. A: First, define the system as the coffee, and the ice or everything else as the surroundings. Then, the change in internal energy for the coffee is given by the following equation:

$$\Delta U_{coffee} = q + w = q + 0 = q$$

Because there is no volume change in the solid or water mixture, the work is zero (W = -PΔV), and ΔU will be equal to the amount of heat transferred. As the coffee hits the ice, it will lose its heat to the ice, which will begin to melt as it absorbs heat.

$$\Delta U_{coffee} = -600 \text{ J and } \Delta U_{ice} = +600 \text{ J}$$

$$-\Delta U_{coffee} = +\Delta U_{ice}$$

40. C: First, find the change in internal energy (ΔU_{gas}) for the system, and then reverse the sign of ΔU_{gas} to find the change in energy for the surroundings, ΔU_{surr}. Based on the First Law of Thermodynamics, the internal energy can be expressed as:

$$\Delta U_{gas} = q_{gas} + W_{gas} = -\Delta U_{surr}$$

The amount of heat that the gas or system loses is 900 J. Note that to find work, use the equation $W = -P\Delta V$.

$$W_{gas} = -P\Delta V = -1.20 \text{ atm}(V_f - V_i)$$

$$V_f = 0.50 \text{ L and } V_i = 1.50 \text{ L}$$

Also, because work in must be in units of Joules, the conversion $1 \ atm \cdot L = 101.3 \ J$ must be included. Because the gas is compressed, work is being done on the system by the surroundings. The system will have gained positive work energy:

$$W_{gas} = -P\Delta V = -1.20 \text{ atm}(0.50 \text{ L} - 1.50 \text{ L}) = -1.20 \text{ atm} \times -1.00 \text{ L} = 1.20 \text{ atm} \cdot \text{L}$$

$$W_{gas} = 1.20 \text{ atm} \cdot \text{L} \times \frac{101.3 \text{ J}}{1 \text{ atm} \cdot \text{L}} = +121.559 \text{ J}$$

The change in internal energy when the gas compresses is:

$$\Delta U_{gas} = -900 \text{ J} + (-P\Delta V) = -900 \text{ J} + 121.559 \text{ J} = -778.44 \text{ J}$$

For the surroundings, the magnitude of value is the same but opposite in sign to the system.

$$\Delta U_{gas} = -\Delta U_{surr}$$

$$\Delta U_{surr} = -\Delta U_{gas} = -(-778 \text{ J}) = +778 \text{ J}$$

41. B: The temperature of ice will increase from -15.0°C to 0.00°C as heat is added. Because the temperature is changing, the sensible heat is given by:

$$q_1 = mc\Delta T = (870.0 \text{ g})(2.03 \text{ J} \cdot \text{g}^{-1} \cdot °\text{C}^{-1})(0.00°\text{C} - (-15.0°\text{C})) = 26491.5 \text{ J}$$

At 0.00°C, water will coexist as a liquid and solid, and as a certain amount of heat is added, the temperature will remain constant. The latent heat is given by the enthalpy of fusion. However, this value must be multiplied by the mass to get units of energy.

$$q_{2,p} = m\Delta H_{fus} = (870.0 \text{ g})(0.334 \text{ kJ} \cdot \text{g}^{-1}) = 290.58 \text{ kJ}$$

Lastly, the amount of heat added to liquid water must be determined when the temperature increases from 0.00 to 15.0°C. The sensible heat is given by:

$$q_3 = mc\Delta T = (870.0 \text{ g})(4.184 \text{ J} \cdot \text{g}^{-1} \cdot °\text{C}^{-1})(15.0 °\text{C} - 0.00°\text{C}) = 54601.2 \text{ J}$$

The total amount of heat added is:

$$q_T = q_1 + q_2 + q_3 = 371,672.7 \text{ J or } 372 \text{ kJ (Correct to three sig figs)}$$

42. E: If the warm piece of moon rock is dropped into the heat bath, then heat moves from the rock to the heat bath:

$$-q_{rock} = +q_{water}$$

The amount of heat absorbed by water with $T_f = 25.0°C$ and $T_i = 22.0°C$ is given by:

$$q_{water} = mc_{water}\Delta T = (100.0\ g)(4.184\ J \cdot g^{-1} \cdot °C^{-1})(25.0\ °C - 22.0°C) = 1255.2\ J$$

Because the amount of heat that is lost by the moon rock to water is known then,

$$+q_{water} = -q_{rock} = -1255.2\ J$$

The specific heat of the moon rock can be found because the final and initial temperatures and amount of heat are given.

$$-q_{rock} = mc_{rock}\Delta T \quad or \quad c_{rock} = \frac{-q_{rock}}{m\Delta T} where\ \Delta T = T_f - T_i$$

The final and initial temperatures for the moon rock are $T_f = 25.0\ °C$ and $T_i = 60.0\ °C$.

$$c_{rock} = \frac{-q_{rock}}{m\Delta T} = \frac{-1255.2\ J}{(1.47\ g)(25.0°C - 60.0\ °C)} = 24.4\ J \cdot g^{-1} \cdot °C^{-1}$$

The mystery moon rock has an unusually high specific heat capacity, which would suggest that it's not of this world. However, it should be noted that no such moon rock with such a relatively high specific heat capacity exists.

43. A: The exothermicity or endothermicity of a reaction can be found by determining the enthalpy of reaction, $\Delta H°_{rxn}$, which is given by the following equation:

$$\Delta H°_{rxn} = \sum n\Delta H°_f\ (products) - \sum m\Delta H°_f\ (reactants)$$

Note that because all the reactants have a $\Delta H°_f$ of zero, this becomes:

$$\Delta H°_{rxn} = \sum n\Delta H°_f\ (products)$$

Because the metal oxide is the only product, then:

$$\Delta H°_{rxn} = n\Delta H°_f(metal\ oxide) = 2\Delta H°_f; \quad where\ n = 2\ from\ the\ equation.$$

The value of $\Delta H°_{rxn}$ for each metal oxide is simply two times $\Delta H°_f$.

$$MgO: \Delta H°_{rxn} = 2 \times (-601.2\ kJ/mol) = -1202.4\ kJ/mol$$

$$CaO: \Delta H°_{rxn} = 2 \times (-635.1\ kJ/mol) = -1270.2\ kJ/mol$$

$$SrO: \Delta H°_{rxn} = 2 \times (-592.0\ kJ/mol) = -1184.0\ kJ/mol$$

$$BaO: \Delta H°_{rxn} = 2 \times (-548.1\ kJ/mol) = -1096.2\ kJ/mol$$

The formation of MgO is the most exothermic because ΔH_{rxn}° is the greatest in magnitude and most negative. Although, computing ΔH_{rxn}° is not necessary in this case because the reactants have an enthalpy of formation of zero and because there is only one product. Therefore, ΔH_{rxn}° can be determined by observing the value of ΔH_f° for each metal oxide. Regarding the extraction of a metal (e.g., MgO to Mg) from its oxide, the most endothermic value corresponds to MgO.

44. C: Because other types of intermolecular forces such as hydrogen bonding and dipole-dipole forces are not included, the correct molecule is relatively nonpolar and will primarily exhibit London dispersion forces. Note that hydrogen bonding is a special type of a dipole-dipole intermolecular force. Choice *B* contains an -OH group, which will exhibit hydrogen bonding, so it's not a correct answer choice. Similarly, Choice *E* is not correct because it contains chlorine, Cl, bonded to carbon, which can form a dipole-dipole interaction with a neighboring hydrogen atom. The remaining choices are hydrocarbons that contain only carbon and hydrogen, where each molecule contains seven carbon atoms. The main difference between Choices *A*, *C*, and *D* is the degree of branching. Choice *A* has a five-carbon chain with two side methyl groups (CH_3). Choice *C* has a six-carbon chain with one methyl side group, and Choice *D* has a five-carbon chain with two methyl side groups. London dispersion forces will be most significant for unbranched chains, and lowest for branched chains dues to the difficulty of forming direct intermolecular or dispersion forces. Because Choice *C* is the most linear or least branched, it would have the greatest dispersion forces.

45. E: Recall that chemical changes involve changes to the molecular structure, whereas physical changes have to do with the appearance of the substance.

For Choice *A*, the condensation of water is represented by:

$$H_2O(g) \leftrightarrow H_2O(l)$$

This process is considered a physical change because there is no change in the identity of the substance. The process is reversible and is a common occurrence. The water vapor, or humidity, in the air tends to condense around cooler objects. If you are wearing glasses and walking from a cold building or car to the outside where it's warm, your glasses will quickly fog up due to condensation. For Choice *B*, the dissolution of table sugar $C_{12}H_{22}O_{11}$ in water will not result in any bond breaking in sugar, but a change in phase. If it did, then your sugary beverage would not taste so good. The process is reversible. If you were to add a tablespoon of sugar into a cup of water, and then boil out the water, you would obtain sugar.

$$C_{12}H_{22}O_{11}\,(s) \overset{H_2O}{\leftrightarrow} C_{12}H_{22}O_{11}\,(aq)$$

When a glass mason jar is broken, which is the same as a glass cup, there are chemical bonds that break; however, the nature of the substance (glass, or SiO_2) is still the same. In other words, the glass is still glass. Therefore, this process is characterized as a physical change. In Choice *D*, mixing sand in water is similar to the dissolution of sugar in water, and it will not change the chemical structure of sand. The process is reversible if you were to evaporate out the water, leaving only sand. In Choice *E*, baking is considered a chemical process for many reasons. For example, most people have smelled the aroma that is associated with baking brownies or a cake. There is also a color change in the batter (eggs, flour, butter, chocolate chips) as its heated. This process is not reversible, which is a key characteristic of a chemical change, because once we make the brownies, there is no way to get the eggs or butter back. New chemical bonds form as the mixture is heated, e.g., the flaky crust and slightly burnt edges around the brownie.

46. C: Entropy is a measure of the disorder, or randomness in a given chemical system and will increase if substances have more room to move around or orient in a given space. Solids will have the least amount of entropy because the molecules or atoms have chemical bonds that are fixed in a given space. Gaseous molecules, elements (e.g., Ar), or molecular elements are substances with the highest entropy because they are not fixed to any one point in space and are continuously moving in a disordered or random fashion. Reaction *i* is a decomposition reaction and has a positive entropy change, $+\Delta S°$, because two of the products are in a gas phase as opposed to the initial solid phase. Reaction *ii* is a synthesis reaction and has a negative entropy change, $-\Delta S°$, because of the two gaseous reactants that transition to a liquid substance. Reaction *iii* is a combustion reaction, where there is no noticeable change in entropy, so from a qualitative point of view, the value of $\Delta S°$ will remain about the same. Three gaseous reactants form three gaseous products. Thus, Choice *C* is the only correct choice.

47. D: The standard change in entropy ($\Delta S°$) and enthalpy of reaction ($\Delta H_{rxn}°$) can be determined by knowing the individual standard entropies ($S°$) and enthalpies of formations ($\Delta H_f°$) at 25.00°C for each reactant and product. However, it's not necessary to compute them because they are already given. The standard Gibbs free energy is given by:

$$\Delta G° = \Delta H° - T\Delta S°$$

The $\Delta H°$ and $\Delta S°$ are already given and are $\Delta H_{rxn}° = -802.3\ kJ/mol$ and $\Delta S° = -4.96\ J/mol \cdot K$, respectively. Note that the entropy change for the given reaction is negative, which is relatively close to zero. To find the temperature at which the reaction is not thermodynamically favorable, let $\Delta G = 0$ and solve for the temperature. Because this is no longer at standard temperature, the ° symbols are dropped.

$$\Delta G = \Delta H - T\Delta S$$

$$0 = \Delta H - T\Delta S$$

$$T\Delta S = \Delta H$$

$$T = \frac{\Delta H}{\Delta S} = \frac{-802.3\ \frac{kJ}{mol}}{-4.96\ J/(mol \cdot K)} = \frac{-802.3\ \frac{kJ}{mol}}{-4.96\ \frac{J}{mol \cdot K} \times \frac{1\ kJ}{1000\ J}} = 1.62 \times 10^5 K$$

The reaction is exothermic at room temperature, so it's expected that at some temperature above 25.00 °C or 298.15 *K*, the reaction will not be spontaneous. According to Le Chatelier's principle, high temperatures will push the reaction to the left. From the Gibbs free energy perspective, the $\Delta H°$ term is much larger than the $-T\Delta S°$ (recall Scenario 3 from our previous table reproduced below). So, the temperature must be relatively high for $-T\Delta S°$ to become dominant.

Scenario	Spontaneity		K	$\Delta G°$	$\Delta H°$	$-T\Delta S°$	$\Delta S°$
3	Spontaneous	(low T)	> 1	$-$	$-$	$+$	$-$
	Nonspontaneous	(high T)	< 1	$+$			

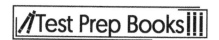

According to Scenario 3, when the temperature reaches 1.62×10^5 K or above, the reaction is no longer thermodynamically favorable (assuming that ΔH and ΔS don't change).

The standard Gibbs free energy $\Delta G°$ at 25.00°C is:

$$\Delta G° = \Delta H° - T\Delta S°$$

$$\Delta G° = \left(-802.3 \, \frac{kJ}{mol}\right) - (273.15 + 25.00 \text{ K}) \left(-4.96 \, \frac{J}{mol \cdot K}\right)\left(\frac{1 \text{ kJ}}{1000 \text{ J}}\right)$$

Note that because the units of the standard Gibbs free energy contain kJ, then the units of entropy must be converted to kJ by multiplying with the appropriate conversion factor (1kJ = 1000 J).

$$\Delta G° = \left(-802.3 \, \frac{kJ}{mol}\right) - (298.15 \, K) \left(-0.00496 \, \frac{kJ}{mol \cdot K}\right)$$

$$\Delta G° = \left(-802.3 \, \frac{kJ}{mol}\right) + 1.4788 \, \frac{kJ}{mol} = -800.821 = -800.8 \, \frac{kJ}{mol}$$

48. A: To make this reaction thermodynamically favorable, the reactions must be coupled to solve for $\Delta G°$, which is similar to how Hess's Law is used to solve for $\Delta H°_{rxn}$. However, in this scenario, the reaction equations are not reversed because of the idea to use a thermodynamically-favorable reaction to drive an unfavorable reaction. For the production of iron by the decomposition of iron (III) oxide, three moles of oxygen are produced. The oxygen species is also found in the second reaction where carbon monoxide reacts with oxygen. However, there is only one mole of oxygen, so the second reaction equation must be multiplied by a factor of three, which will triple the value of $\Delta G°$ from $-514.4 \, kJ$ to $1543 \, kJ$. The coupled, or combined, reactions are:

$$2Fe_2O_3(s) \rightarrow 4 \, Fe(s) + 3O_2(g); \quad \Delta G° = +1487 \text{ kJ}$$

$$\underline{3 \times (2CO(g) + O_2(g) \rightarrow 2CO_2(g); \quad \Delta G° = 3 \times (-514.4 \text{ kJ})}$$

$$2Fe_2O_3(s) + 6CO(g) + \cancel{3O_2(g)} \rightarrow 4 \, Fe(s) + \cancel{3O_2(g)} + 6CO_2(g)$$

$$2Fe_2O_3(s) + 6CO(g) \rightarrow 4 \, Fe(s) + 6CO_2(g)$$

$$\Delta G° = +1487 \text{ kJ} + (-1543 \text{ kJ}) = -56 \text{ kJ}$$

Note that oxygen acts as an intermediate and is not shown in the final reaction equation. In practice, the production and reduction of iron from iron ore is carried out in a blast furnace. So, using a Yeti® cooler is likely not the best reaction vessel.

49. A: To obtain the net chemical equation, some reaction stages may have to be reversed and multiplied. Because the overall reaction contains H_2O_2 on the products' side, then the first reaction stage 1) must be reversed to place H_2O_2 on the products' side. Consequently, the sign of ΔH_1 is reversed.

$$1.) \, 2H_2O(l) + O_2(g) \rightarrow 2H_2O_2(l); \, \Delta H_1 = +196.0 \text{ kJ}$$

The second reaction stage does not have to be reverses because hydrogen and oxygen gas are on the reactants' side as they are in the overall reaction equation. However, the second equation must be multiplied by a factor of two to cancel out the water intermediate. The value of ΔH_2 is multiplied by a factor of two:

$$2.)\ 2H_2(g) + \frac{2}{2}O_2(g) \rightarrow 2H_2O(l); \Delta H_2 = -2 \times 285.8\ kJ$$

The final first and second reaction stages are:

$$1.)\ 2H_2O(l) + O_2(g) \rightarrow 2H_2O_2(l); \Delta H_1 = +196.0\ kJ$$

$$2.)\ 2H_2(g) + \frac{2}{2}O_2(g) \rightarrow 2H_2O(l); \Delta H_2 = -2 \times 285.8\ kJ = -571.6\ kJ$$

$$Net\ reaction:\ 2H_2(g) + 2O_2(g) \rightarrow 2H_2O_2(l); \Delta H_1 + \Delta H_2 = \Delta H = -375.6\ kJ$$

The last equation shows the correct order of the reactants and products but differs by a factor of two compared to the original reaction. The chemical equation can be reduced by dividing each coefficient by two. Dividing by a factor of two must also be done to the enthalpy of reaction:

$$\frac{1}{2} \times \left(2H_2(g) + 2O_2(g) \rightarrow 2H_2O_2(l) \right); \Delta H = \frac{1}{2} \times -375.0\ kJ$$

$$Overall\ reaction:\ H_2(g) + O_2(g) \rightarrow H_2O_2(l); \Delta H = -187.8\ kJ$$

An alternative method for completing this problem requires reversing the order of Stage 1 and multiplying it by -½.

$$\frac{1}{2} \times \left(2H_2O(l) + O_2(g) \rightarrow 2H_2O_2(l) \right); \Delta H_1 = \frac{1}{2} \times (196.0\ kJ) = 98.00\ kJ$$

Reaction Stage 2 remains unchanged as shown below:

$$1.)\ H_2O(l) + \frac{1}{2}O_2(g) \rightarrow H_2O_2(l); \Delta H_1 = 98.00\ kJ$$

$$2.)\ H_2(g) + \frac{1}{2}O_2(g) \rightarrow H_2O(l); \Delta H_2 = -285.8\ kJ$$

$$Overall\ reaction:\ H_2(g) + O_2(g) \rightarrow H_2O_2(l); \Delta H_1 + \Delta H_2 = \Delta H = -187.8\ kJ$$

50. E: For the reaction to run in the forward direction and products to be formed, the reaction quotient, Q, should be less than the equilibrium constant, K. For this reaction, K = 15, so Q should be less than 15. The equation to calculate Q for this reaction is:

$$Q = \frac{[H_2S]^2}{[H_2]^2[S_2]}$$

Plugging in the given concentrations, Q = 9 / (36 x $[S_2]$). Plugging in 15 for Q to find the concentration of S_2 at equilibrium, you find that S_2 = 1.67 M at equilibrium. Higher concentrations of S_2 decrease the value of Q, which would make the reaction run in the forward direction. Lower concentrations increase the value of Q and favor the formation of reactants. Therefore, only Choice E is correct.

51. A: For the reaction aA + bB → cC + dD , the equilibrium constant, K, is calculated using the equation:

$$K = \frac{[C]^c[D]^d}{[A]^a[B]^b}$$

The initial concentrations do not factor into the calculation of K. Using the equilibrium concentrations and plugging them into the equation yields: K = ((2)(2)) / (3) = 1.33, which makes Choice *A* the correct answer. If the initial concentrations were plugged in instead of the equilibrium concentrations, the answer would come out to zero, which is Choice *B*; this is incorrect. If the reactant was incorrectly used as the numerator of the equation instead of the denominator, the answer would come out to 0.75, which is Choice *E*.

52. C: Particulate diagrams are useful visual representations of chemical reactions. They give a pictorial representation of the reactant and product molecules. They show the proportion of reactants needed to form the products according to the stoichiometry of the chemical equation, Choice *A*. They also show how the reactant elements dissociate and then re-bond to form the product molecules, Choices *B* and *D*. They also depict the ratio of reactants and products at equilibrium, Choice *E*. Because it is a pictorial representation, particulate diagrams do not indicate the value of the equilibrium constant for the reaction, Choice *C*, so this is the correct answer.

53. B: Le Chatelier's principle states that if equilibrium of a reaction is disturbed, the reaction will react in a manner to restore equilibrium of the system. This reaction would run in reverse if heat was taken away from the system. The gaseous products would become the solid and liquid reactants. There would be an increase in the concentration of C and H_2O, making Choice *B* the correct answer. Choices *A* and *C* indicate an increase in the product concentrations, which is incorrect. Choices *D* and *E* have a mix of reactant and product molecules, which would not occur since the reaction would have to run in either the forward or reverse direction, so both choices are incorrect.

54. D: When properties of a chemical reaction system change—such as changes in concentration, volume, pressure, and temperature—the reaction reacts to restore equilibrium. The only property that changes the value of the equilibrium constant, K, is a change in temperature, Choice *D*. Changes in concentration, volume, and pressure all change the value of the reaction quotient, Q, but do not change the value of K, making Choices *A*, *B*, *C*, and *E* incorrect.

55. C: Finding the base ionization constant for this reaction requires a few calculations. When the pK_a and pK_b of the reaction are added together, they always equal 14. The pK_b of the reaction is calculated as:

$$pK_b = -log\,Kb$$

To find the value of pK_b:

$$pK_a + pKb = 14$$

$$pK_b = 14 - (-6.11)$$

$$pK_b = 20.11$$

To find the value of K_b:

$$pK_b = 20.11 = -\log Kb$$

$$K_b = -(10^{20.11}) = 7.7 \times 10^{-21}$$

Choice *A* is the water dissociation constant, K_w. Choice *D* is equal to the K_a of the reaction, which is calculated by pK_a = -log K_a. Choice *E* is equal to the pK_b of the reaction.

56. A: At the halfway equivalence point of a chemical reaction, the pH of the solution is equal to the pK_a of the reaction. Therefore, if the pK_a of the reaction is 4, the pH would be 4 at the halfway equivalence point, making Choice *A* the correct answer. The Henderson-Hasselbalch equation is:

$$pH = pK_a + \log \frac{[A^-]}{[HA]}$$

Because A⁻ and Br⁻ here, and HA and HBr here have the same concentration, the log of their quotient is equal to zero, making the pH equal to the pK_a. The solution does not have a neutral pH at the halfway equivalence point, so Choice *B* is incorrect. Choices *D* and *E* are both basic pH values, so would not make sense for the acidic chemical reaction.

57. D: Conjugate acid-base pairs are formed from one molecule donating a proton and another accepting it or one molecule accepting a proton and another donating it. The base accepts a proton, whereas the acid donates the proton. In this reaction, the ammonia molecule, NH_3, accepts a proton and becomes NH_4^+. Therefore, the NH_3 is the base and the NH_4^+ is the conjugate acid, making Choice *D* correct. The H_2O acts as the acid because it donates its proton and becomes OH⁻, which becomes the conjugate base, Choice *A*. The other Choices, *B*, *C*, and *E*, do not form acid-base pairs.

58. B: Q, the reaction quotient, indicates the ratio of products and reactants at any given moment of the reaction. Here, because Q is greater than K, there are more products than reactants present, and the reaction would be driven in the reverse direction, so AgCl would be precipitated, making Choice *B* correct and Choice *D* incorrect. The reaction is not at equilibrium, Choice *A*, because Q and K are not equal. The concentration of Cl- at equilibrium can be calculated and is 1.33×10^{-5} M, so Choice *C* is incorrect. Q is not less than K, so Choice *E* is incorrect.

59. C: Chemical reactions run spontaneously when there is a negative change in free energy for the reaction, which means that $\Delta G°$ has a negative value, as indicated in Choice *C*. The reaction is exergonic, and the bonds that are formed between the products are stronger than the ones that are broken between the reactants. Choices *A*, *B*, *D*, and *E* all describe reactions in which the change in free energy, $\Delta G°$, would be positive, and the reaction would not be spontaneous. Endergonic reactions absorb energy from the environment, and have a K value of less than one.

60. D: The reaction is favored to run in the forward direction until equilibrium is reached. If an influx of product, in this case ammonia, is added to the system, the reaction will reverse to form more reactants to reach equilibrium concentrations again, so Choice *D* is correct. The reaction does not indicate that it is affected by heat, so Choice *A* is incorrect. Adding either hydrogen or nitrogen, Choices *B* and *C*, would both drive the reaction in the forward direction so that more product is formed. Adding equal proportions of reactants and products would not affect the direction of the reaction, so Choice *E* is incorrect.

61. D: Equal sharing of electrons is correct. In water, the electronegative oxygen sucks in the electrons of the two hydrogen atoms, making the oxygen slightly negatively-charged and the hydrogen atoms slightly positively-charged. This unequal sharing is called "polarity." This polarity is responsible for the slightly-positive hydrogen atoms from one molecule being attracted to a slightly-negative oxygen in a different molecule, creating a weak intermolecular force called a hydrogen bond, so A and B are true. Choice C is also true, because this unique hydrogen bonding creates intermolecular forces that literally hold molecules with low enough kinetic energy (at low temperatures) to be held apart at "arm's length" (really, the length of the hydrogen bonds). This makes ice less dense than liquid, which explains why ice floats, a very unique property of water. Choice D is the only statement that is false, so it is the correct answer.

62. A: The metric prefix for 10^{-3} is "milli." 10^{-3} is $1/10^3$ or $1/1000$ or 0.001. If this were grams, 10^{-3} would represent 1 milligram or $1/1000$ of a gram. For multiples of 10, the prefixes are as follows: deca = 10^1, hecta = 10^2, kilo = 10^3, mega = 10^6, giga = 10^9, tera = 10^{12}, peta = 10^{15}, and so on. For sub-units of 10, the prefixes are as follows: deci=10^{-1}, centi=10^{-2}, milli=10^{-3}, micro=1^{-6}, nano=10^{-9}, pico=10^{-12}, femto=10^{-15}, and so on. Metric units are usually abbreviated by using the first letter of the prefix (except for micro, which is the Greek letter mu, or μ). Abbreviations for multiples greater than 10^3 are capitalized.

63. D: Isotones are atoms of different elements that have the same number of neutrons. Isotones will have different mass numbers, as will isotopes, which are atoms of the same element that have a different number of neutrons. Isobars are atoms that have the same number of nucleons (protons and neutrons), and therefore, they must have the same mass number.

64. B: Choices A and D both suggest a different number of protons, which would make a different element. It would no longer be a sodium atom if the proton number or atomic number were different, so those are both incorrect. An atom that has a different number of electrons is called an ion, so Choice C is incorrect as well.

65. B: They form covalent bonds. If nonmetals form ionic bonds, they will fill their electron orbital (and become an anion) rather than lose electrons (and become a cation), due to their smaller atomic radius and higher electronegativity than metals. Choice A is, therefore, incorrect. There are some nonmetals that are diatomic (hydrogen, oxygen, nitrogen, and halogens), but that is not true for all of them; thus, Choice D is incorrect. Organic compounds are carbon-based due to carbon's ability to form four covalent bonds. In addition to carbon, organic compounds are also rich in hydrogen, phosphorus, nitrogen, oxygen, and sulfur, so Choice C is incorrect as well.

66. B: The basic unit of matter is the atom. Each element is identified by a letter symbol for that element and an atomic number, which indicates the number of protons in that element. Atoms are the building block of each element and are comprised of a nucleus that contains protons (positive charge) and neutrons (no charge). Orbiting around the nucleus at varying distances are negatively-charged electrons. An electrically-neutral atom contains equal numbers of protons and electrons. Atomic mass is the combined mass of protons and neutrons in the nucleus. Electrons have such negligible mass that they are not considered in the atomic mass. Although the nucleus is compact, the electrons orbit in energy levels at great relative distances to it, making an atom mostly empty space.

67. A: Nuclear reactions involve the nucleus, and chemical reactions involve electron behavior alone. If electrons are transferred between atoms, they form ionic bonds. If they are shared between atoms, they form covalent bonds. Unequal sharing within a covalent bond results in intermolecular attractions, including hydrogen bonding. Metallic bonding involves a "sea of electrons," where they float around non-specifically, resulting in metal ductility and malleability, due to their glue-like effect of sticking neighboring atoms together. Their metallic bonding also contributes to electrical conductivity and low specific heats, due to electrons' quick response to charge and heat, given to their mobility. Their floating also results in metals' property of luster as light reflects off the mobile electrons. Electron movement in any type of bond is enhanced by photon and heat energy investments, increasing their likelihood to jump energy levels. Valence electron status is the ultimate contributor to electron behavior as it determines their likelihood to be transferred or shared.

68. A: Electrons give atoms their negative charge. Electron behavior determines their bonding, and bonding can either be covalent (electrons are shared) or ionic (electrons are transferred). The charge of an atom is determined by the electrons in its orbitals. Electrons give atoms their chemical and electromagnetic properties. Unequal numbers of protons and electrons lend either a positive or negative charge to the atom. Ions are atoms with a charge, either positive or negative.

69. B: On the periodic table, the elements are grouped in columns according to the configuration of electrons in their outer orbitals. The groupings on the periodic table give a broad view of trends in chemical properties for the elements. The outer electron shell (or orbital) is most important in determining the chemical properties of the element. The electrons in this orbital determine charge and bonding compatibility. The number of electron shells increases by row from top to bottom. The periodic table is organized with elements that have similar chemical behavior in the columns (groups or families).

70. B: We can use the following equation to determine the volume of concentrated acid that is needed:

$$M_{initial}V_{initial} = M_{final}V_{final}$$

The final volume and molarity are 350 mL and 2.00 M HNO_3, which represents the final diluted concentration. The initial molarity represents the more concentrated acid, 5.00 M HNO_3. Solving for the initial volume gives:

$$V_{initial} = \frac{M_{final}V_{final}}{M_{initial}} = \frac{2.00 \text{ M} \times 350 \text{ mL}}{5.00 \text{ M}} = 140 \text{ mL of } 5.00 \text{ M } HNO_3$$

Note that units of mL can be used because the molarity units cancel. Choices C and D do not represent the correct volumes. To prepare 350 mL of a 2.00 M HNO_3 solution, add 140 mL of 5.00 M HNO_3 into a large graduated cylinder, and dilute with water up to the 350 mL marker or add 210 mL of water to make a 2.00 M solution of HNO_3.

71. B: The law of conservation of mass states that matter cannot be created or destroyed, but that it can change forms. This is important in balancing chemical equations on both sides of the arrow. Unbalanced equations will have an unequal number of atoms of each element on either side of the equation and violate the law.

72. D: Solids all increase solubility with Choices A, B, and C. Powdered hot chocolate is an example to consider. Heating (A) and stirring (B) make it dissolve faster. Regarding Choice C, powder is in chunks that collectively result in a very large surface area, as opposed to a chocolate bar that has a very small relative surface area. The small surface area to volume area ratio dramatically increases solubility. Decreasing the solvent (most of the time, water) will decrease solubility.

73. C: A solution is the term for a homogeneous mixture. A solution contains a solute (particle) dissolved in a solvent (usually a liquid, such as water). Solutions have the smallest solutes of the mixtures,

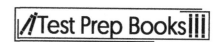

dissolving very easily, and the solute is spread out evenly, or homogeneous. A colloid has medium-sized particles that are somewhat evenly spread out, but their major difference from solutions is that they have the Tyndall effect. Because their particles are larger, they will reflect light that will appear as a beam (Tyndall). The sun's rays are an example of Tyndall; the light is reflecting off of the large gas particles in the atmosphere. A suspension has very large particles that actually settle, creating a heterogeneous mixture.

74. A: When salt is added to water, it increases its boiling point. This is an example of a colligative property, which is any property that changes the physical property of a substance. This particular colligative property of boiling point elevation occurs because the extra solute dissolved in water reduces the surface area of the water, impeding it from vaporizing. If heat is applied, though, it gives water particles enough kinetic energy to vaporize. This additional heat results in an increased boiling point. Other colligative properties of solutions include the following: their melting points decrease with the addition of solute, and their osmotic pressure increases (because it creates a concentration gradient that was otherwise not there).

75. C: Pressure has little effect on the solubility of a liquid solution because liquid is not easily compressible; therefore, increased pressure won't result in increased kinetic energy. Pressure increases solubility in gaseous solutions, since it causes them to move faster.

76. B: Nonpolar molecules have hydrophobic regions that do not dissolve in water. Oils are nonpolar molecules that repel water. Polar molecules combine readily with water, which is, itself, a polar solvent. Polar molecules are hydrophilic or "water-loving" because their polar regions have intermolecular bonding with water via hydrogen bonds. Some structures and molecules are both polar and nonpolar, like the phospholipid bilayer. The phospholipid bilayer has polar heads that are the external "water-loving portions" and hydrophobic tails that are immiscible in water. Polar solvents dissolve polar solutes, and nonpolar solvents dissolve nonpolar solutes. One way to remember these is "Like dissolves like."

77. A: A catalyst increases the rate of a chemical reaction by lowering the activation energy. Enzymes are biological protein catalysts that are utilized by organisms to facilitate anabolic and catabolic reactions. They speed up the rate of reaction by making the reaction easier (perhaps by orienting a molecule more favorably upon induced fit, for example). Catalysts are not used up by the reaction and can be used over and over again.

78. C: Increasing temperature increases the rate of reactions due to increases in the kinetic energy of atoms and molecules. This increased movement results in more collisions between reactants (with each other as well as with enzymes), which is the cause of the rate increase.

79. C: The lower subscript to the left of the chemical symbol indicates the atomic number (Z), which is equal to the number of protons. There are 88 protons in radium and also 88 electrons because the number of electrons must equal to the number of protons in a neutral atom. The upper left superscript indicates the mass number (A), which is the sum of the protons and neutrons. To find the number of neutrons use the following equation:

$$\text{neutron number } (N) + \text{atomic number } (Z) = \text{mass number } (A)$$

$$N = A - Z$$

$$N = 226 - 88 = 138$$

80. A: Ionic bonding is the result of electrons that have been transferred between atoms. When an atom loses one or more electrons, a cation, or positively-charged ion, is formed. An anion, or negatively-charged ion, is formed when an atom gains one or more electrons. Ionic bonds are formed from the attraction between a positively-charged cation and a negatively-charged anion. The bond between sodium and chlorine in table salt, or sodium chloride (NaCl), is an example of an ionic bond.

81. A: Oxidation is when a substance loses electrons in a chemical reaction, and reduction is when a substance gains electrons. Any element by itself has a charge of 0, as iron and oxygen do on the reactant's side. In the ionic compound formed, iron has a +3 charge, and oxygen has a -2 charge. Because iron had a zero charge that then changed to +3, it means that it lost three electrons and was oxidized. Oxygen, which gained two electrons, was reduced.

82. C: Fission occurs when heavy nuclei are split and is currently the energy source that fuels power plants. Fusion, on the other hand, is the combining of small nuclei and produces far more energy, and it is the nuclear reaction that powers stars like the sun. Harnessing the extreme energy released by fusion has proven impossible so far, which is unfortunate since its waste products are not radioactive, while waste produced by fission typically is.

83. A: Alpha decay involves a helium particle emission (with two neutrons). Beta decay involves emission of an electron or positron, and gamma is just high-energy light emissions.

84. B: Choice *A* is false because it is possible to have a very strong acid with a pH between 0 and 1. Choice *C* is false because the pH scale is from 0 to 14, and while -1 is outside the usual scale, it would indicate a very strong acid (10M hydronium ions in solution), not a weak one. Choice *D* is false because a solution with a pH of 2 has ten times fewer hydronium ions than a solution with pH of 1.

85. C: Gamma is the lightest radioactive decay with the most energy, and this high energy is toxic to cells. Due to its weightlessness, gamma rays are extremely penetrating. Alpha particles are heavy and can be easily shielded by skin. Beta particles are electrons and can penetrate more than an alpha particle because they are lighter. Beta particles can be shielded by plastic.

86. C: 2:9:10:4. These are the coefficients that follow the law of conservation of matter. The coefficient times the subscript of each element should be the same on both sides of the equation.

87. D: The reaction quotient will need to be determined and compared to the equilibrium constant to determine if a precipitate will form. First, write a balanced equation and calculate the amount of each reactant from their concentrations and initial volumes:

$$2\,AgNO_3(aq) + Na_2SO_4(aq) \rightarrow Ag_2SO_4(s) + 2\,NaNO_3(aq)$$

$$mol\ AgNO_3 = 0.100\ L \times 0.05\ \frac{mol\ AgNO_3}{L} = 5.0 \times 10^{-3}\ mol\ AgNO_3 = 5.0 \times 10^{-3}\ mol\ Ag^+$$

$$mol\ Na_2SO_4 = 0.0500\ L \times 0.05\ \frac{mol\ Na_2SO_4}{L} = 2.5 \times 10^{-3}\ mol\ Na_2SO_4 = 2.5 \times 10^{-3}\ mol\ SO_4^{2-}$$

The total volume after adding sodium sulfate is 150 mL or 0.150 L. The molarity of each ion is:

$$[Ag^+] = \frac{5.0 \times 10^{-3} \text{ mol AgNO}_3}{0.150 \text{ L}} = 0.0333 \text{ M}$$

$$[SO_4^{2-}] = \frac{2.5 \times 10^{-3} \text{ mol Na}_2SO_4}{0.150 \text{ L}} = 0.0166 \text{ M}$$

Now write out the dissociation of silver sulfate and set up an ICE table:

$$Ag_2SO_4(s) \rightleftharpoons 2\,Ag^+(aq) + SO_4^{2-}(aq)$$

	$[Ag^+]$	$[SO_4^{2-}]$
Initial	0	0
Change	$+2S$	$+S$
Equilibrium	$2S$	S

Expressed in terms of molar solubility, the reaction quotient for the dissociation of silver sulfate is:

$$Q = [Ag^+]^2[SO_4^{2-}] = (2S)^2(S) = 4S^3$$

Now calculate Q:

$$Q = [Ag^+]^2[SO_4^{2-}] = (0.0333)^2(0.0166) = 1.85 \times 10^{-5}$$

Because $Q > K_{sp}$, this indicates that there will be a precipitate. The concentration of silver and the sulfate ion are both in excess. This can be proven by comparing the solution to the ion concentrations at equilibrium. Since $K_{sp} = 6.9 \times 10^{-15}$ for silver sulfate, solving for its molar solubility at equilibrium would give:

$$[SO_4^{2-}] = S = 1.2 \times 10^{-5} \text{M and } [Ag^+] = 2S = 2.4 \times 10^{-5} \text{ M}$$

For the molar solubility in the given solution, use the calculated reaction quotient $Q = 1.9 \times 10^{-5}$,

$$[SO_4^{2-}] = S = 1.7 \times 10^{-2} \text{ M and } [Ag^+] = 2S = 3.3 \times 10^{-2} \text{ M}$$

These concentrations are much greater, so the ions will be present in excess and form a precipitate of silver sulfate.

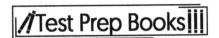

88. B: The nitrogen atom has the orbital and electron configuration shown below.

N: $1s^2 2s^2 2p^3$

The atomic number of nitrogen is seven, which means there are seven electrons for a neutral nitrogen atom. Nitrogen is found in the p-block (group 3A to 8A elements with p orbitals) within Group 5A in the periodic table. There are five valence electrons (group 5A), and three of those electrons are found in the p subshell. Counting from left to right (3A to 5A) in the p block indicates there will be three electrons dispersed in the p subshell. Note that subshell and orbital don't necessarily have the same meaning for the p subshell. There are three p orbitals in one p subshell, which contains three electrons. This p subshell belongs to the L shell and is denoted 2p. There is another subshell in the L shell called the 2s subshell, which contains only one orbital and holds a maximum of two electrons. Therefore, there are five electrons in the L shell for the nitrogen atom, which is shown by the orbital diagrams. The sum of the superscripts given by the electron configuration also gives the total number of electrons in nitrogen.

89. D: Water's unique properties are due to intermolecular hydrogen bonding. These forces make water molecules "stick" to one another, which explains why water has unusually high cohesion and adhesion (sticking to each other and sticking to other surfaces). Cohesion can be seen in beads of dew. Adhesion can be seen when water sticks to the sides of a graduated cylinder to form a meniscus. The stickiness to neighboring molecules also increases surface tension, providing a very thin film that light things cannot penetrate, which is observed when leaves float in swimming pools. Water has a low freezing point, not a high freezing point, due to the fact that molecules have to have a very low kinetic energy to arrange themselves in the lattice-like structure found in ice, its solid form.

90. A: Salts are formed from compounds that use ionic bonds. Disulfide bridges are special bonds in protein synthesis that hold the protein in their secondary and tertiary structures. Covalent bonds are strong bonds formed through the sharing of electrons between atoms and are typically found in organic molecules like carbohydrates and lipids. London dispersion forces are fleeting, momentary bonds that occur between atoms that have instantaneous dipoles but quickly disintegrate.

91. C: As in the last question, covalent bonds are special because they share electrons between multiple atoms. Most covalent bonds are formed between the elements H, F, N, O, S, and C, while hydrogen bonds are formed nearly exclusively between H and either O, N, or F of other molecules. Covalent bonds may inadvertently form dipoles, but this does not necessarily happen. With similarly electronegative atoms like carbon and hydrogen, dipoles do not form, for instance. Crystal solids are typically formed by substances with ionic bonds like the salts sodium iodide and potassium chloride.

92. A: Gases like air will move and expand to fill their container, so they are considered to have an indefinite shape and indefinite volume. Liquids like water will move and flow freely, so their shapes change constantly, but do not change volume or density on their own. Solids change neither shape nor volume without external forces acting on them, so they have definite shapes and volumes.

93. A: The product of the addition reaction is ethyl bromide. Choice *B* is incorrect because no atom or group from ethene is substituted by the attacking atoms (i.e. hydrogen or bromide). Choice *C* is incorrect because no atom or group is eliminated from ethene in the reaction. Choice *D* is incorrect because polymerization involves bonding of similar molecules to form a larger molecule.

94. C: The correct answer is 2-methyl-1-butanol, which has a chiral C atom with 4 different groups/atoms. See the structures below:

a.	b.	c.	d.

$$\text{a.} \quad CH_3-CH_2-\underset{\underset{CH_3}{|}}{\overset{\overset{OH}{|}}{C}}-CH_3$$

$$\text{b.} \quad CH_3-CH_2-\overset{\overset{OH}{|}}{CH}-CH_3$$

$$\text{c.} \quad CH_3-CH_2-\underset{\underset{CH_3}{|}}{\overset{\overset{H}{|}}{C}}-CH_2OH$$

$$\text{d.} \quad CH_3-\overset{\overset{CH_3}{|}}{C}=CH-CH_2OH$$

95. D: All the compounds in the list, including formaldehyde (H—CHO), trimethyl acetaldehyde [$(CH_3)_3C$—CHO], and benzaldehyde (C_6H_5—CHO), lack an α-hydrogen, and therefore can undergo a Cannizzaro reaction.

96. C: Clemmensen reduction converts carbonyl group (C=O) into -CH_2 groups, causing the formation of a saturated hydrocarbon. In Choices *A*, *B*, and *D*, carbonyl compounds are converted to hydrocarbons, and thus followed a Clemmensen reduction. According to Clemmensen reaction, acetic acid is reduced to form ethane, but not ethanol.

97. D: Choices *A* and *B* are positive electrophiles, while Choice *C* is a neutral electrophile. Choice *D* is a neutral nucleophile.

98. C: The number of enantiomers of an organic compound is 2^n, where "n" refers to the number of chiral C atoms in the molecule. Glucose has four chiral C atoms, and therefore the number of possible optical isomers would be $2^4 = 16$.

99. C: See the numbering of C-chain below and review IUPAC numbering section. The longest C-chain should be selected for numbering and naming as follows:

$$\overset{7}{CH_3}-\overset{6}{CH_2}-\overset{5}{CH_2}-\overset{4}{\underset{\underset{CH_3}{|}}{C}}\overset{\overset{CH_3}{|}}{}-\overset{3}{CH_2}-\overset{2}{\overset{\overset{CH_3}{|}}{CH}}-\overset{1}{CH_3}$$

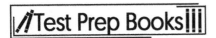

100. A: This is an addition reaction:

$$CH_3 - CH = CH_2 \quad + \quad Br_2 \quad \longrightarrow \quad CH_3 - CHBr - CH_2Br$$

Propene   1, 2 - dibromopropane

101. B: Aspirin is an acetylated *ester* of salicylic acid. Therefore, the conversion of aspirin to salicylic acid is *best* described as an ester hydrolysis. See the reaction below.

The conversion of aspirin to salicylic acid via ester hydrolysis

$OCOCH_3$ OH

COOH   COOH

        Ester hydrolysis

$+ H_2O \quad \longrightarrow \quad + CH_3COOH$

Acetylsalicylic acid   Salicylic acid   Acetic acid

Asprin

102. A: For further review, see the aldol condensation section.

103. D: Sucrose, lactose, and maltose are disaccharides. Mannose is a monosaccharide.

104. B: Electrophilic addition refers to an addition reaction in which an electrophile (electron attracting species) is added to a molecule, mostly carrying unsaturated bonds.

105. C: This molecule has an asymmetric C atom attached with 4 different atoms/groups. Therefore, it would constitute isomers that are optically active.

106. D: Ethene, diethyl ether, and acetaldehyde could be converted to ethanol by appropriate chemical reactions that follow:

Forming Ethanol From Various Compounds

i.   From ethanol by hydration

$$CH_2=CH_2 + H_2SO_4 \longrightarrow CH_3-CH_2 \xrightarrow[\text{Heat}]{H_2O} CH_3-CH_2-OH$$

Ethene   Ethyl hydrogen sulphate   Ethanol

ii.   From diethyl ether by hydrolysis

$$C_2H_5-O-C_2H_5 + H_2O \longrightarrow 2\,CH_3-CH_2-OH$$

Diethyl ether   Ethanol

iii.   From acetaldehyde by reduction

$$CH_3-\overset{\overset{\displaystyle O}{\|}}{C}-H + 2[H] \xrightarrow{LiAlH_4} CH_3-CH_2-OH$$

Acetaldehyde   Ethanol

107. C: The general formula for an alkane, alkene, and alkyne are C_nH_{2n+2}, C_nH_{2n}, and C_nH_{2n-2}, respectively.

108. B: Nitrogen trifluoride is similar to ammonia (NH_3) and will have a molecular geometry that is trigonal pyramidal about the nitrogen atom. However, the electron-group geometry will be tetrahedral because there is a lone pair electron around the nitrogen atom, which will occupy space like a chemical bond. Three bonds and a lone pair will minimize repulsion by pointing or aligning to the corners of a tetrahedron.

109. C: A Brønsted-Lowry base is any substance capable of accepting a proton, or H^+. Choices *A* and *D* are incorrect because they both have dissociable hydrogens, meaning they would likely donate a hydrogen in an aqueous solution. Choice *B*, carbon dioxide (CO_2), would not be considered a base on its own because it is nonpolar and does not normally accept a proton in the presence of water. Choice *C*, ammonia (NH3) is the only choice that acts as a base in aqueous solution, since it is more likely to accept a proton from water, forming ammonium (NH_4^+) and hydroxide ions.

110. A: The neutrons and protons make up the nucleus of the atom. The nucleus is positively charged due to the presence of the protons. The negatively charged electrons are attracted to the positively charged nucleus by the electrostatic or Coulomb force; however, the electrons are not contained in the nucleus. The positively charged protons create the positive charge in the nucleus, and the neutrons are electrically neutral, so they have no effect. Radioactivity does not directly have a bearing on the charge of the nucleus.

111. B: The conversion from Celsius to Fahrenheit is $°F = \frac{9}{5}(°C) + 32$. Substituting the value for °C gives $°F = \frac{9}{5}(35) + 32$, which yields 95 °F. The other choices do not apply the formula correctly and completely.

112. C: When dissolved into pure water, sodium hydroxide (NaOH) dissociates into Na^+ and OH^- ions. However, in a solution containing potassium hydroxide, there will be some hydroxide ions already present. Since there are already OH^- ions, the common ion effect takes place, reducing the effective solubility of NaOH, so Choice B is incorrect. Choice A is incorrect because if there's too much NaOH, it might not be able to completely dissociate. Choice D is incorrect because some amount of NaOH will still dissolve, but potentially less than in pure water. Therefore, C is the correct choice.

113. D: The correct set of quantum numbers is determined according to the number of electrons for a given atom. The possible values of l and m_l are given by the formulas $l = n - 1$ and $m_l = 2l + 1$. For example, in Choice A, the possible quantum numbers are:

$$n = 2$$

$$l = n - 1 = 2 - 1 = 1$$

$$m_l = 2l + 1 = 2(1) + 1 = 3 \quad \{-1, 0, +1\}$$

Choices A and B contain an allowed set of quantum numbers. Choice C also shows a unique set of quantum number belonging to $n = 1$. However, for Choice D, $m_l = -2$ is not possible because $l = 1$.

114. C: The question is asking to find the number of oxygen atoms (not molecular oxygen, O_2). Use the following unit conversion roadmap to determine the number of oxygen atoms.

$$\text{liters of } CO_2 \rightarrow \text{moles of } CO_2 \rightarrow \text{moles of O} \rightarrow \text{\# of O atoms}$$

$$0.500 \text{ L } CO_2 \times \frac{1 \text{ mol } CO_2}{22.4 \text{ L } CO_2} \times \frac{2 \text{ mol O atoms}}{1 \text{ mol } CO_2} \times \frac{6.022 \times 10^{23} \text{ O atoms}}{1 \text{ mol O atoms}} = 2.69 \times 10^{22} \text{ O atoms}$$

Note that the molar volume is used to convert liters of gas to moles. If the substance were a solid (e.g., dry ice or solid carbon dioxide), then one would use its molar mass to convert to moles. Then, using Avogadro's number, the moles of a substance can be converted to the number of gas molecules. However, there are two moles of O atoms per one mole of carbon dioxide, so this stoichiometric relationship must be accounted for in the equation above. Therefore, Choice A is incorrect. Choices B and D are not correct because those answers refer to molecular oxygen.

115. A: To find the mass of caffeine, use the following roadmap conversion:

$$\text{molecule of caffeine} \xrightarrow{\text{use } N_A} \text{moles of caffeine} \xrightarrow{\text{use molar mass}} \text{mass of caffeine}$$

The molar mass of caffeine is approximately 194.2 g/mol.

$$1 \text{ molecule caffeine } \times \frac{1 \text{ mol}}{6.022 \times 10^{23} \text{ molecules}} \times \frac{194.2 \text{ g caffeine}}{1 \text{ mol caffeine}} \times \frac{1 \text{ kg}}{1000 \text{ g}} = 3 \times 10^{-25} \text{ kg caffeine}$$

Choice *C* would have been a correct answer if the units were in grams. If the question asked for grams of one caffeine molecule, then formula above would have neglected the last conversion formula giving a value of 3×10^{-22} g.

116. A: The compound containing nitrogen and oxygen has an empirical formula of N_xO_y where x and y represent the whole number atom rations. If there are 1.51 g of oxygen, then the mass of nitrogen is:

$$\text{mass of nitrogen} = \text{mass of compound} - \text{mass of oxygen} = 2.04 \text{ g} - 1.51 \text{ g} = 0.53 \text{ g N}$$

The mass of each substance can now be converted to moles.

$$\text{Mass of nitrogen: } 0.53 \text{ g O} \times \frac{1 \text{ mol N}}{14.01 \text{ g N}} = 3.78 \times 10^{-2} \text{ mol N}$$

$$\text{Mass of oxygen: } 1.51 \text{ g O} \times \frac{1 \text{ mol O}}{16.00 \text{ g O}} = 9.44 \times 10^{-2} \text{ mol O}$$

To obtain x and y, divide by the smallest molar value between O and N, which is 3.78×10^{-2}.

$$x = \frac{3.78 \times 10^{-2} \text{ mol N}}{3.78 \times 10^{-2} \text{ mol}} = 1 \text{ for N}$$

$$y = \frac{9.44 \times 10^{-2} \text{ mol O}}{3.78 \times 10^{-2} \text{ mol}} = 2.5 \text{ for O}$$

The y term is not a whole number, so each term must be multiplied by two giving $x = 2$ and $y = 5$. Therefore, the empirical formula is N_2O_5.

117. C: The formula mass of the compound can be determined through the following computation:

$$7.61 \times 10^{20} \text{ molecules } (SO_2)_n \times \frac{1 \text{ mol}}{6.022 \times 10^{23} \text{ molecules}} = 1.26 \times 10^{-3} \text{ mol } (SO_2)_n$$

$$\text{molar mass} = \frac{0.162 \text{ g } (SO_2)_n}{1.26 \times 10^{-3} \text{ mol } (SO_2)_n} = 129 \text{ g/mol}$$

The value of n can then be calculated dividing the molar mass by the molar mass of the empirical formula.

$$n = \frac{\text{formula or molar mass (g mol}^{-1})}{\text{empirical formula mass (g mol}^{-1})}$$

$$n = \frac{129 \text{ g mol}^{-1}}{64.07 \text{ g mol}^{-1}} \approx 2$$

Therefore, using the value of n, the molecular formula can also be found:

$$(SO_2)_n = (SO_2)_2 = S_2O_4$$

118. D: To find the mass of oxygen, covert the molar amount of sucrose to the mass in grams, then multiply by the percentage of oxygen.

$$1.50 \text{ mol } C_{12}H_{22}O_{11} \times \frac{342.296 \text{ g } C_{12}H_{22}O_{11}}{1 \text{ mol } C_{12}H_{22}O_{11}} = 513 \text{ g } C_{12}H_{22}O_{11}$$

$$\text{mass } \% \text{ O} = \frac{176.0 \text{ g O}}{342.296 \text{ g } C_{12}H_{22}O_{11}} \times 100\% = 51.42\%$$

Now multiply the mass of the sample and the fraction of oxygen in the sample.

$$\text{mass of oxygen in sample} = \frac{51.42\%}{100.0\%} \times \text{mass of sucrose} = 0.5142 \times 513 \text{ g} = 264 \text{ g O}$$

119. B: The maximum amount of carbon dioxide created from the reaction, according to its stoichiometry, is the theoretical yield. To find the theoretical yield, determine which reactant produces the least amount of carbon dioxide using the stoichiometry from the reaction.

$$3 \text{ mol CO} \times \frac{2 \text{ mol } CO_2}{2 \text{ mol CO}} = 3 \text{ mol } CO_2$$

$$1 \text{ mol } O_2 \times \frac{2 \text{ mol } CO_2}{1 \text{ mol } O_2} = 2 \text{ mol } CO_2$$

Oxygen is the limiting reactant because it produces fewer moles of carbon dioxide. The theoretical yield is only two moles of carbon dioxide. Carbon monoxide is the excess reactant because it is not entirely consumed and would produce one mole of carbon dioxide if the gas reacted to completion.

120. A: First determine the limiting reactant and theoretical yield of carbon monoxide.

$$2 \text{ mol } CO_2 \times \frac{2 \text{ mol CO}}{1 \text{ mol } CO_2} = 4 \text{ mol CO}$$

$$3 \text{ mol C} \times \frac{2 \text{ mol CO}}{1 \text{ mol C}} = 6 \text{ mol CO}$$

The limiting reactant is carbon dioxide, which means that the theoretical yield is four moles of carbon monoxide. Carbon (charcoal) is the excess reactant and produces two extra moles of carbon monoxide.

121. D: To find the correct theoretical yield of barium sulfate, determine the limiting reactant. The molar masses of aluminum sulfate and barium chloride are 342.14 g/mol and 208.23 g/mol.

$$3.52 \text{ g } Al_2(SO_4)_3 \times \frac{1 \text{ mol } Al_2(SO_4)_3}{342.17 \text{ g } Al_2(SO_4)_3} \times \frac{3 \text{ mol } BaSO_4}{1 \text{ mol } Al_2(SO_4)_3} = 3.09 \times 10^{-2} \text{ mol } BaSO_4$$

$$4.06 \text{ g } BaCl_2 \times \frac{1 \text{ mol } BaCl_2}{208.23 \text{ g } BaCl_2} \times \frac{3 \text{ mol } BaSO_4}{3 \text{ mol } BaCl_2} = 1.95 \times 10^{-2} \text{ mol } BaSO_4$$

It is not necessary to convert moles of barium sulfate to mass to find the limiting reactant. Barium chloride is the limiting reactant because it produces fewer moles of barium sulfate. To find the theoretical yield of barium sulfate in grams, convert the smaller molar amount of barium sulfate to grams.

$$1.95 \times 10^{-2} \text{ mol BaSO}_4 \times \frac{233.40 \text{ g BaSO}_4}{1 \text{ mol BaSO}_4} = 4.55 \text{ g BaSO}_4$$

On an important note, it's useful to perform the calculation in one operation to avoid any potential rounding error: For example, the previous calculations may be performed in one operation:

$$4.06 \text{ g BaCl}_2 \times \frac{1 \text{ mol BaCl}_2}{208.23 \text{ g BaCl}_2} \times \frac{3 \text{ mol BaSO}_4}{3 \text{ mol BaCl}_2} \times \frac{233.40 \text{ g BaSO}_4}{1 \text{ mol BaSO}_4} = 4.55 \text{ g BaSO}_4$$

In other words, if possible, rounding should be done at the end of a calculation. If the theoretical yield of barium sulfate in moles was asked for, then 1.95×10^{-2} mol $BaSO_4$ is acceptable.

122. C: The following equation gives the percent yield:

$$\text{percent yield} = \frac{\text{actual yield (experiment)}}{\text{theoretical yield}} \times 100\% = \frac{3.96 \text{ g}}{4.55 \text{ g}} \times 100\% = 87.0\%$$

Note that the percent yield is given as a percentage and that the theoretical yield, found by using stoichiometry, is always greater than the actual yield. Experimentally, it can be difficult to collect all the barium sulfate. Consider the scenario where you are baking a cake. Not all the batter (eggs, flour, etc.) goes into the cooking pan, as some of it gets left behind in the mixing bowl.

123. C: The balanced chemical equation is:

$$Al_2(SO_4)_3(aq) + 3 \text{ BaCl}_2(aq) \rightarrow 3 \text{ BaSO}_4(s) + 2 \text{ AlCl}_3(aq)$$

For simplification, rounding is shown in each step of the following calculation:

$$3.52 \text{ g Al}_2(SO_4)_3 \times \frac{1 \text{ mol Al}_2(SO_4)_3}{342.17 \text{ g Al}_2(SO_4)_3} \times \frac{3 \text{ mol BaSO}_4}{1 \text{ mol Al}_2(SO_4)_3} = 3.09 \times 10^{-2} \text{ mol BaSO}_4$$

$$4.06 \text{ g BaCl}_2 \times \frac{1 \text{ mol BaCl}_2}{208.23 \text{ g BaCl}_2} \times \frac{3 \text{ mol BaSO}_4}{3 \text{ mol BaCl}_2} = 1.95 \times 10^{-2} \text{ mol BaSO}_4$$

Aluminum sulfate is the excess reactant because it produces a greater molar amount of barium sulfate. One method to determine the excess amount is to perform the following calculations:

$$3.09 \times 10^{-2} \text{ mol BaSO}_4 - 1.95 \times 10^{-2} \text{ mol BaSO}_4 = 1.14 \times 10^{-2} \text{ excess moles of BaSO}_4$$

Now convert the excess moles of barium sulfate to grams of aluminum sulfate:

$$1.14 \times 10^{-2} \text{ excess moles of BaSO}_4 \times \frac{1 \text{ mol Al}_2(SO_4)_3}{3 \text{ mol of BaSO}_4} \times \frac{342.17 \text{ g Al}_2(SO_4)_3}{1 \text{ mol Al}_2(SO_4)_3} = 1.30 \text{ g Al}_2(SO_4)_3$$

Note that if rounding were done at the very last step, a value of 1.30 g of $Al_2(SO_4)_3$ would have also been obtained. Alternatively, the mass of barium chloride could have been used to calculate the limiting mass of aluminum sulfate. The difference between the two values would give the excess amount of aluminum sulfate.

$$4.06 \text{ g BaCl}_2 \times \frac{1 \text{ mol BaCl}_2}{208.23 \text{ g BaCl}_2} \times \frac{1 \text{ mol Al}_2(\text{SO}_4)_3}{3 \text{ mol BaCl}_2} \times \frac{342.17 \text{ g Al}_2(\text{SO}_4)_3}{1 \text{ mol Al}_2(\text{SO}_4)_3} = 2.22 \text{ g Al}_2(\text{SO}_4)_3$$

$$\text{excess mass of Al}_2(\text{SO}_4)_3 = \text{initial mass of Al}_2(\text{SO}_4)_3 - 2.22 \text{ g} = 3.52 \text{ g} - 2.22 \text{ g} = 1.30 \text{ g Al}_2(\text{SO}_4)_3$$

124. B: Choices *A*, *B*, and *D* are common ways to express the concentration of a solute in a solution. The molality (m) is defined as the moles of solute per kilogram of solvent.

$$m = \frac{\text{moles of solute}}{\text{kilograms of solvent}}$$

The molarity (M) is moles of solute per liter of solution.

$$M = \frac{\text{moles of solute}}{\text{liters of solution}}$$

Note that molality and molarity use the terms solvent and solution in the denominator. The solvent, typically water, refers to the mass of that solvent only. For molarity, the volume of the solution (solute and solvent) is used. For solutes that are present in small amounts within a solution, parts per mass, e.g., parts per billion (ppb), parts per million (ppm), and percent mass can be used. The general formula is defined as follows:

$$\text{parts per mass} = \frac{\text{mass of solute}}{\text{mass of solution}} \times \text{factor}$$

If the factor is 100, then it's called mass percentage (%). If the factor is a million, 10^6, then the term is called parts per million (ppm). If the factor is a billion (10^9), then the term is called parts per billion (ppb). Choice *C* is called the molar volume and is defined as the volume that is taken up or occupied by an ideal gas standard temperature and pressure. Although the units are similar (liters per mole) the molar volume is primarily used for gases, and the molarity applies to the concentration of a solute in a solution. Therefore, Choice *B* is the correct answer.

$$\text{molar volume} = \frac{V \text{ (volume of gas in liters)}}{n \text{ (moles of gas)}}$$

125. A: The mass of silver(I) chloride can be converted to a molar amount, which can then be equated to the known molality (m).

$$250.0 \text{ g AgCl} \times \frac{\text{mol AgCl}}{143.32 \text{ g AgCl}} = 1.744 \text{ moles AgCl}$$

The molality (m) is given by:

$$m = \frac{\text{moles of solute}}{\text{kilograms of solvent}}$$

The solvent in this case is water, and the solute is silver(I) chloride. The mass of the solvent can be found by rearrangement:

$$m = 0.35 \text{ mol kg}^{-1} = \frac{1.744 \text{ moles AgCl}}{\text{kilograms of water}}$$

$$\text{kilograms of water} = \frac{1.744 \text{ mol AgCl}}{0.35 \text{ mol kg}^{-1}} = 5.0 \text{ kg water}$$

Choices *A* and *D* represent answers with the correct units. Choice *D* is incorrect because it gives the incorrect mass of the solvent. Choices *C* and *B* are incorrect because the units of mass should be in kilograms.

126. B: We can use the following equation to determine the volume of concentrated acid that is needed:

$$M_{\text{initial}} V_{\text{initial}} = M_{\text{final}} V_{\text{final}}$$

The final volume and molarity are 350 mL and 2.00 M HNO_3, which represents the final diluted concentration. The initial molarity represents the more concentrated acid, 5.00 M HNO_3. Solving for the initial volume gives:

$$V_{\text{initial}} = \frac{M_{\text{final}} V_{\text{final}}}{M_{\text{initial}}} = \frac{2.00 \text{ M} \times 350 \text{ mL}}{5.00 \text{ M}} = 140 \text{ mL of } 5.00 \text{ M } HNO_3$$

Note that units of mL can be used because the molarity units cancel. Choices *C* and *D* do not represent the correct volumes. To prepare 350 mL of a 2.00 M HNO_3 solution, add 140 mL of 5.00 M HNO_3 into a large graduated cylinder, and dilute with water up to the 350 mL marker or add 210 mL of water to make a 2.00 M solution of HNO_3.

127. C: The endpoint refers to a point where a titration is stopped due to a color change (from the addition of an acid-base indicator), indicating equivalence has been reached. Equivalence occurs when the concentrations of hydronium ions and hydroxide ions are equal, $[H^+] = [OH^-]$. Based on this, the molarity of the diprotic acid can be found as follows:

$$\text{mol KOH} = 1.500 \text{ M KOH} \times 35.00 \text{ mL KOH} \times \frac{1 \text{ L}}{1000 \text{ mL}} = 5.250 \times 10^{-2} \text{ mol KOH}$$

Then the moles of the base must be converted to moles of acid (equivalence). However, because the acid is diprotic, the stoichiometric relationship must be taken into account.

$$5.250 \times 10^{-2} \text{ mol KOH} \times \frac{1 \text{ mol } H_2SO_4}{2 \text{ mol KOH}} = 2.626 \times 10^{-2} \text{ mol } H_2SO_4$$

Finally, the concentration of the acid is given by the following equation:

$$\text{M } H_2SO_4 = \frac{2.626 \times 10^{-2} \text{ mol } H_2SO_4}{15.00 \text{ mL} \times \frac{1 \text{ L}}{1000 \text{ mL}}} = 1.750 \text{ M } H_2SO_4$$

Note that a volume of 15.00 mL is used because this is the initial volume of the acid corresponding to a concentration of 1.750 M. The volume is not the combined volume of the acid and base because that volume represents a neutral solution.

128. C: The physical process is exothermic because liquid water has a kinetic energy greater than solid ice. Because the average kinetic energy is proportional to the temperature, as water freezes heat will flow out of the system.

$$H_2O(l) \rightarrow H_2O(s) + heat$$

The reaction is exothermic because heat is released from the molecular system. Consequently, because heat is lost by the chemical system $(-q_{system})$, then heat must be released or gained by the surroundings $(+q_{surroundings})$. The surroundings include the plant and air around it. If a plastic bag is placed over the plant, it will be able to trap or absorb the heat, thereby keeping the plant from completely freezing.

129. B: The formation of each compound is exothermic, so Choices C and D could not be correct choices.

$$Li + Br \rightarrow LiBr + heat \quad \Delta H_f^\circ = -351.2 \text{ kJ/mol}$$

$$Li + Cl \rightarrow LiCl + heat \quad \Delta H_f^\circ = -408.6 \text{ kJ/mol}$$

Choice B is correct because the amount of heat produced by the formation of lithium chloride is greater in magnitude compared to lithium bromide.

130. D: Note that the enthalpy of reaction for the combustion of methane is –802.3 kJ/mol. The problem is asking how much heat (in kJ) is produced from 1.50 kg, so it is expected that the value will be relatively large. To determine how much heat is produced, use the following conversion roadmap:

$$\text{kilograms of } CH_4 \xrightarrow{\frac{1000 \text{ g}}{1 \text{ kg}}} \text{grams of } CH_4 \xrightarrow{\frac{1 \text{ mole } CH_4}{16.042 \text{ g } CH_4}} \text{moles of } CH_4 \xrightarrow{\frac{-802.3 \text{ kJ}}{1 \text{ mole } CH_4}} \text{kilojoules}$$

The last conversion factor is important because it relates the stoichiometric relationship for moles and energy.

$$1.50 \text{ kg } CH_4 \times \frac{1000 \text{ g}}{1 \text{ kg}} \times \frac{1 \text{ mol } CH_4}{16.042 \text{ g } CH_4} \times \frac{-802.3 \text{ kJ}}{1 \text{ mol } CH_4} = -7.50 \times 10^4 \text{ kJ}$$

Choices B and C are not correct because these values indicate heat is absorbed. Choice A is not correct because $\Delta H_{rxn}^\circ = -802.3$ kJ per mole of methane. Choice D is correct because this value indicates that the combustion of 1.50 kg of methane produces -7.50×10^4 kJ.

131. A: Use the following equation to find ΔH_{rxn}°.

$$\Delta H_{rxn}^\circ = \sum n \Delta H_f^\circ(\text{products}) - \sum n \Delta H_f^\circ(\text{reactants})$$

$$\Delta H_{rxn}^\circ = \left[1 \times \overbrace{-1675.7 \frac{kJ}{mol}}^{Al_2O_3} + 2 \times \overbrace{0 \frac{kJ}{mol}}^{Fe} \right] - \left[2 \times \overbrace{0 \frac{kJ}{mol}}^{Al} + 1 \times \overbrace{-824.2 \frac{kJ}{mol}}^{Fe_2O_3} \right] = -851.5 \text{ kJ/mol}$$

The reaction is known as a thermite reaction and is exothermic. The value of –851.5 kJ is also correct (energy released per mole unit) because the equation says that for every two moles of aluminum that reacts, –851.5 kJ is produced. Or, for every 1 mole of iron(III) oxide that reacts, –851.5 kJ is produced.

132. C: The heat of reaction equation to determine the amount of heat that is lost from the metal and transferred to the water bath.

$$q_{\text{water}} = m \times C_{\text{s, water}} \times \Delta T$$

$$q_{\text{water}} = 100.0 \text{ g} \times 4.184 \text{ J/(g °C)} \times T_{\text{f}} - T_{\text{i}}$$

$$q_{\text{water}} = 100.0 \text{ g} \times 4.184 \text{ J/(g °C)} \times (37.0 \text{ °C} - 25.0\text{°C}) = +5.02 \times 10^3 \text{ J}$$

Note that 100.0 g of water is equivalent to 100.0 mL of water because the density is 1 g/mL. The initial temperature used should be that of water. Because of the amount of heat that is absorbed, the heat of reaction should be positive for water but negative for the metal.

$$+q_{\text{water}} = -q_{\text{metal}}$$

The heat of reaction is used again, but this time to find the specific heat capacity of the metal. Note that metal loses -5.02×10^3 J.

$$q_{\text{metal}} = m \times C_{\text{s, metal}} \times \Delta T$$

$$C_{\text{s, metal}} = \frac{q_{\text{metal}}}{m\Delta T} = \frac{-5.02 \times 10^3 \text{ J}}{43.0 \text{ g} \times (37.0 \text{ °C} - 100.0 \text{ °C})} = 1.85 \text{ J/(g °C)}$$

The final temperature will be that of the water bath; thermal equilibrium is reached at that temperature. Because the initial temperature is greater than the final one, the ΔT will be negative. However, the negative signs will cancel out and give a positive specific heat (it must be positive).

133. D: Entropy is a measure of disorder or randomness in a system. The contraction of a gas either due to cooling or compression is an example where the change in entropy is negative $(-\Delta S)$. As the volume decreases, the gas molecules/particles will occupy fewer spaces. The randomness decreases. The freezing of water is an example of where water molecules that were free to occupy different positions within a liquid now become fixed and ordered in the form of a solid and fixed crystal lattice. There is more order (less disorder), meaning that the change in entropy is negative $(-\Delta S)$.

The reaction of solid calcium oxide and carbon dioxide gas to form solid calcium carbonate is an example of a system that becomes more ordered. The products do not contain a gas which can occupy different points in space. The entropy change is negative $(-\Delta S)$. Adding sugar to water is an example where the entropy change is positive $(+\Delta S)$. The molecular structure of sugar does not break down, rather the crystal structure of table sugar breaks apart and dissolves in water. The sugar molecules are free to move around in solution and are not restrained to fixed positions. The disorder increases.

134. A: The table below indicates that when both the enthalpy and entropy change are positive, then the reaction will be spontaneous only when the temperature increases (Scenario 4, last row). Initially, at low temperatures, the positive ΔH term has a greater than the $-T\Delta S$ term, which makes the Gibbs free energy positive. However, as the temperature increases, the $-T\Delta S$ term becomes more negative and makes a more negative contribution than ΔH, which will make the Gibbs free energy more negative.

ΔG Sign conventions for different signs of ΔH and ΔS							
Scenario	Spontaneity		K	ΔG	ΔH	−TΔS	ΔS
1	Spontaneous		> 1	−	−	−	+
2	Nonspontaneous		< 1	+	+	+	−
3	Spontaneous	(low T)	> 1	−	−	+	−
	Nonspontaneous	(high T)	< 1	+			
4	Nonspontaneous	(low T)	< 1	+	+	−	+
	Spontaneous	(high T)	> 1	−			

135. C: Based on the table displaying free energy sign conventions, assuming that $\Delta G \approx \Delta G°$, the reaction would correspond to Scenario 3. The entropy of reaction is negative because there is more order as two molecules of a gas form one molecule. The enthalpy of reaction is negative and exothermic. If the temperature is low enough, the magnitude of the $-\Delta H$ term will be greater than the positive $T\Delta S$ term, and the reaction will be spontaneous. However, as the temperature is increased, the magnitude of the $+T\Delta S$ term will outweigh the $-\Delta H$ term, which will make the Gibbs free energy positive and nonspontaneous.

136. D: The figure below shows several orbital diagrams and quantum numbers corresponding to different electrons.

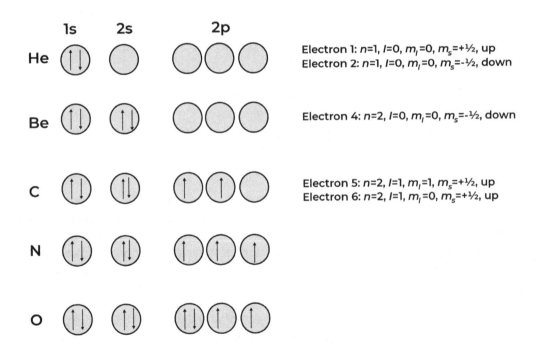

Recall that the Aufbau principle explains how electrons are added to orbitals, such that repulsion is minimized. Electrons are filled in the first three p orbitals with the same spin quantum number (i.e.,

$m_s = +\frac{1}{2}$). If a seventh electron were added to carbon, it would have the following set of quantum numbers:

$$n = 2, l = 1, m_l = +1, m_s = +\frac{1}{2}$$

Note that each answer choice has $n = 2$, meaning that only electrons are considered in the L shell or s and p subshells. The $l = 1$ quantum number is referring to the 2p subshell. Choices A and C could not be possible choices because each has $l = 0$, which corresponds to electrons in the 2s subshell. The eighth electron, based on the Aufbau principle, should be added to the p subshell. Choice B represents electron five (shown in the figure above) and has $m_s = +\frac{1}{2}$. Therefore, the answer choice must be D:

$$n = 2, l = 1, m_l = -1, m_s = -\frac{1}{2}$$

Note that, the magnetic quantum number ($m_l = -1$) will be the same as electron 5, but the spin quantum number must be $m_s = -\frac{1}{2}$ based on the Pauli exclusion principle.

137. A: Choices B, C, and D all represent possible electron configurations for nitrogen. However, Choice A does not represent a possible set of quantum number because the magnetic quantum number $m_l = -2$ is not an allowed value within the 2p subshell. The only permitted values when $l = 1$ are $m_l = -1$, 0, and $+1$. Choices B and C correspond to electrons in the 1s and 2s subshells, while Choice D refers to an electron in the 2p subshell.

138. B: The formal charges for each structure are shown below, according to the following formula:

$$\text{formal charge (FC)} = \text{\# of valence } e^- - \left(\frac{1}{2}\text{\# of bonding } e^- + \text{\# of nonbonding } e^-\right)$$

I	II	III

$$:\!\ddot{N}\!=\!N\!=\!\ddot{O}\!: \qquad :\!N\!\equiv\!N\!=\!\ddot{\ddot{O}}\!: \qquad :\!\ddot{\ddot{N}}\!-\!N\!\equiv\!O\!:$$

Structure I:

$$\text{formal charge for outer nitrogen atom} = 5 - \left(\frac{1}{2}(4) + 4\right) = -1$$

$$\text{formal charge for cental nitrogen atom} = 5 - \left(\frac{1}{2}(8) + 0\right) = +1$$

$$\text{formal charge for oxygen atom} = 6 - \left(\frac{1}{2}(4) + 4\right) = 0$$

Structure II:

$$\text{formal charge for outer nitrogen atom} = 5 - \left(\frac{1}{2}(6) + 2\right) = 0$$

$$\text{formal charge for central nitrogen atom} = 5 - \left(\frac{1}{2}(8) + 0\right) = +1$$

$$\text{formal charge for oxygen atom} = 6 - \left(\frac{1}{2}(2) + 6\right) = -1$$

Structure III:

$$\text{formal charge for outer nitrogen atom} = 5 - \left(\frac{1}{2}(2) + 6\right) = -2$$

$$\text{formal charge for central nitrogen atom} = 5 - \left(\frac{1}{2}(8) + 0\right) = +1$$

$$\text{formal charge for oxygen atom} = 6 - \left(\frac{1}{2}(6) + 2\right) = +1$$

All structures are neutrally charged. Structure II or Choice B is the most likely structure because the oxygen atom, the most electronegative, bears the negative charge, while the central nitrogen contains a positive charge. Choice C includes a positive charge on the oxygen atom, which makes the structure more unstable. Choice A has a structure with a negative charge on the outer nitrogen atom, the least electronegative atom. Choice D is also incorrect because it lists the formal charges incorrectly.

139. C: Choices A, B, and D are all true. In Choice A, the A to B phase transition corresponds to sublimation, a solid to gas transition. In Choice B, the change from C to B represents a liquid to gas transition called evaporation. In choice D, the A to C transition is a solid to liquid transition called melting. Choice C is incorrect because although the slope of the provided phase diagram is positive, the slope of the solid-liquid equilibrium of water is negative. If the slope were positive for the phase diagram of water, then ice skating would be difficult or impossible. Because the solid-liquid equilibrium line has a negative slope, the pressure of the skates over the ice causes a solid to liquid phase transition, melting a small portion of ice under the skates, and thus reducing friction with the solid ice. However, if the slope was positive, at 0 °C, increasing the pressure would not result in a phase transition.

140. A: To determine the freezing point depression, first calculate the molality of the solution.

$$m = \frac{\text{moles of solute}}{\text{kilograms of solvent}} = \frac{1.00 \times 10^3 \text{ g } C_2H_6O_2 \times \frac{1 \text{ mol } C_2H_6O_2}{62.07 \text{ g } C_2H_6O_2}}{5.00 \times 10^3 \text{ g } H_2O \times \frac{1 \text{ kg}}{1000 \text{ g}}} = 3.22 \frac{\text{mol}}{\text{kg}}$$

Solving for the freezing point depression gives:

$$\Delta T_f = m \times K_f = 3.22 \text{ mol kg}^{-1} \times 1.86 \frac{°C}{\text{mol kg}^{-1}} = 5.99 \text{ °C}$$

The actual freezing point can be found by expanding the left-hand side of the previous equation, and accounting for the negative sign.

$$\Delta T_f = T_{\text{solvent}} - T_{\text{solution}} = 0.00 \text{ °C} - 5.99 \text{ °C} = -5.99 \text{ °C}$$

Note that 0.00 °C corresponds to the freezing point of pure water or the solvent.

141. C: The following form of the Arrhenius equation can be used to find the factor or ratio at which the rate constant increases.

$$\ln\frac{k_2}{k_1} = \frac{E_a}{R}\left(\frac{1}{T_1} - \frac{1}{T_2}\right)$$

The ratio or factor is given by the following rearrangement of the previous equation:

$$\frac{k_2}{k_1} = e^{\frac{E_a}{R}\left(\frac{1}{T_1} - \frac{1}{T_2}\right)}$$

Note that $E_a = 125$ kJ/mol, $R = 8.314$ J/(mol K), $T_1 = 25.0 + 273.15$ K, and $T_2 = 50.0 + 273.15$ K.

First, evaluate the term within the exponent:

$$\frac{E_a}{R}\left(\frac{1}{T_1} - \frac{1}{T_2}\right) = \frac{125 \text{ kJ mol}^{-1}}{8.314 \text{ J mol}^{-1}\text{ K}^{-1}} \times \frac{1000 \text{ J}}{1 \text{ kJ}} \times \left(\frac{1}{298.15 \text{ K}} - \frac{1}{323.15 \text{ K}}\right) = 3.901$$

Now evaluate the following expression:

$$\frac{k_2}{k_1} = e^{3.901} = 49.5$$

Increasing the temperature to 50.0 °C increases the rate by nearly 50 times the original rate.

142. B: Use the following equations to determine the time it takes for the element X to decay to 35.0 percent of its original amount:

$$\ln[B]_t = -kt + \ln[B]_0 \text{ and } t_{1/2} = \frac{0.693}{k}$$

First, find the rate constant:

$$k = \frac{0.693}{t_{1/2}} = \frac{0.693}{7.95 \text{ days}} = 0.087170 \text{ days}^{-1}$$

Now solve for t using the rearranged equation above.

$$\ln[B]_t = -kt + \ln[B]_0$$

$$\ln[B]_t - \ln[B]_0 = -kt$$

$$t = -\ln\left(\frac{[B]_t}{[B]_0}\right)\frac{1}{k}$$

The terms $[B]_t$ and $[B]_0$ represent the amount at time t (35%) and the initial amount at time zero. For simplification, suppose that $[B]_0 = 1$, then $[B]_t = 0.350([B]_0) = 0.350$. Regardless of the amount, the term within the natural logarithm will be 0.350.

$$t = -\ln\left(\frac{0.350([B]_0)}{[B]_0}\right)\frac{1}{0.087170 \text{ days}^{-1}} = 12.0 \text{ days}$$

143. D: The overall equilibrium expression will be the product of each constant or expression. First, determine the equilibrium constant expression for the first reaction, followed by the second one.

$$2\,CO(g) + O_2(g) \rightleftharpoons 2\,CO_2$$

$$K_1 = \frac{[CO_2]^2}{[CO]^2[O_2]}$$

$$CO_2(g) + C(s) \rightleftharpoons 2\,CO(g)$$

$$K_1 = \frac{[CO]^2}{[CO_2][C]}$$

Now multiply each equilibrium constant to obtain the overall K value.

$$K_1 \times K_2 = K_{overall} = \frac{[CO_2]^2}{[CO]^2[O_2]} \times \frac{[CO]^2}{[CO_2][C]} = \frac{[CO_2]}{[O_2][C]}$$

This expression could have also been obtained if both equations from above were combined.

$$2\,CO(g) + O_2(g) \rightleftharpoons 2CO_2$$

$$CO_2(g) + C(s) \rightleftharpoons 2\,CO(g)$$

$$O_2(g) + C(s) \rightleftharpoons CO_2(g) \quad K_{overall} = \frac{[CO_2]}{[O_2][C]}$$

144. D: The equilibrium expression of the reaction is:

$$K = \frac{1}{[SO_2]}$$

The solid reactant and products are not included in the expression and will not impact the equilibrium constant or direction of equilibrium. Therefore, Choices A and B are not correct. A decrease in SO_2 pressure at equilibrium means that the reaction must shift to the right because fewer gas particles exert fewer forces over an area and lead to lower pressure. If the volume of the reaction vessel, containing the chemical system, was decreased, then the equilibrium will shift to the right where there are no gaseous products. In contrast, in Choice C, increasing the volume of the reaction vessel will initially decrease the pressure, but when equilibrium is reestablished, the reaction will shift to the left to produce more SO_2. Heat can be considered a product of the reaction. If heat is removed from the system, such as chilling the reaction vessel, then the reaction will shift to the right, reducing the overall pressure. If the temperature is reduced, then heat is removed, and the reaction will proceed to the right. At equilibrium, the pressure of SO_2 will be lower.

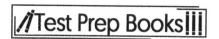

145. B: First set up the equilibrium constant expression for the reaction of ammonia in water.

$$\overbrace{NH_3(aq)}^{\substack{\text{Brønsted–}\\\text{Lowry}\\\text{base}}} + \overbrace{H_2O(l)}^{\substack{\text{Brønsted–}\\\text{Lowry}\\\text{acid}}} \rightleftharpoons \overbrace{OH^-(aq)}^{\substack{\text{conjugate}\\\text{base}}} + \overbrace{NH_4^+(aq)}^{\text{conjugate acid}}$$

$$K_b = \frac{[OH^-][NH_4^+]}{[NH_3]} = 1.8 \times 10^{-5}$$

Set up an ICE table to formulate an expression in terms of x or the hydroxide concentration.

	$[NH_3]$	$[OH^-]$	$[NH_4^+]$
Initial	0.35	0	0
Change	$-x$	$+x$	$+x$
Equilibrium	$0.35 - x$	x	x

$$K_b = \frac{(x)(x)}{(0.35 - x)} = \frac{x^2}{0.35 - x} = 1.8 \times 10^{-5}$$

Assuming x is much smaller than 0.35, then:

$$1.8 \times 10^{-5} = \frac{x^2}{0.35}$$

$$x = \sqrt{1.8 \times 10^{-5} \times 0.35} = 0.0025$$

Applying the 5 percent rule or substituting into the original equation gives:

$$\frac{0.0025}{0.35} \times 100\% = 0.71\% \qquad \frac{(0.0025)^2}{0.35 - (0.0025)} = 1.8 \times 10^{-5}$$

The value of x is reasonable because its less than 5 percent with respect to the initial concentration. The pOH can be found from the hydroxide concentration:

$$0.0025 = [OH^-]$$

$$pOH = -\log(0.0025) = 2.6$$

The pH of ammonia can then be found using the pOH:

$$pOH + pH = 14.00$$

$$pH = 14.00 - pOH = 14.00 - 2.6 = 11.4$$

146. A: First set up the equilibrium constant expression.

$$K_c = \frac{[HI]^2}{[H_2][I_2]}$$

Note that the concentration of each reactant is $1.00 \text{ mol}/1.0 \text{ L} = 1.0$ M. Next, set up an ICE table.

	$[H_2]$	$[I_2]$	$[HI]$
Initial	1.0	1.0	0
Change	$-x$	$-x$	$+2x$
Equilibrium	$1.0 - x$	$1.0 - x$	$2x$

Then, substitute into the equilibrium expression:

$$49.7 = \frac{(2x)^2}{(1.0 - x)(1.0 - x)} = \frac{(2x)^2}{(1.0 - x)^2}$$

The right-hand side of the equation is a perfect square. Taking the square root of both sides gives:

$$\sqrt{49.7} = \sqrt{\frac{(2x)^2}{(1.0 - x)^2}}$$

$$\pm 7.0498 = \frac{2x}{1.0 - x}$$

Both positive and negative values must be considered. Rearranging the equation gives:

Case 1 (+7.0498)

$$x_1 = 0.78$$

Case 2 (−7.0498)

$$x_2 = 1.4$$

The second value (1.4) is not a possible choice because it's greater than the initial concentration, so the first value (0.78) is the correct answer. The equilibrium concentrations are:

$$[H_2] = [I_2] = 1.0 - 0.78 = 0.22 \text{ M}$$

$$[HI] = 2(0.78) = 1.6 \text{ M}$$

147. C: The radioactive decay process describes electron capture, because it's the only type of decay that involves electron absorption.

$$^{92}_{44}\text{Ru} + ^{0}_{-1}\text{e} \rightarrow ^{92}_{43}\text{Tc}$$

Ruthenium-92 will accept an electron from its inner orbitals, transforming one of its protons to a neutron. Consequently, the atomic mass will decrease from 44 to 43, while the mass number remains the same. Choices A and B are not correct because the mass number changes, and because ruthenium does not decompose to rhodium. Instead, the atom that has an atomic number of 43 is technetium. Choice D is incorrect because electron capture does not produce an alpha particle.

148. B: First find the N/Z ratio of Pb-212, which has $Z = 82$ (see exact values in the periodic table). The number of neutrons is $A - Z = 212 - 82 = 130$. The N/Z ratio is $130/82 = 1.59$. Based on the valley of stability, it lies slightly outside the shaded region. Because the N/Z ratio is too high, beta decay will occur, and a neutron will be converted to a proton.

$$^{212}_{82}Pb \rightarrow {}^{212}_{83}Bi + {}^{0}_{-1}e$$

The number of neutrons in bismuth-212 is $A - Z = 212 - 83 = 129$, which is one less than the parent nuclide. The N/Z ratio is $129/83 = 1.55$.

149. B: Choice A is incorrect because MnO_4^- is reduced in the reaction:

$$\underset{\underset{\overbrace{MnO_4^-}}{x=+7}}{x+4(-2)=-1} \quad \rightarrow \quad \underset{\underset{\overbrace{MnO_2}}{x=+4}}{x+2(-2)=0}$$

The oxidation state of manganese changes from Mn^{+7} to Mn^{+4}, which indicates a reduction took place. Because manganese is reduced, it acts as an oxidizing agent, making Choice B the correct choice. Iodide changes from I^- to I_2, meaning that its oxidation state changes from -1 to 0, indicating oxidation took place. The iodide ion, therefore, acts as the reducing agent.

150. B: The aluminum half-reaction has a relatively more negative potential compared to nickel. Therefore, aluminum will act as the anode and proceed in the reverse direction, thereby repelling electrons and becoming oxidized in the process. In contrast, nickel will act as the cathode and accept electrons because it has a more positive potential.

$$\text{Cathode (reduction): } Ni^{2+}(aq) + 2e^- \rightarrow Ni(s) \quad E^\circ = -0.23 \text{ V}$$

$$\text{Anode (oxidation): } Al(s) \rightarrow Al^{3+}(aq) + 3e^- \quad E^\circ = +1.66 \text{ V}$$

Because aluminum is oxidized, the sign of E° changes from negative to positive. Each equation must be multiplied by a factor such that the net equation will have an equal number of electrons on each side.

$$3 \times \left(Ni^{2+}(aq) + 2e^- \rightarrow Ni(s)\right) \quad E^\circ = -0.23 \text{ V}$$

$$2 \times \left(Al(s) \rightarrow Al^{3+}(aq) + 3e^-\right) \quad E^\circ = +1.66 \text{ V}$$

The spontaneous net equation is:

$$3Ni^{2+}(aq) + 6e^- \rightarrow 3Ni(s) \quad E^\circ = -0.23 \text{ V}$$

$$2Al(s) \rightarrow 2Al^{3+}(aq) + 6e^- \quad E^\circ = +1.66 \text{ V}$$

$$3Ni^{2+}(aq) + 2Al(s) \rightarrow 3Ni(s) + 2Al^{3+}(aq) \quad E^\circ_{cell} = -0.23 \text{ V} + 1.66 \text{ V} = +1.43 \text{ V}$$

Note that the electrode cell potentials, E°, for each half-cell reaction were not multiplied by a factor. The reason is that E° is an intensive property and does not depend on the molar amount. E° remains the same regardless of the factor that is used to multiply the equation.

151. A: First calculate the molar concentration (units of mol/L) of silver chloride from its solubility. The molar mass of silver chloride is 143.32 g/mol.

$$1.9 \times 10^{-3} \frac{g}{L} = \frac{1 \text{ mol AgCl}}{143.32 \text{ g AgCl}} = 1.326 \times 10^{-5} \text{ mol/L}$$

First, write out the dissociation of silver chloride in water:

$$AgCl(s) \rightleftharpoons Ag^+(aq) + Cl^-(aq)$$

Now set up an ICE table and solve for K_{sp}:

	$[Ag^+]$	$[Cl^-]$
Initial	0	0
Change	$+S$	$+S$
Equilibrium	S	S

The K_{sp} of silver chloride can be then found from the concentration of dissolved silver chloride:

$$K_{sp} = [Ag^+][Cl^-]$$

$$K_{sp} = (S)(S) = (S)^2 = (1.326 \times 10^{-5})^2 = 1.758 \times 10^{-10} \approx 1.8 \times 10^{-10}$$

152. C. The equilibrium constant can be used to predict the extent and direction of the reaction. For example, a large K value ($> 10^3$) would indicate that the reaction proceeds strongly to the right, toward the products. By knowing the initial concentrations and equilibrium constant, an ICE table can be used to help set up a mathematical expression of the equilibrium expression in terms of the equilibrium concentrations. There is no information on the kinetics or how long it will take a reaction to proceed within the equilibrium constant

Dear Customer,

We would like to start by thanking you for purchasing this chemistry study guide. We hope that we exceeded your expectations. With that said, if you found something that you feel was not up to your standards, please send us an email and let us know.

Thanks Again and Happy Testing!
Product Development Team
info@studyguideteam.com

FREE Test Taking Tips Video/DVD Offer

To better serve you, we created videos covering test taking tips that we want to give you for FREE. **These videos cover world-class tips that will help you succeed on your test.**

We just ask that you send us feedback about this product. Please let us know what you thought about it—whether good, bad, or indifferent.

To get your **FREE videos**, you can use the QR code below or email freevideos@studyguideteam.com with "Free Videos" in the subject line and the following information in the body of the email:

      a. The title of your product

      b. Your product rating on a scale of 1-5, with 5 being the highest

      c. Your feedback about the product

If you have any questions or concerns, please don't hesitate to contact us at info@studyguideteam.com.

Thank you!

Made in the USA
Columbia, SC
12 May 2023